4th International Conference on Stem Cell Engineering 2014

Coronado, California, USA
16-19 March 2014

ISBN: 978-1-5108-1762-3

Printed from e-media with permission by:

Curran Associates, Inc.
57 Morehouse Lane
Red Hook, NY 12571

Some format issues inherent in the e-media version may also appear in this print version.

Printed by Curran Associates, Inc. (2016)

For permission requests, please contact AIChE
at the address below.

AIChE
120 Wall Street, FL 23
New York, NY 10005-4020

Phone: (800) 242-4363
Fax: (203) 775-5177

www.aiche.org

Additional copies of this publication are available from:

Curran Associates, Inc.
57 Morehouse Lane
Red Hook, NY 12571 USA
Phone: 845-758-0400
Fax: 845-758-2634
Email: curran@proceedings.com
Web: www.proceedings.com

TABLE OF CONTENTS

STEM CELL TISSUE ENGINEERING & MORPHOGENESIS

REGENERATIVE THERAPIES

BIOPROCESSING & BIOMANUFACTURING STRATEGIES

MICROTECHNOLOGIES & HIGH THROUGHPUT SCREENING

COMPUTATIONAL STEM CELL BIOLOGY

Decellularized Extracellular Matrix Based Scaffolds for Regenerative Medicine

Christman, K.,

Originally presented on: 3/17/2014 0:00:00 - 0:20:00

The extracellular matrix (ECM) is nature's scaffold, and in recent years, researchers have isolated these scaffolds for tissue engineering applications by removing all of the cellular components, a process called decellularization. These scaffolds are known to promote cell influx, regeneration, and healing in a variety of tissues, and their degradation products have angiogenic, chemoattractant, and antimicrobial properties, as well as promote cell migration and proliferation. By removal of the cellular antigens, these scaffold are considered biocompatible, and xenogeneic sources can be used. While these scaffolds retain the native ECM structure, they are not amenable to minimally invasive, injectable procedures or for cell culture on defined 2D surfaces and 3D structures. We have developed a variety of injectable ECM derived materials that self-assemble to form porous, nanofibrous scaffolds once injected in vivo or brought to physiological conditions *in vitro*. These materials can also be used as substrate coatings for 2D cell culture. These ECM based scaffolds have been shown to increase tissue specific differentiation and maturation of a variety of progenitor and stem cells *in vitro*, and are showing promise *in vivo* in several tissues including the myocardium, skeletal muscle, and adipose tissue. This talk will cover the progress with these materials in both *in vitro* and *in vivo* stem cell engineering.

Sticky Science:
the Role of Adhesion and Mechanics in Regulating Stem Cell Fate

Adam J. Engler
University of California, San Diego
Sanford Consortium for Regenerative Medicine
http://ecm.ucsd.edu

17 March 2014
4th Intl. Conference on Stem Cell Eng'g

UCSD

Sanford Consortium FOR REGENERATIVE MEDICINE

The Cell in Context

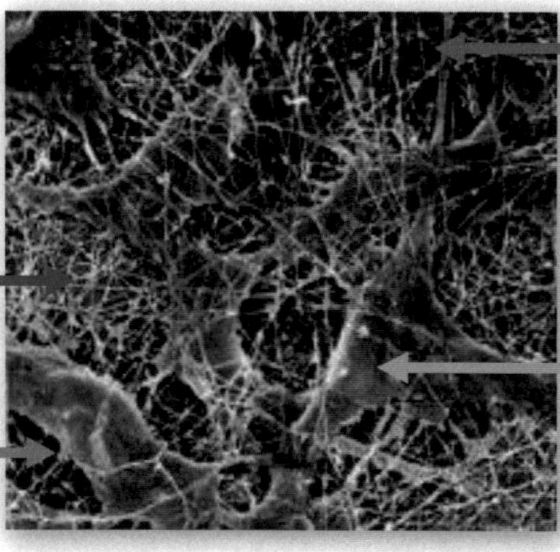

Alberts (4th Edition)

Biological, Chemical, and Mechanical Interfaces: "Building Blocks" for Cells

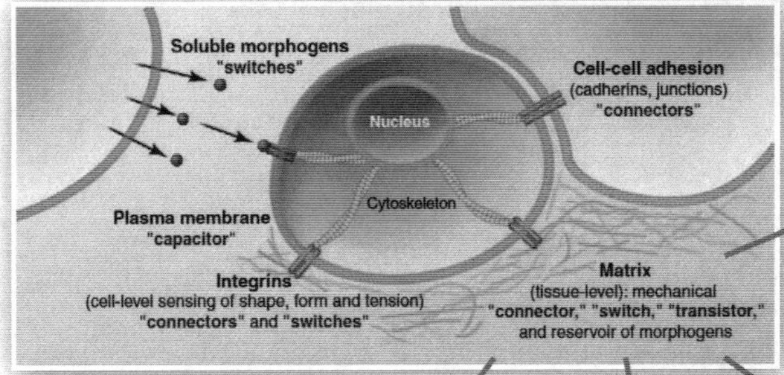

Engler et al, *Science* (2009)

Cell Shape

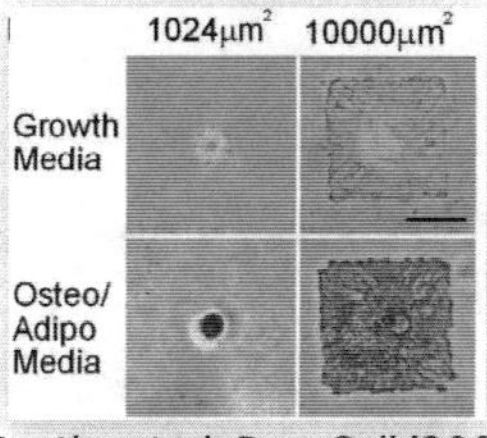

McBeath, et al *Dev Cell* (2004)
Fu, et al *Nat Methods* (2010)

Matrix Ligand

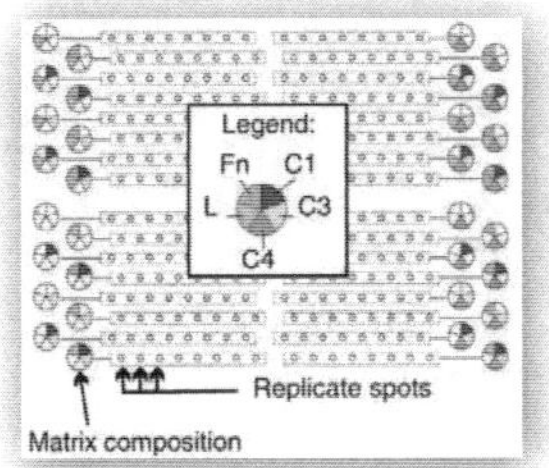

Ott, et al *Nature Med* (2008)
Flaim, et al *Nature Meth* (2005)

Topography

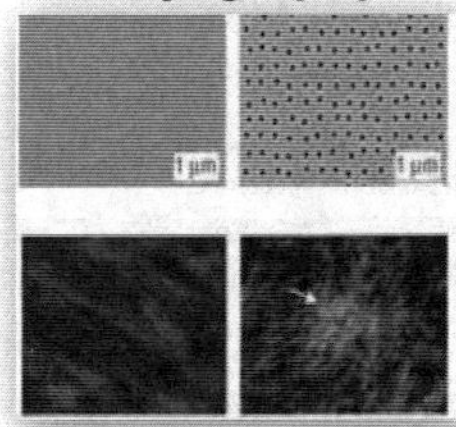

Dalby, et al *Nature Mat* (2007)
Oh, et al *PNAS* (2009)

Matrix Elasticity

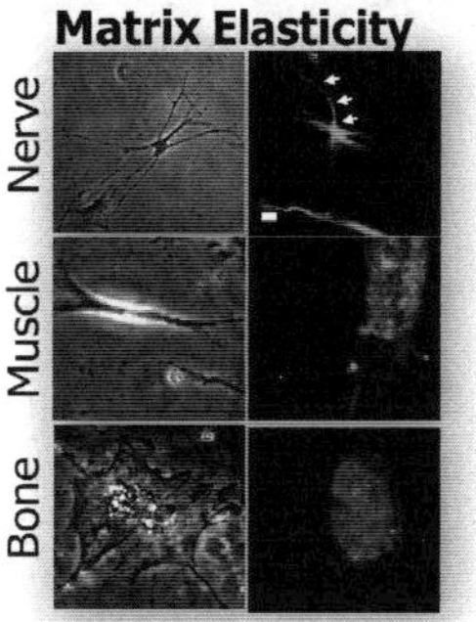

Engler, et al *Cell* (2006)
Gilbert, et al *Cell* (2010)

Regularly-spaced vs. Disordered adhesions

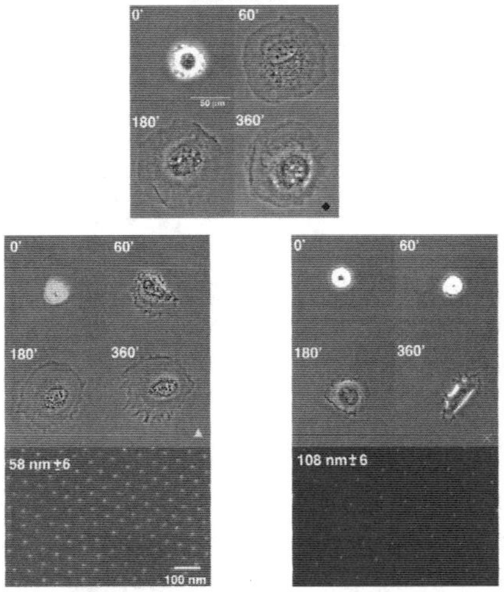

58 nm

108 nm

Cavalcanti-Adam et al, *Biophys J* (2007)

Huang et al, *Nano Lett* (2009)

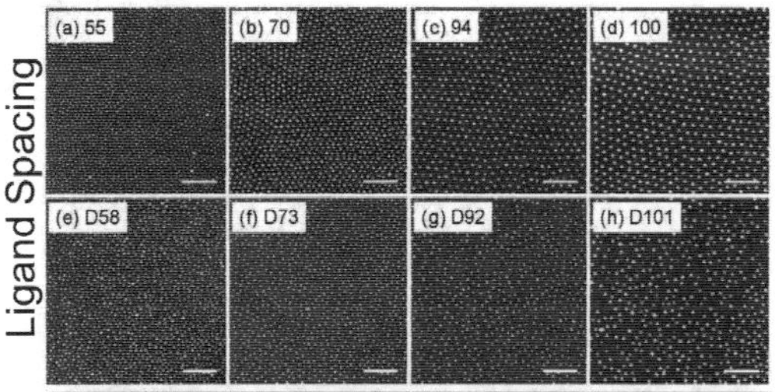

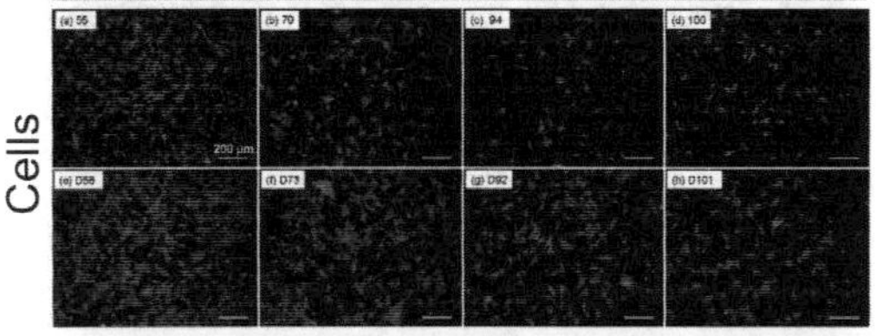

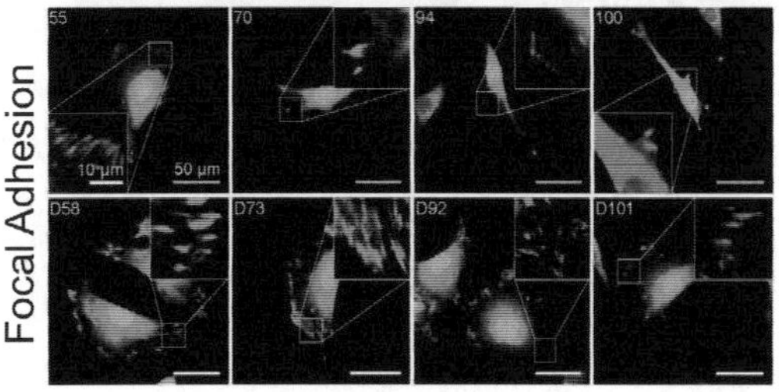

Extracellular Matrix is naturally adhesive, but it has "Adhesive Heterogeneity"

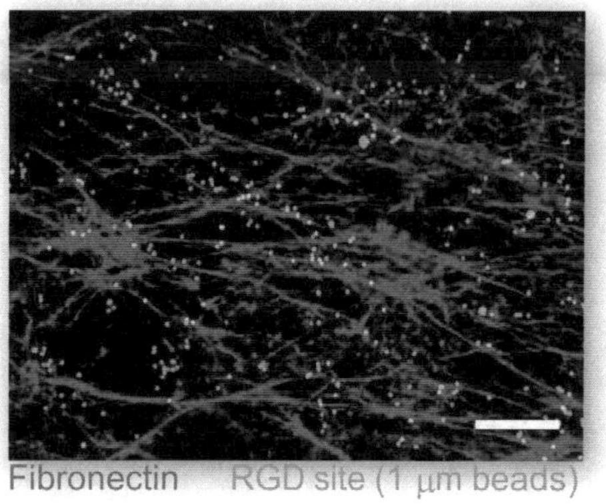

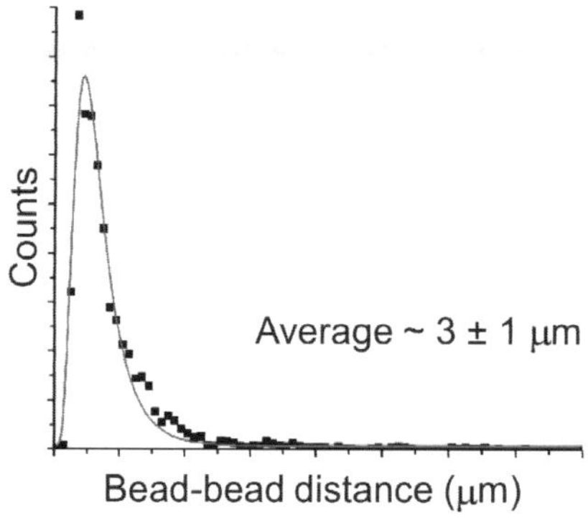

Cell-Instructive Nano-Patterned Soft Materials

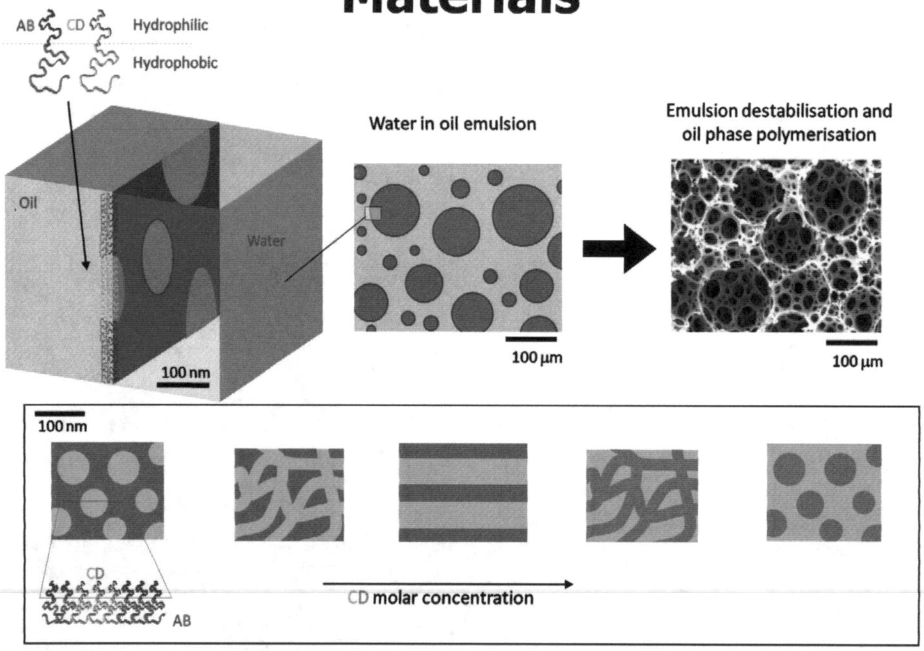

Giuseppe Battaglia (University College London)

Imaging Fluorescently-Labeled Nano-Domains

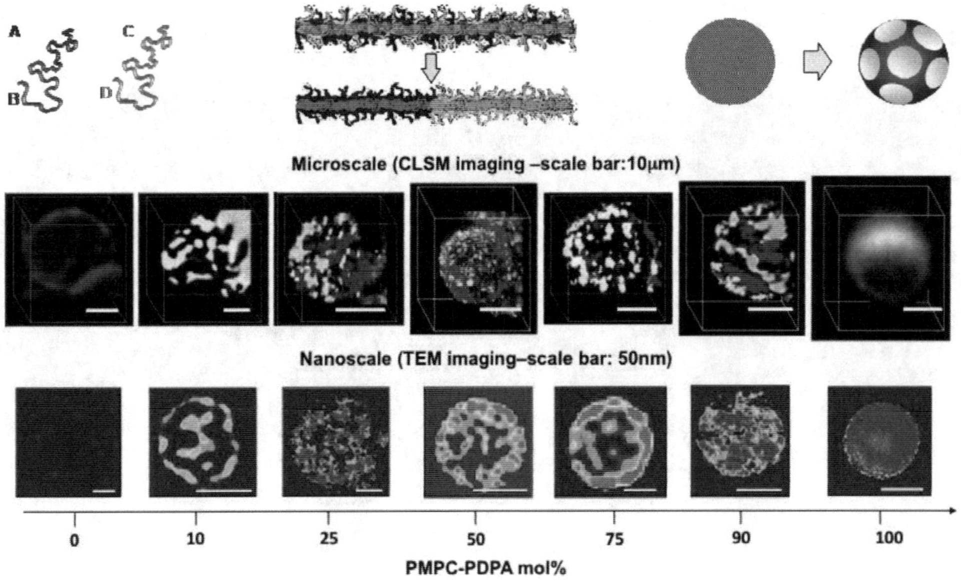

AB: PEO16-PBO25
CD: PMPC25-PDPA75

Lo Presti et al, *ACS Nano* (2011); Battaglia et al, *Small* (2011)

Cell-Instructive Nano-Patterned Soft Foams: Structure

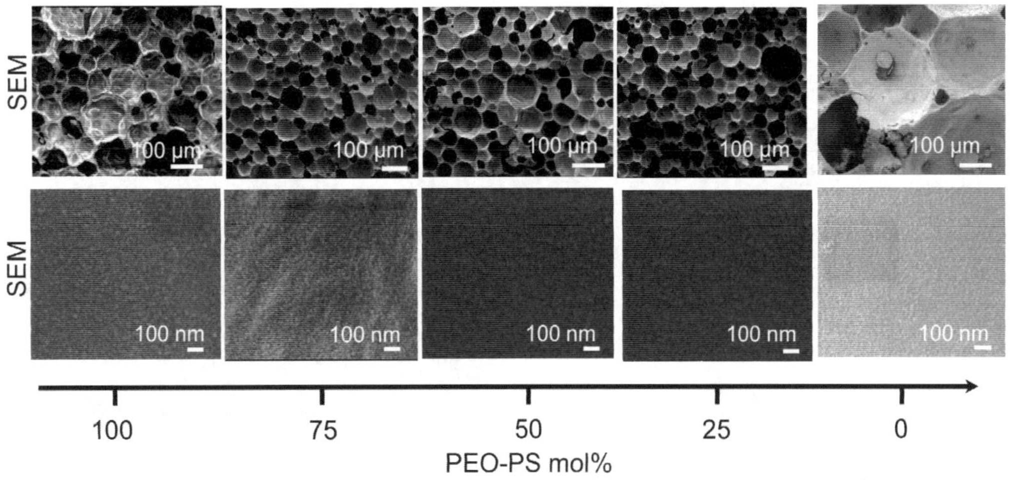

Priya Viswanathan
Somyot Chirasatitsin

AB: PEO33-PS25
CD: PAA16-PS25

Viswanathan, Chriasatitsin et al, *JACS* (2012)

Cell-Instructive Nano-Patterned Soft Foams: Nano-Domains

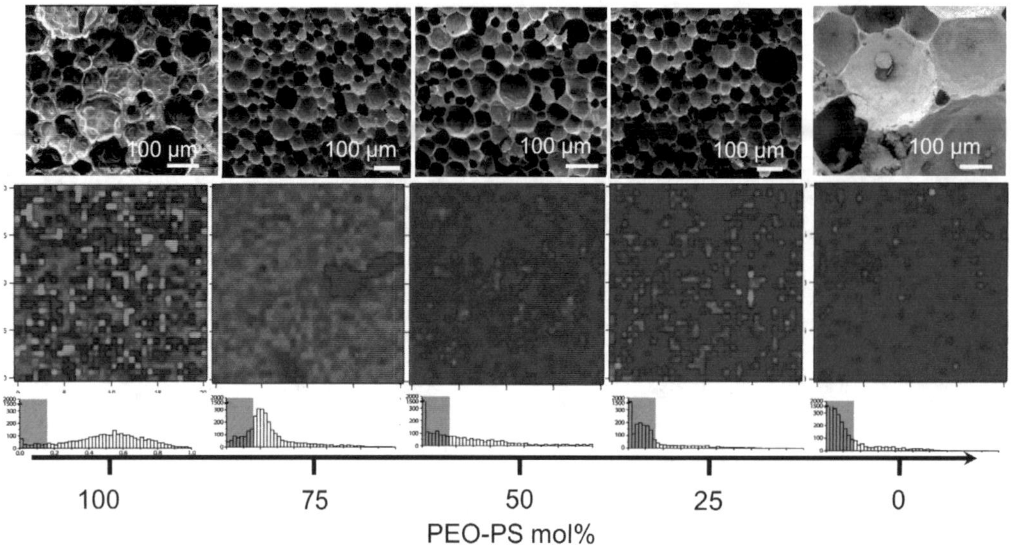

AB: PEO33-PS25
CD: PAA16-PS25

Viswanathan, Chriasatitsin et al, *JACS* (2012)

Cell-Instructive Nano-Patterned Soft Foams: Nano-Domains

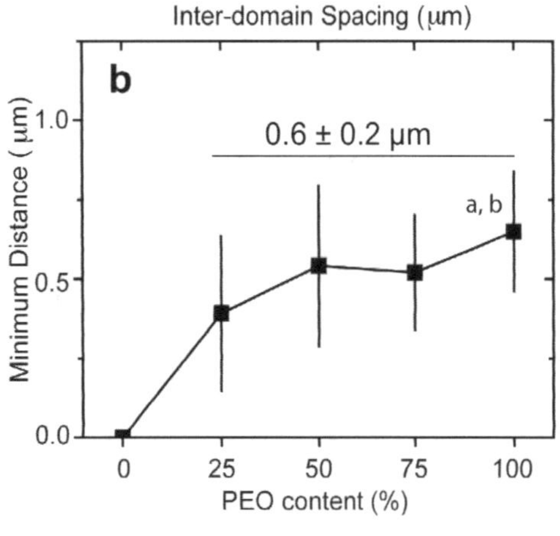

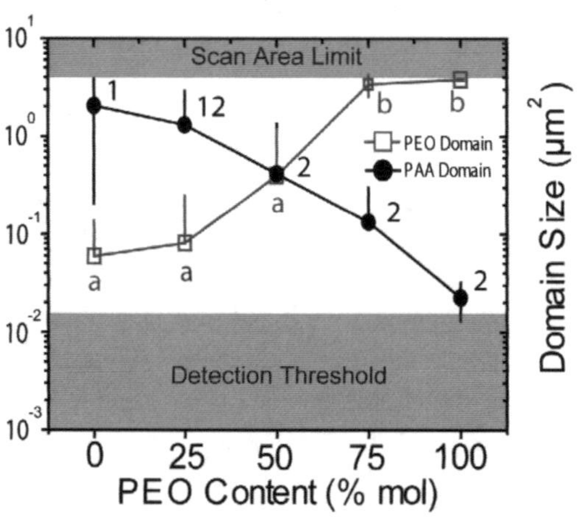

AB: PEO33-PS25
CD: PAA16-PS25

Viswanathan, Chriasatitsin et al, *JACS* (2012)

Cell-Instructive Nano-Patterned Soft Foams: Fibronectin

PEO75

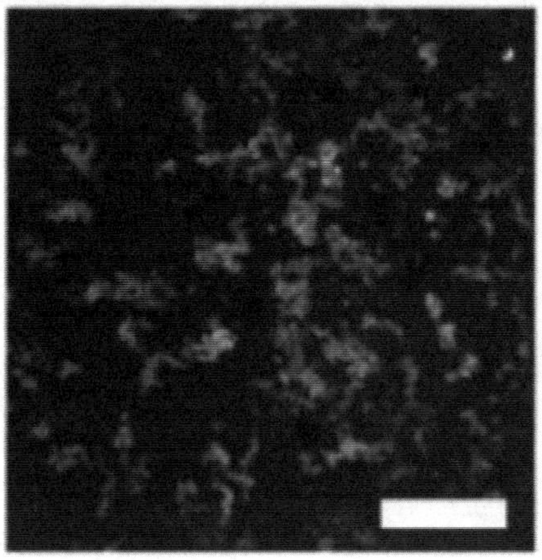

PEO25

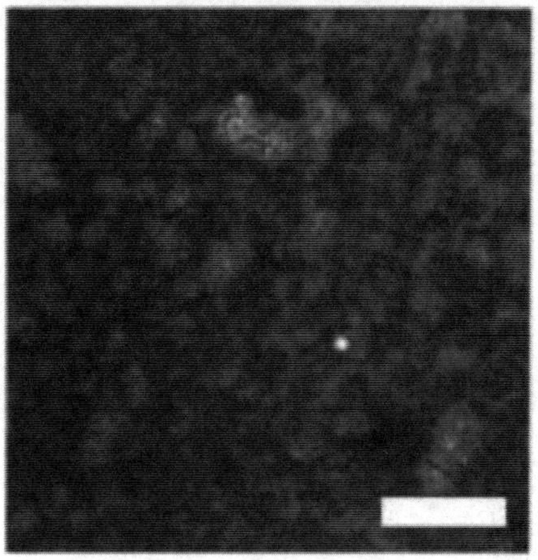

AB: PEO33-PS25
CD: PAA16-PS25

Viswanathan, Chriasatitsin et al, *JACS* (2012)

Marrow-Derived Adult Human Stem Cells

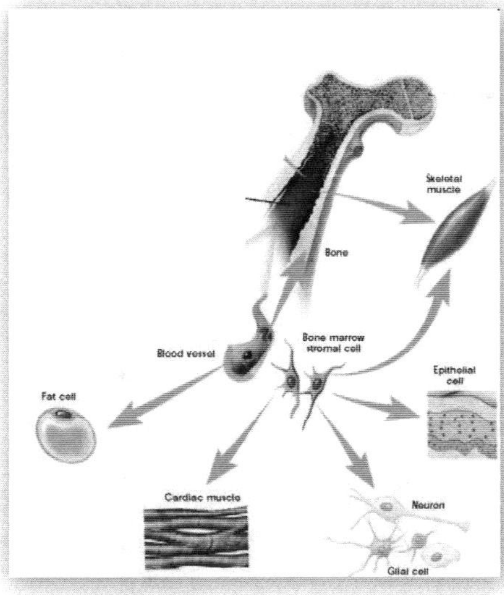

Mesenchymal Stem Cell Expression Profile

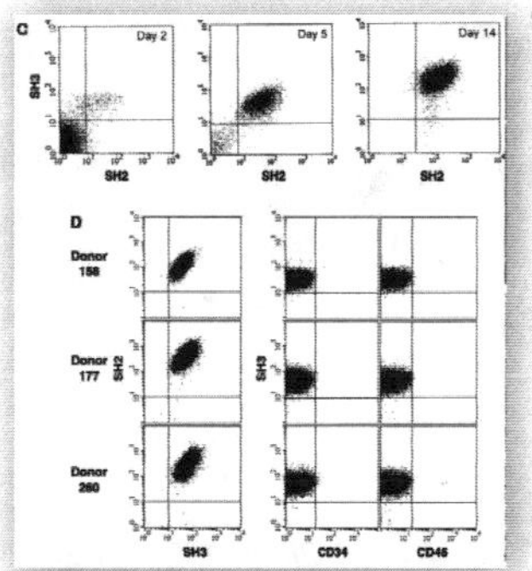

Mesenchymal Stem Cell Differentiation

(via chemical cues)

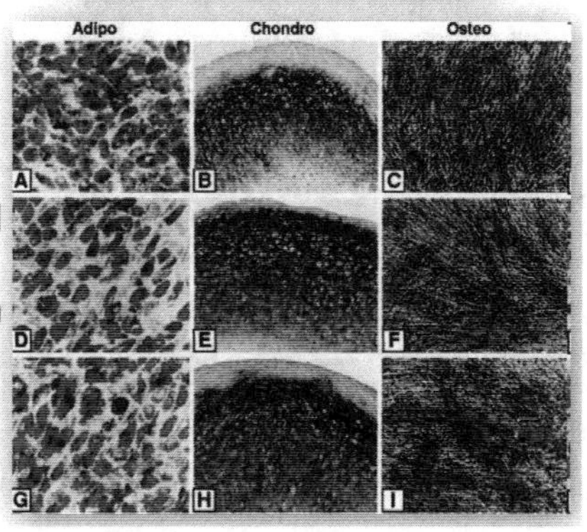

Pittenger et al., *Science* 1999

Cell Adhesion to Foams is Composition-Dependent

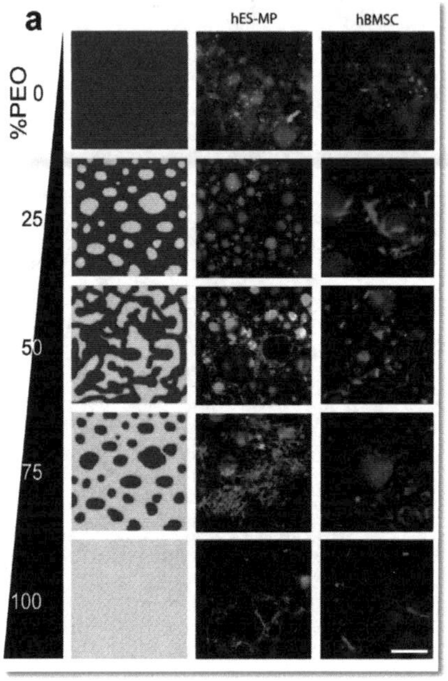

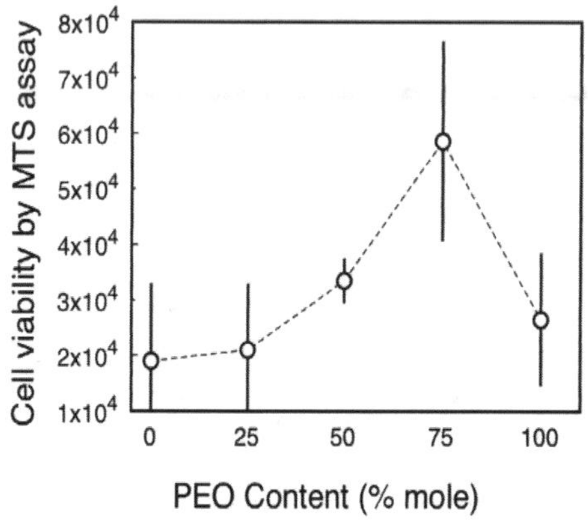

Viswanathan, Chriasatitsin et al, *JACS* (2012)

Cell Differentiation is Foam Composition-Dependent

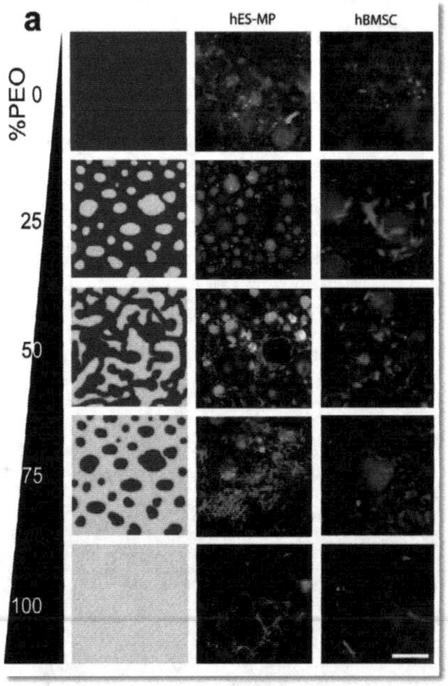

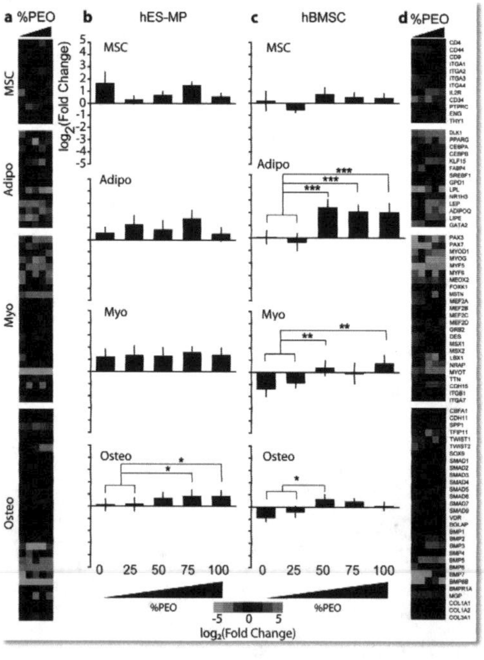

Cell Differentiation is Foam Composition-Dependent

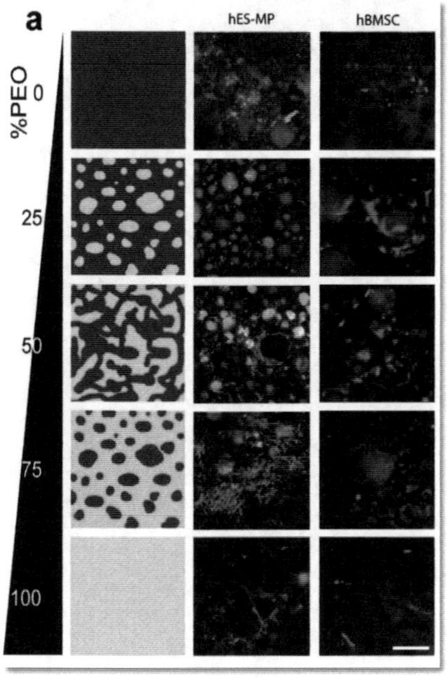

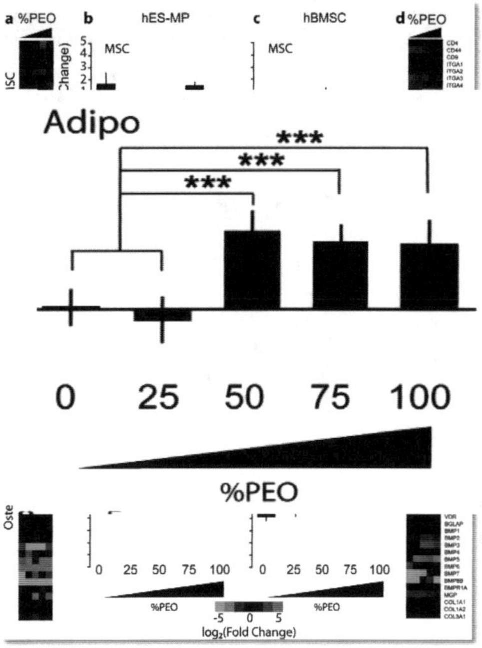

Composition-Dependent Lineages Reflect the Presence of Nano-domains

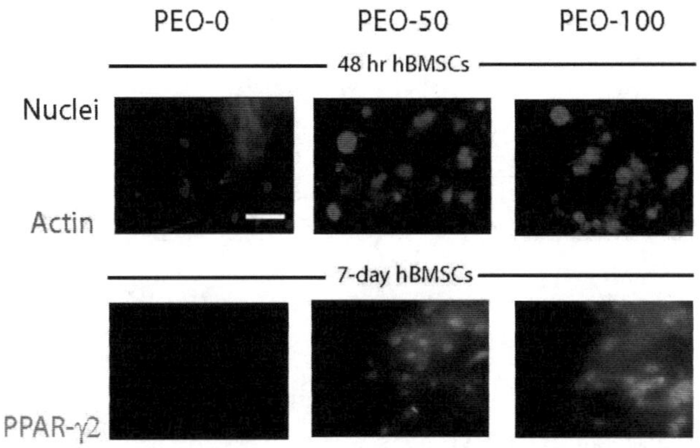

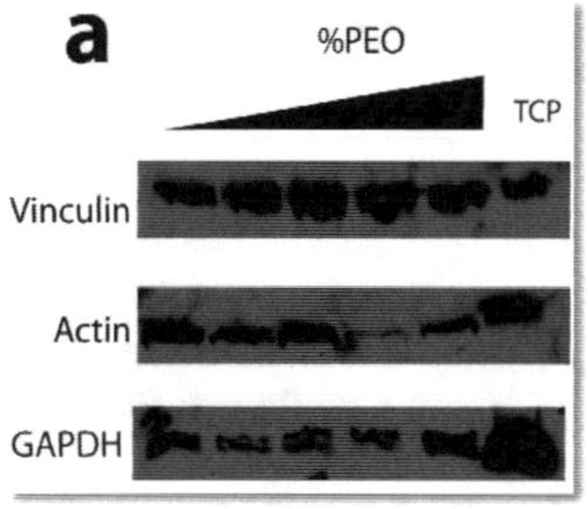

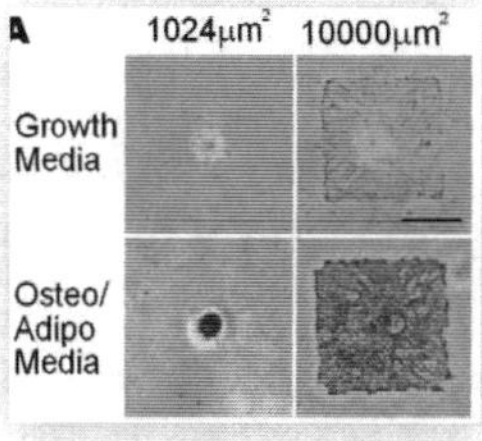

Adhesive Heterogeneity "Take Home Messages"

- High internal phase emulsions (HIPEs) with interface de-mixing create foams with nano-domains (~0.1 μm^2) that are spaced apart with the approximately the same heterogeneity as native matrix (36%)

Adhesive Heterogeneity "Take Home Messages"

- Stem cell adhesion is maximal on foams with nano-domains

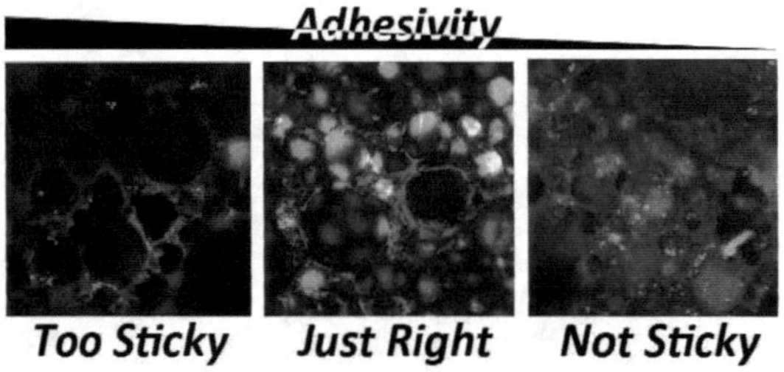

- Stem cell differentiation:
  - Adult stem cells express less contractile lineages in a nano-domain–dependent manner (adipo) and remain adhesive but unspread

Tissue Cells Interact With Their Substrate Mechanically

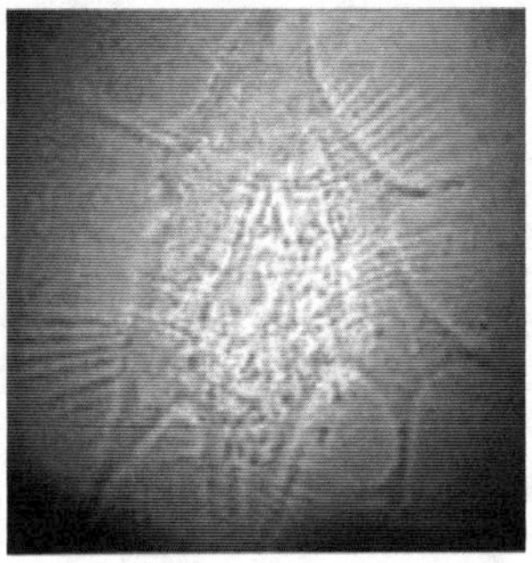

Wrinkling of a thin silicone film (A.K. Harris et al, Science 1980)

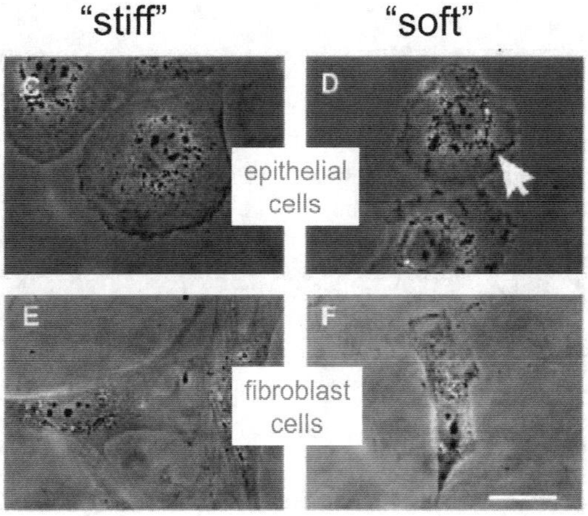

Changes in cell spreading (Pelham and Wang, PNAS 1997)

Ligand Matters in Differentiation

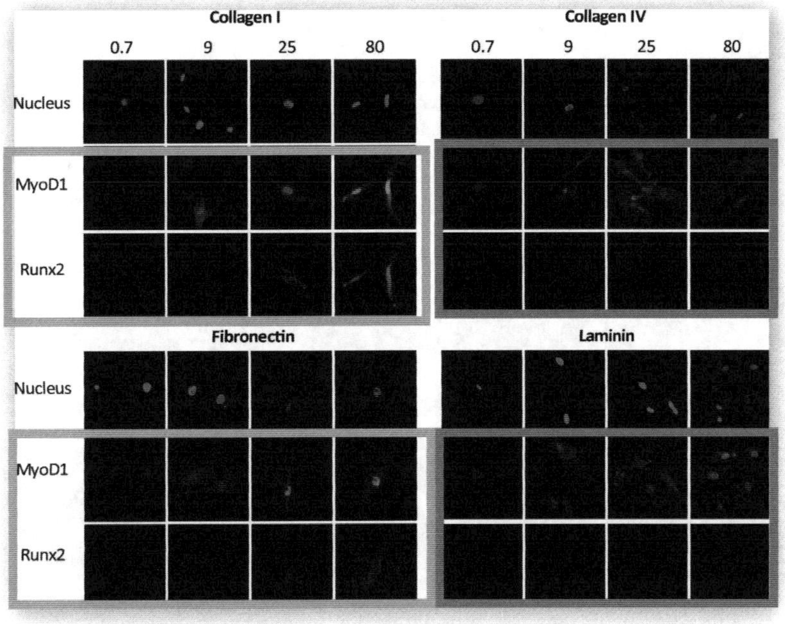

Rowlands et al., *AJP: Cell Physiol* 2008

<u>Matrisome</u>: cell must remain differentiated despite different matrix proteins (DMD vs. WT)

Ln(DMD/WT)	
2.46	Decorin
1.67	Type VI Collagen, alpha 2
1.10	Type VI Collagen, alpha 1
0.77	Type I Collagen, alpha 2
0.72	Type I Collagen, alpha 1
-0.10	Lumican
-0.12	Type V Collagen, alpha 2
-0.31	Type V Collagen, alpha 1
-0.32	Type II Collagen, alpha 1
-0.34	Type III Collagen, alpha 1
-0.34	Fibulin 2
-0.79	Fibulin 5
-2.47	Laminin, gamma 1
-2.71	Nidogen
-2.99	Type IV collagen, alpha 2
-3.06	Laminin, beta 2

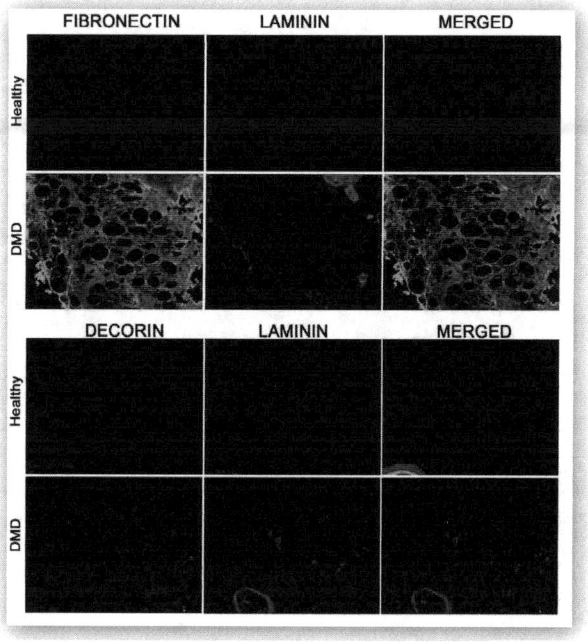

Stem Cell Fate Regulation: Matrix Ligand Tethering vs. Stiffness

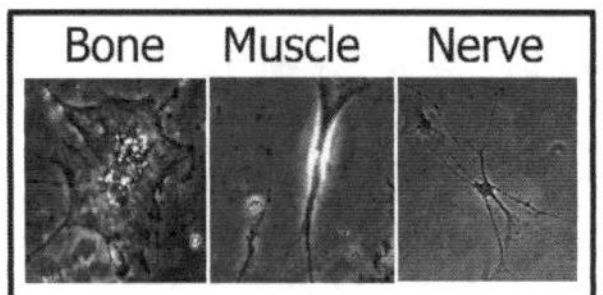

Engler et al, *Cell* 2006

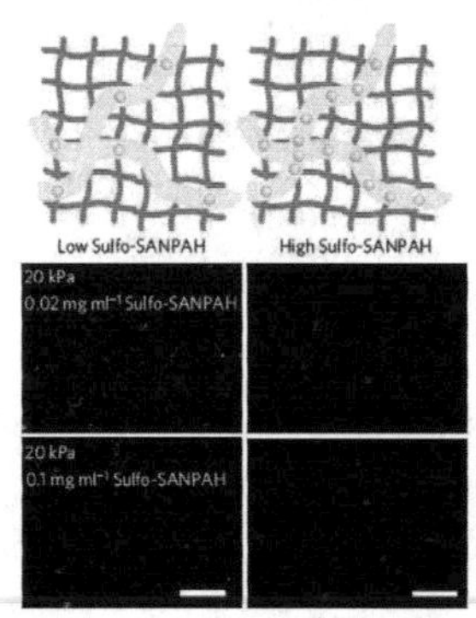

Trappmann et al, *Nat. Materials* 2012

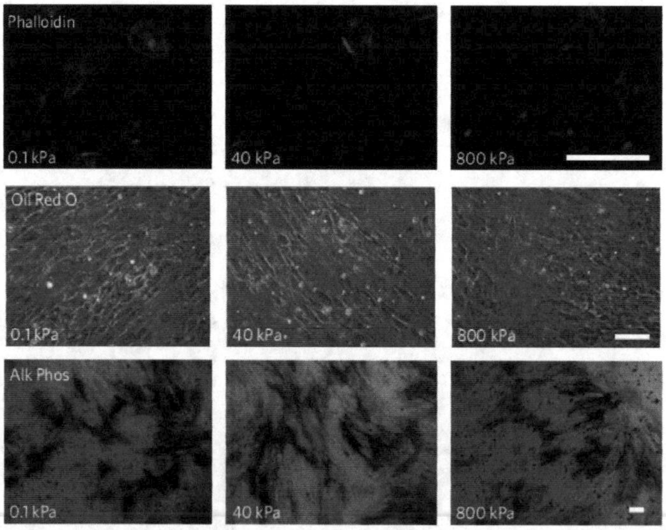

Stem Cell Fate Regulation: Matrix Ligand Tethering

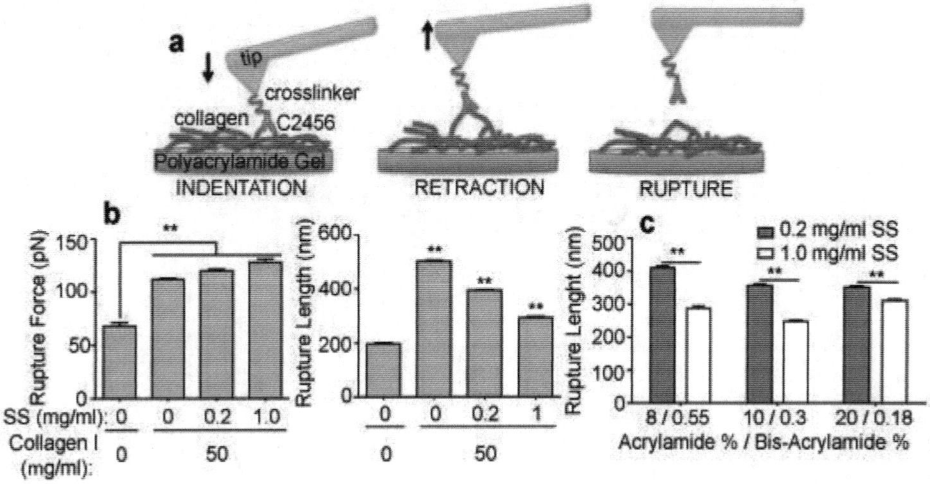

Stem Cell Fate Regulation: Matrix vs. Tether Deformation

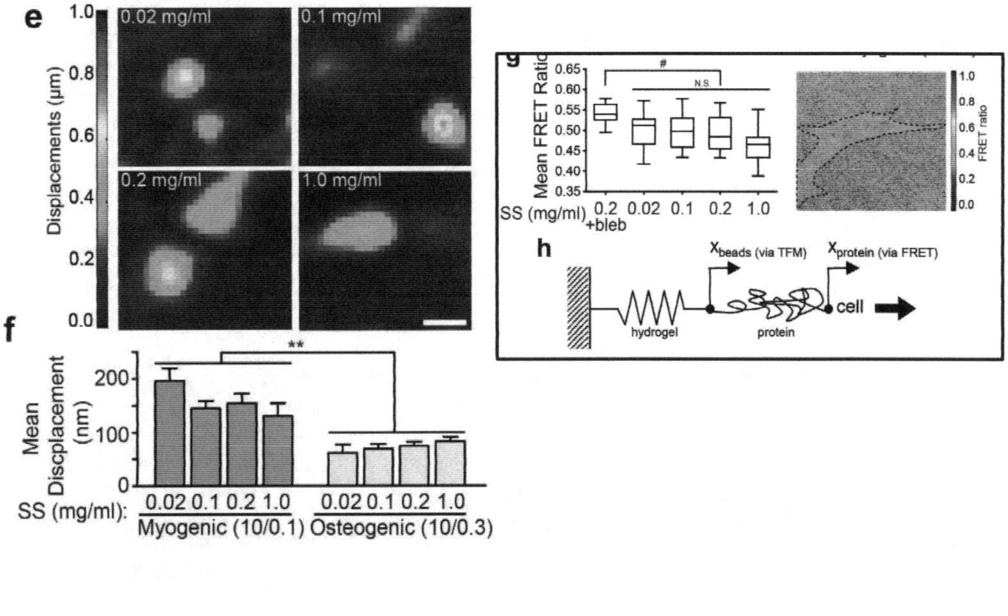

Stem Cell Fate Regulation: Matrix-based Differentiation is stiffness, not tether dependent

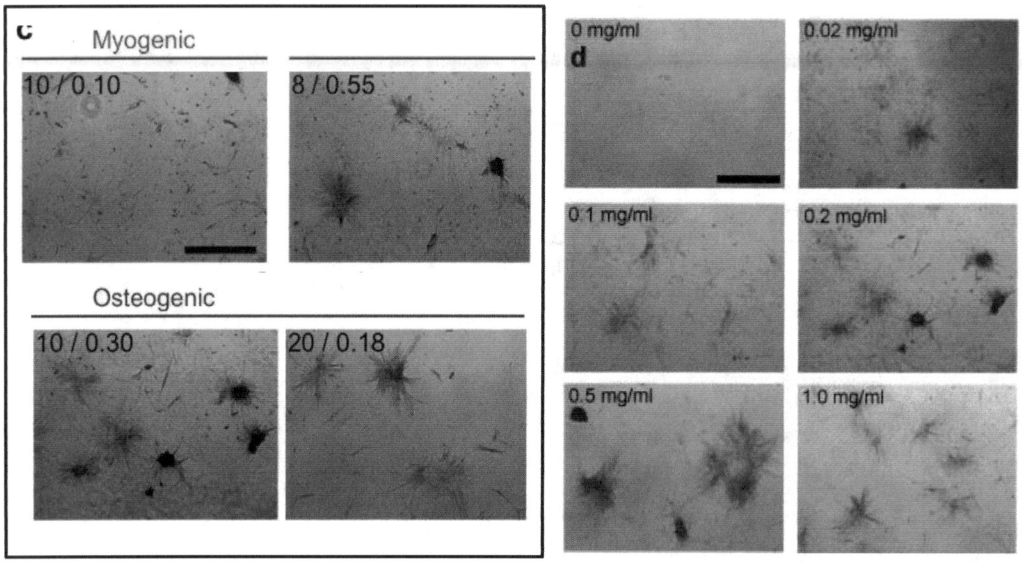

Even in tether-less systems, stiffness induces differentiation

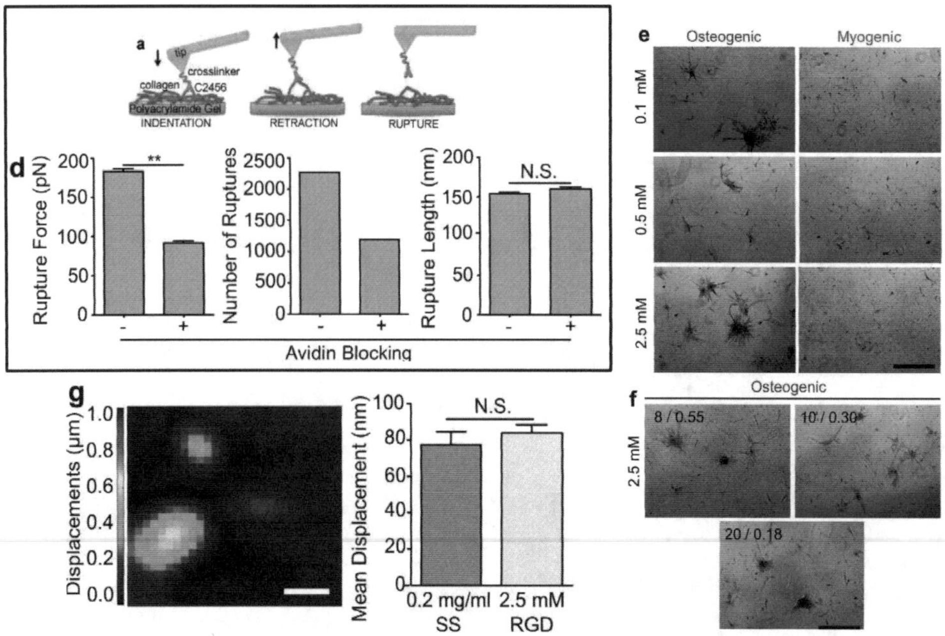

Matrix Stiffness vs. Tethering "Take Home Messages"

- AFM:
  - Differences in tether length either with changing linkage density or modulating pore site
- Cells:
  - Deform their underlying substrate hundreds of nanometers to microns (TFM)
  - Cells do not differentially pull tethers (FRET)
- Differentiation:
  - No observable difference from tethering (outside of no linkage at all) either by tethered or tether-less systems
- Suggest that strain intracellular sensors interpret deformation differences.

Vinculin Knockout Requires MAPK1 site in the Head Domain

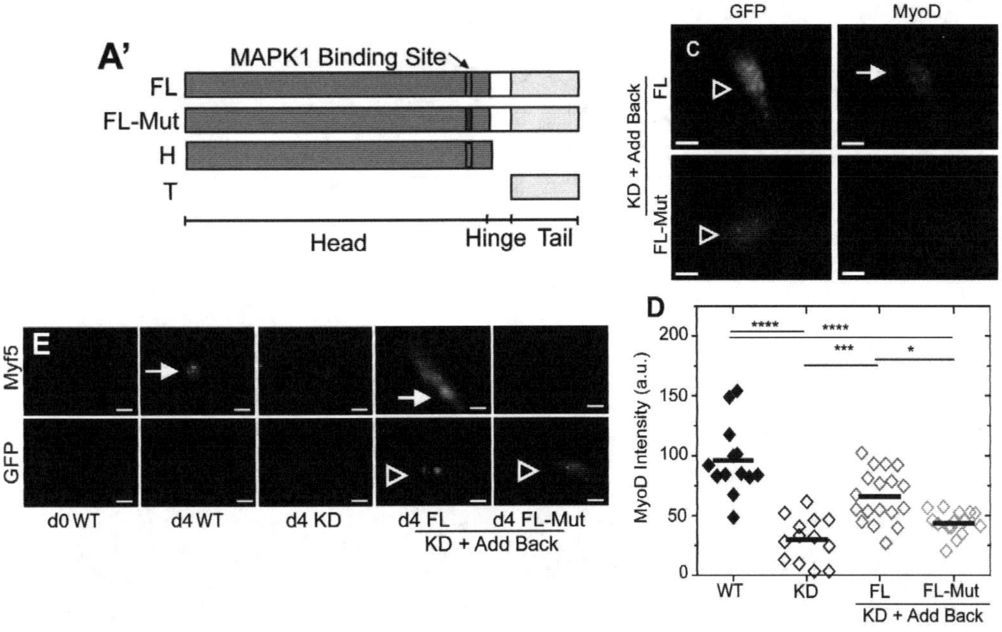

Holle et al, *Stem Cells* 2013

Acknowledgements

Current: (*Lab Manager*) **Jerry Tuler**; (*Postdocs*) Alexander Fuhrmann, **Gretchen Meyer**; (*PhD Students*) Ludovic Vincent, **Gaurav Kaushik**, Hermes Taylor-Weiner, Jessica Wen, Matthew Ondeck, **Ayla Sessions**, **Kelsey Thomas**, Aditya Kumar, **Michael Gibbons**; (Masters) Jacquelyn Schaefer; (*Undergraduates*) **Andrew Lee**, Neerja Ravi

Past: (*Postdocs*) **Yu Suk Choi**; (*PhD Students*) **Somyot Chirasatitsin**, Jennifer Young, Andrew Holle; (*Masters Students*) Xinyi Tang, Jennifer Kiang, **Alice Spenlehauer**; (*Undergraduates*) Justin Tse, Deepthi Vijayraghavan, Nikhil Joshi, **Sarah Richardson**, Kyle Kretchmer

Collaborators: (*UCSD*) Shu Chien, **Marek Dobke**, **Sam Ward**, Juan Carlos del Alamo, Alex Zambon, Karen Christman, Robert Ross, **Jing Wang**; (*UCSF*) Thea Tlsty; (*Stanford*) Joseph Wu; (*Johns Hopkins*) **Anthony Cammarato**; (*Sanford-Burnham*) **Rolf Bodmer**, **Alessandra Sacco**; (*SDSU*) **Sandy Bernstein**; (*UCL, UK*) **Giuseppe Battaglia**; (*NTU, Singapore*) Lay Poh Tan

Funding:
NIH: DP02 OD006460 (New Innovator Award), R01 HL103566, P30 AR061303, R01 AG045428
NSF: GRFPs
DoD: BC121670P1 (IDEA Award)
Human Frontier Science Program: RGY0064/2010
CIRM: Bridges Program
Muscular Dystrophy Association: 241665

NANOMATERIALS, MECHANOBIO, ASCs, FLIES

Biological, Chemical, and Mechanical Interfaces: "Building Blocks" for Cells

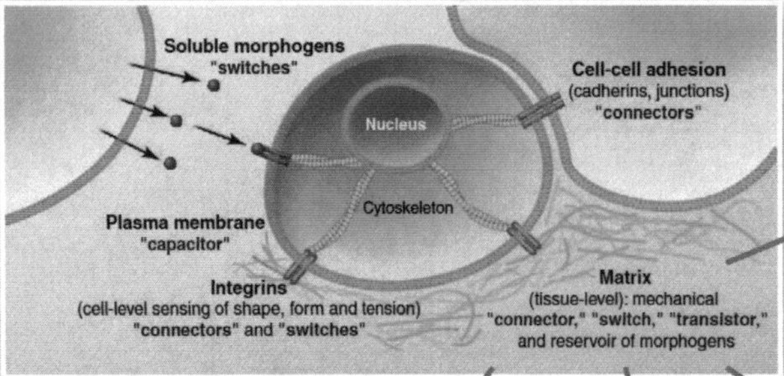

Engler et al, *Science* (2009)

Cell Shape

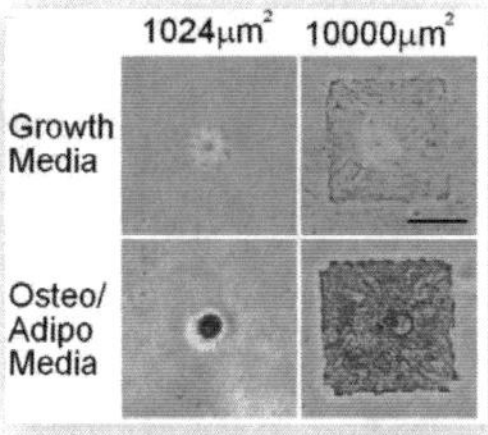

McBeath, et al *Dev Cell* (2004)
Fu, et al *Nat Methods* (2010)

Matrix Ligand

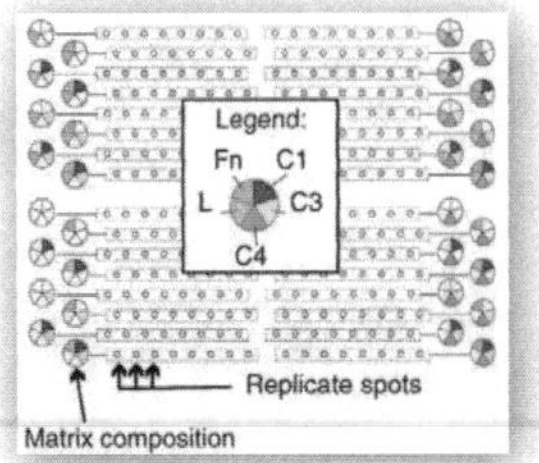

Ott, et al *Nature Med* (2008)
Flaim, et al *Nature Meth* (2005)

Topography

Dalby, et al *Nature Mat* (2007)
Oh, et al *PNAS* (2009)

Matrix Elasticity

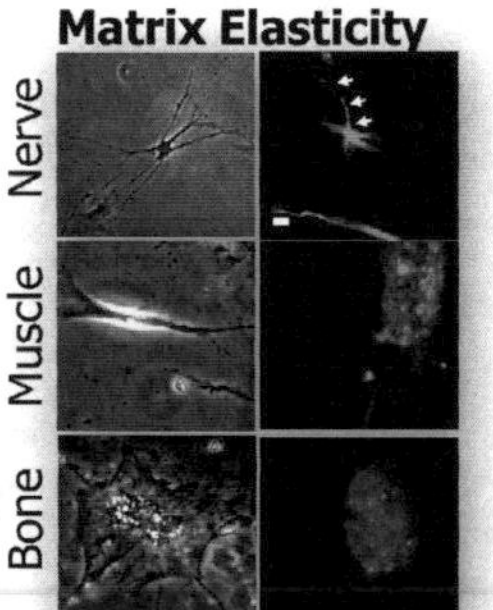

Engler, et al *Cell* (2006)
Gilbert, et al *Cell* (2010)

Force As a Regulator of Mesodermal Differentiation

Taby Ahsan, Ph.D.

Burk-Kleinpeter Professor
Department of Biomedical Engineering
Tulane University

www.tulane.edu/~tahsan

Role of Physical Cues in Stem Cell-based Therapies

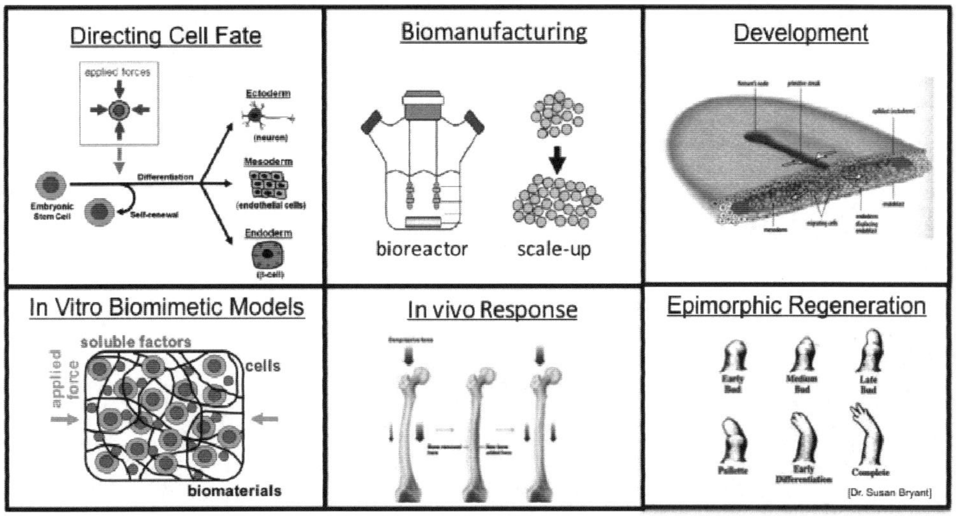

Mechanical Environment *in vivo*

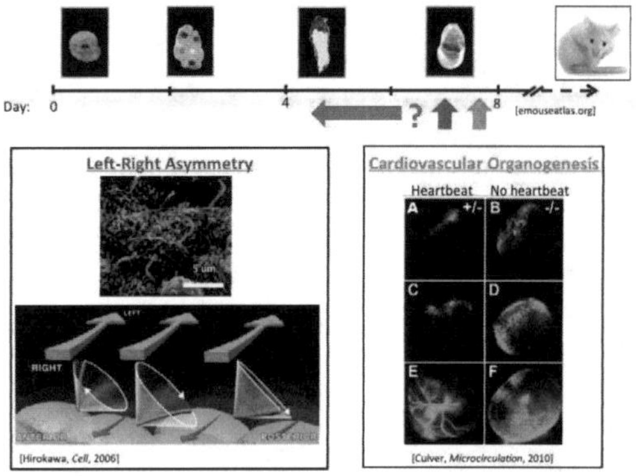

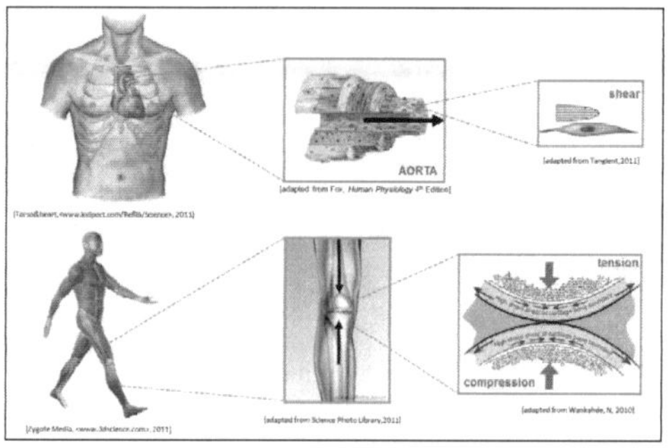

During development, organogenesis and left-right asymmetry have been shown to be shown to depend on mechanical factors.

Endothelial cells and chondrocytes, of the mesodermal lineage, are exposed to a dynamic mechanical environment in adult physiology.

Deformations During Early Development

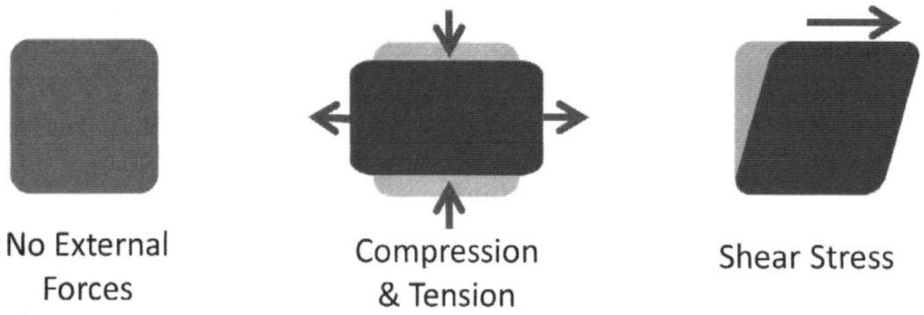

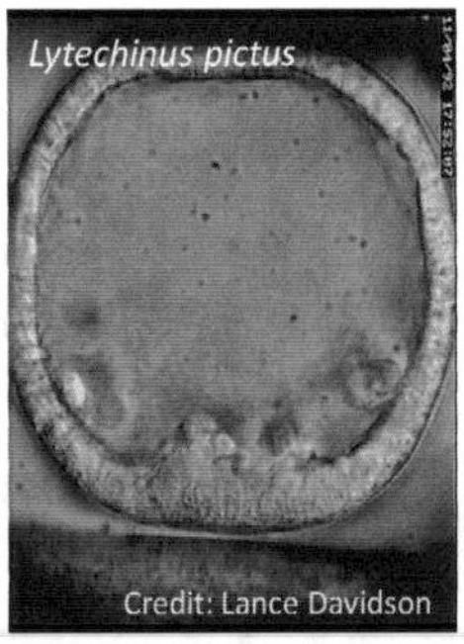

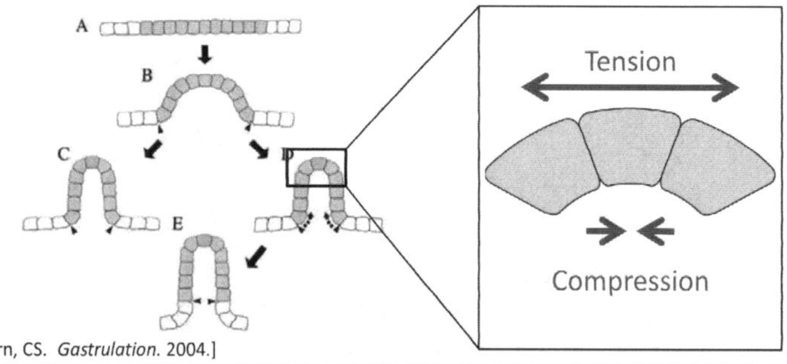

[Stern, CS. *Gastrulation.* 2004.]

Germ Lineage Specification

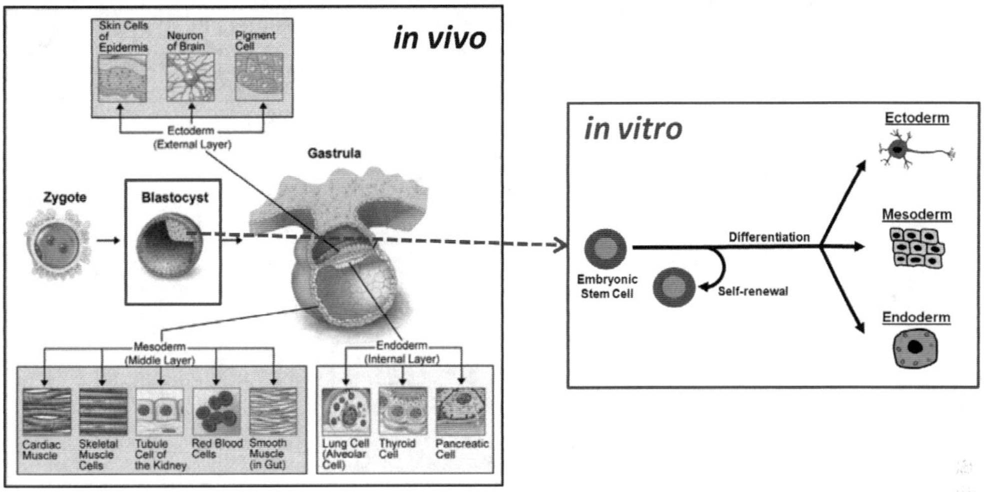

[ncbi.nlm.nih.gov/About/primer/genetics_cell]

2D Mechanical Modulation *In Vitro*

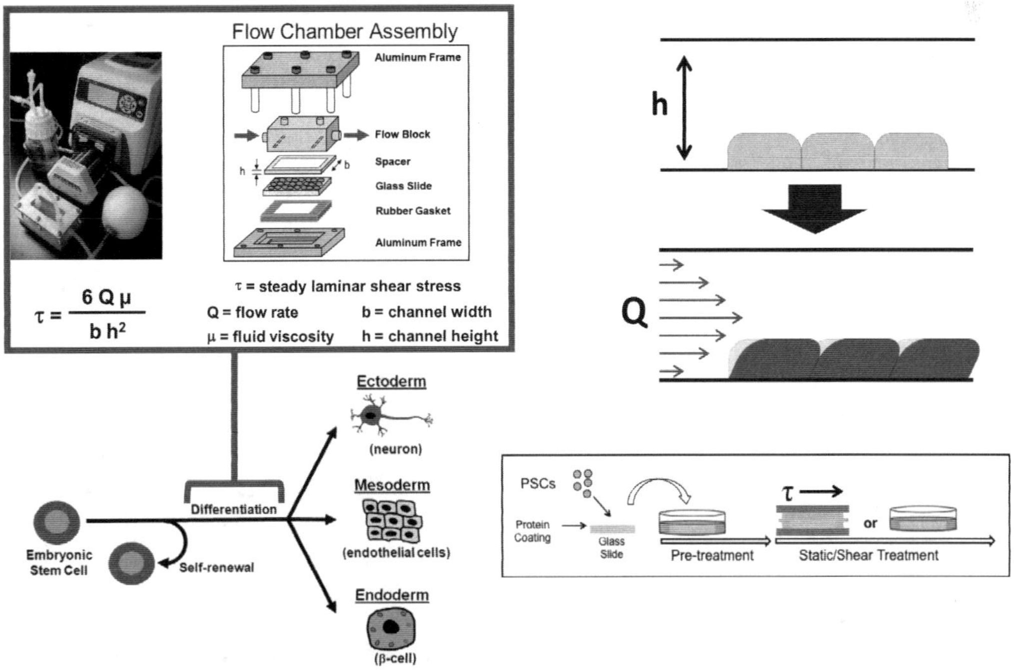

Effects of Applied Shear Stress on Germ Specification

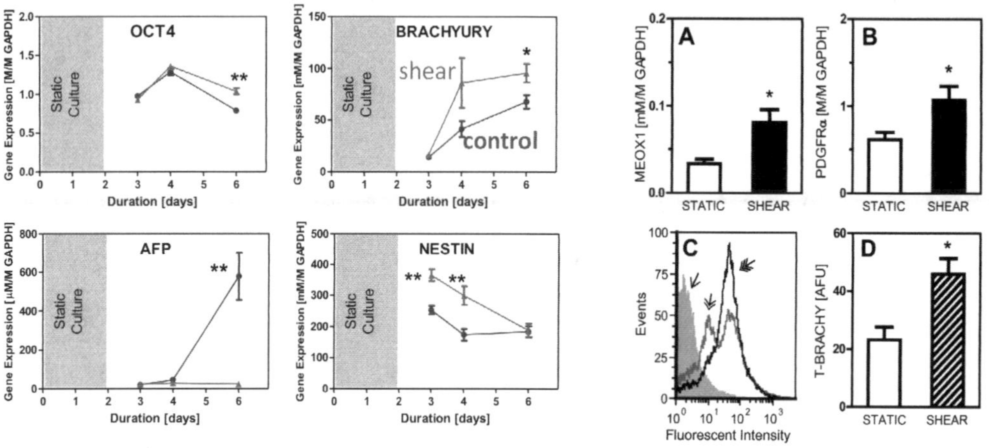

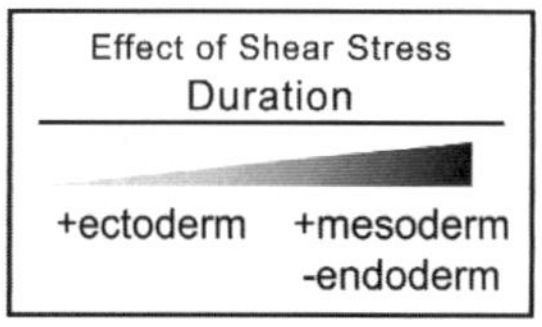

Shear stress applied to ESCs for 4 days induces mesodermal differentiation.

[Wolfe et al, *Integrative Biology* 2012]

Shear Stress for Generating Mesodermal Phenotypes

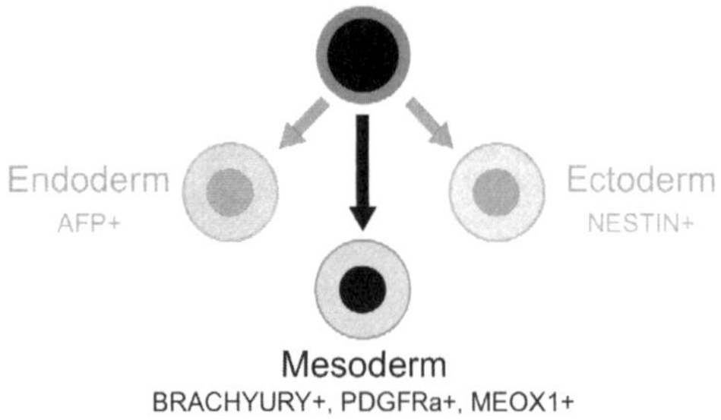

Shear stress promotes mesodermal specification.

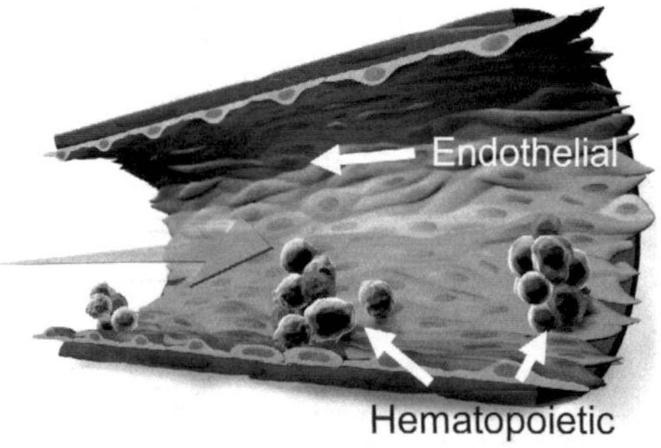

[Adamo & Garcia-Cardena, Dev Biol, 2012]

Can applied shear stress promote downstream phenotypes?

Effects of Treatment Duration and Protein Substrate

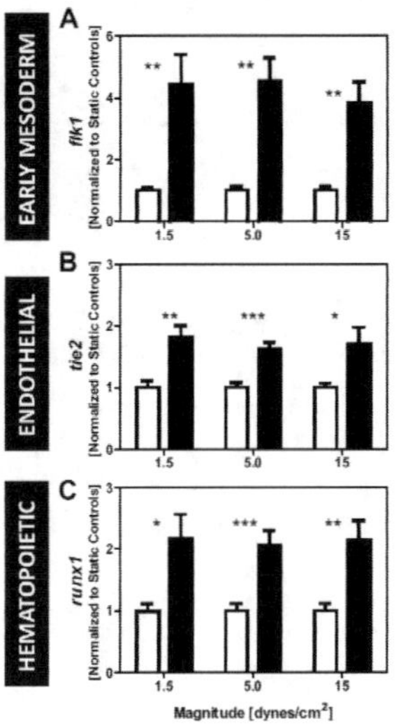

Applied shear stress promotes early endothelial and hematopoietic differentiation.

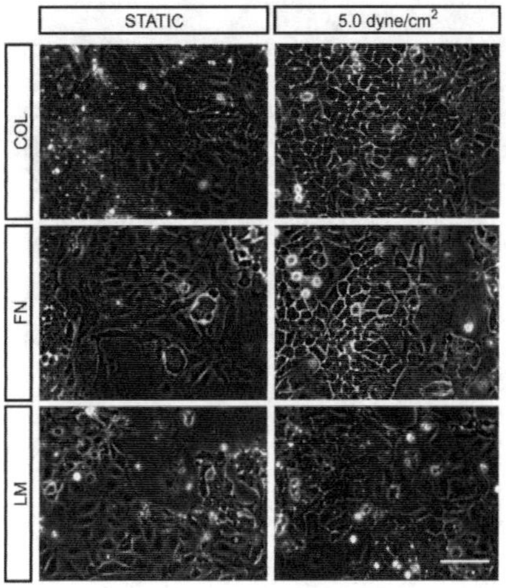

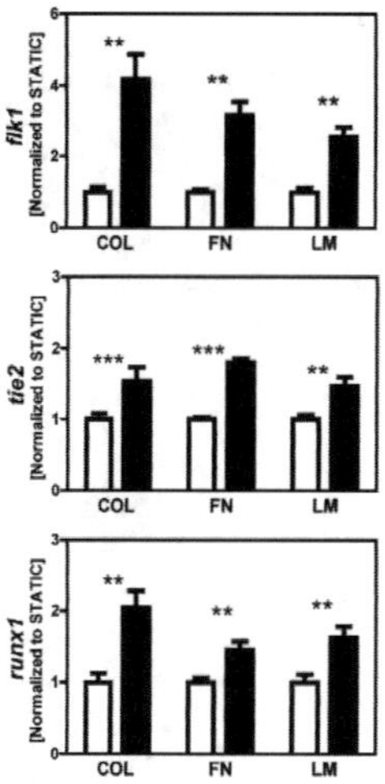

The response to shear stress is consistent across a number of protein substrates, including collagen type IV, fibronectin, and laminin.

[Wolfe & Ahsan, *Biotech Bioeng* 2013]

Moving towards Generating Functional Phenotypes

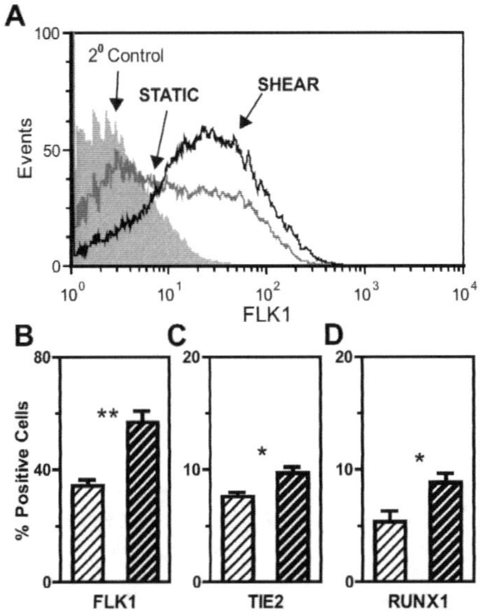

Shear stress applied for 4 days promotes protein expression of endothelial and hematopoietic markers.

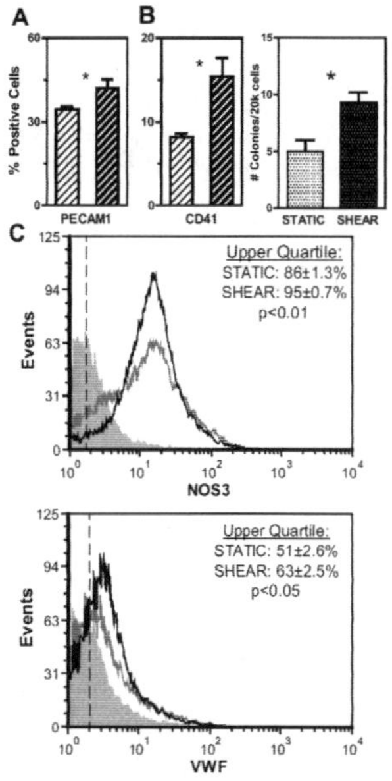

Further maturation can also promote functional markers.

[Wolfe & Ahsan, *Biotech Bioeng* 2013]

Change in Mechano-sensitivity with Differentiation

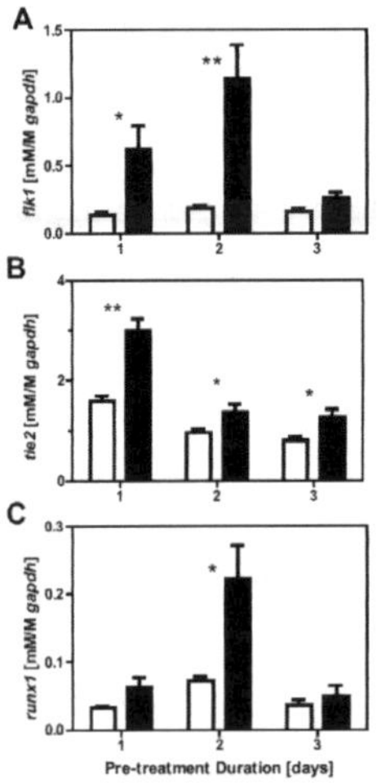

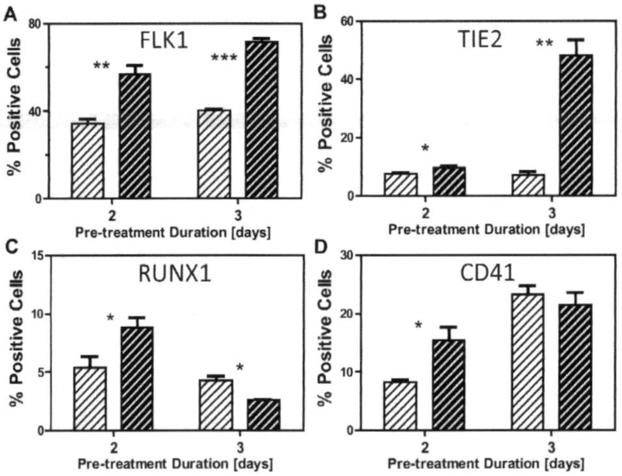

After 3 days of initial differentiation, shear stress still promotes an endothelial phenotype but no longer hematopoietic differentiation.

[Wolfe & Ahsan, *Biotech Bioeng* 2013]

Cytoskeleton & Shear Stress Response

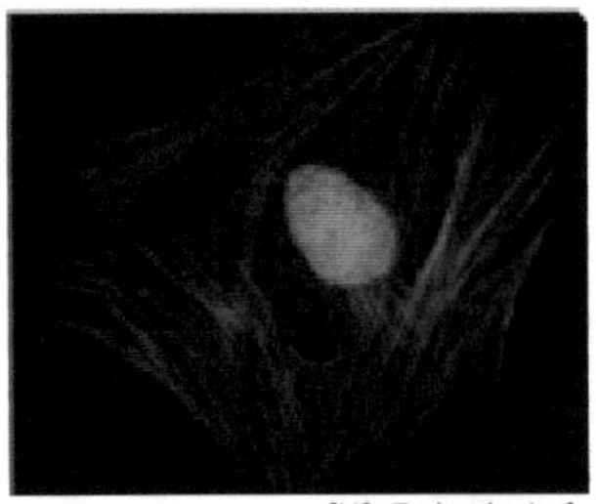

[Life Technologies]

ENDOTHELIAL CELLS

Shear stress alters signal transduction and gene & protein expression.

[Fisher, Am J Phys Lung Cell Mol Phys '01]
[Chien, Hypertension '98]

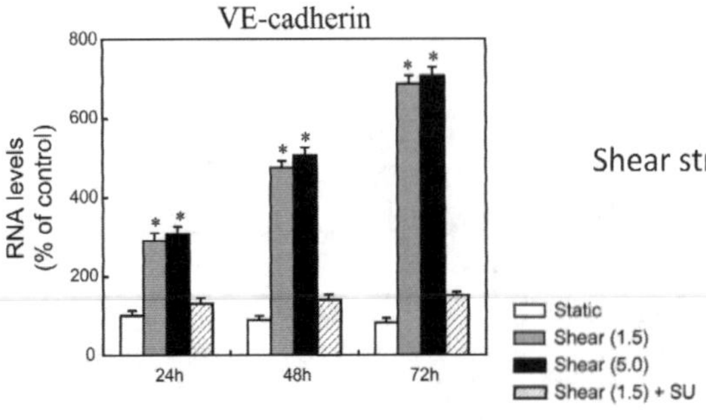

FLK1+ PROGENITOR CELLS

Shear stress effects are also magnitude dependent.

[Yamamoto, Am J Phy Ht Circ Phy '05]

Changes in the Cytoskeleton with ESC Differentiation

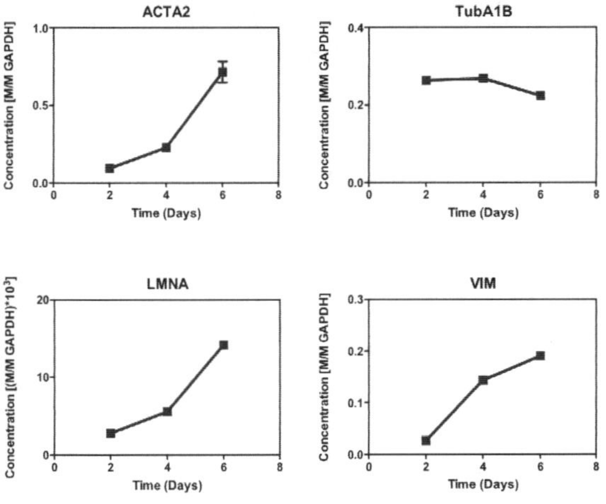

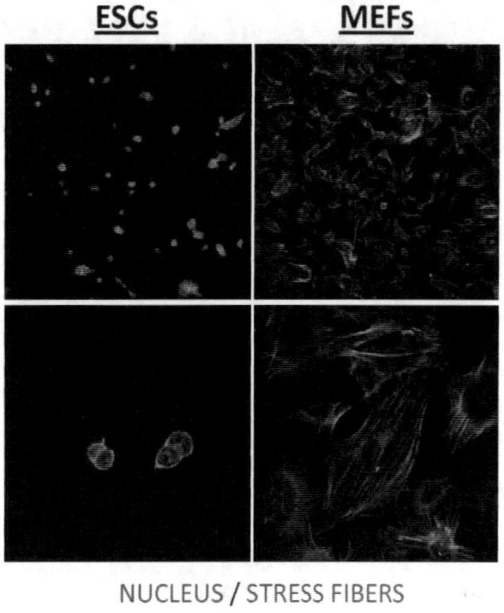

Increases in cytoskeletal expression in mouse ESCs with differentiation.

Relative differences in protein expression of phalloidin with MEFs.

Force Transfer Across the Cell Membrane

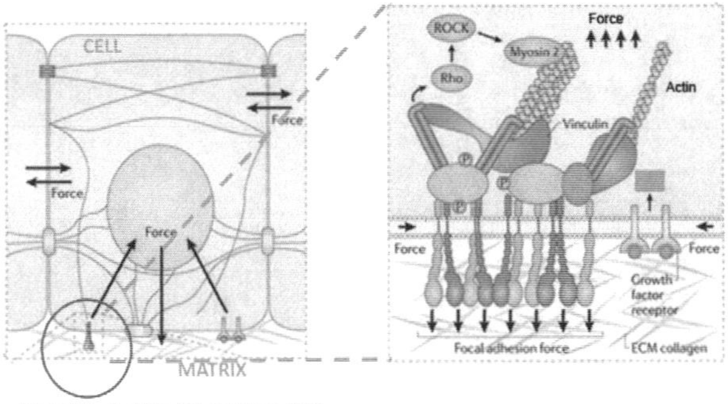

[adapted from F. Michor, *Nature Reviews*, 2011]

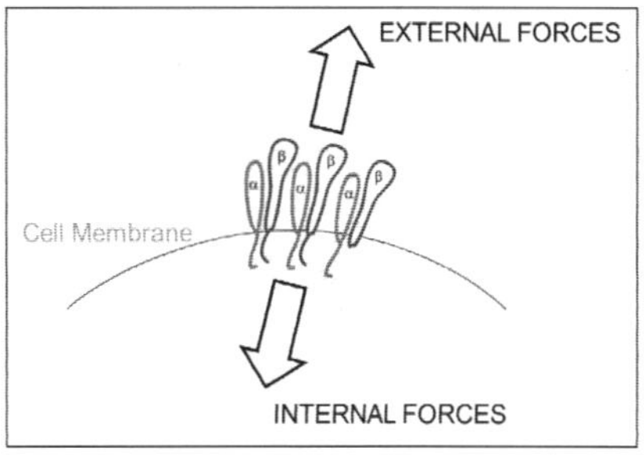

- Equilibrium across cell boundaries is reached through the balance of externally applied force and internally generated force
- Force balance likely includes transmembrane proteins, such as integrins.

Quantifying Protein Expression in Heterogenous Stem Cell Populations

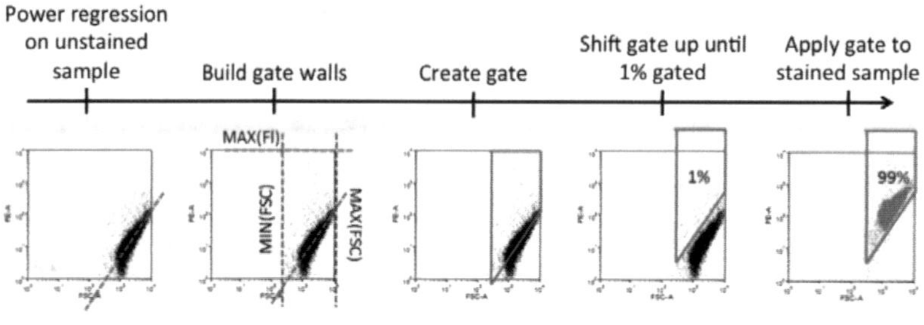

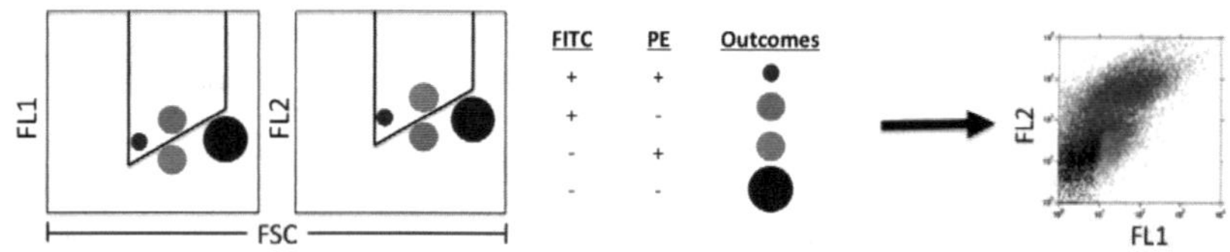

Changes in Integrin Expression with Differentiation

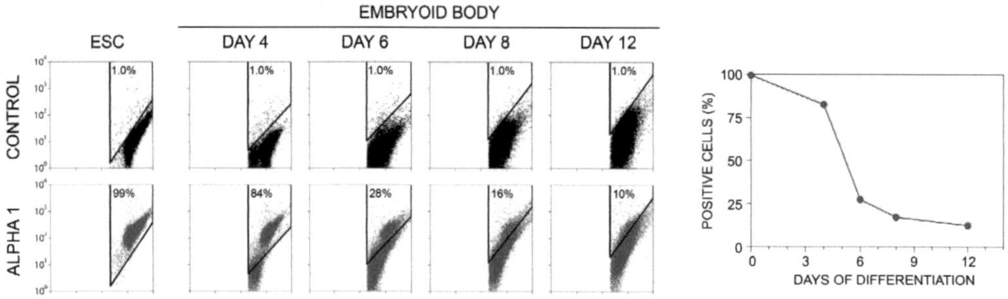

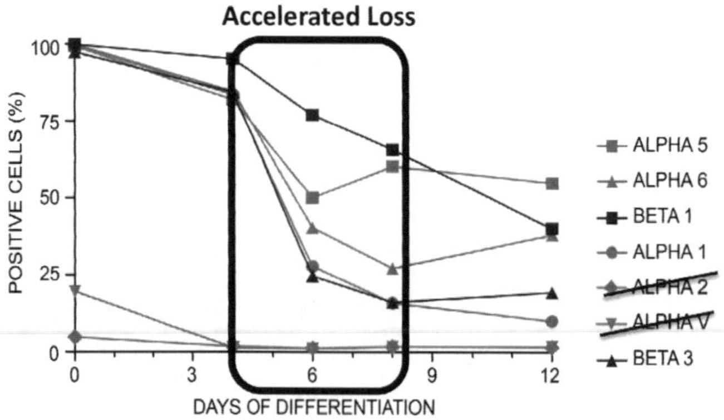

Integrin Expression in Subpopulations

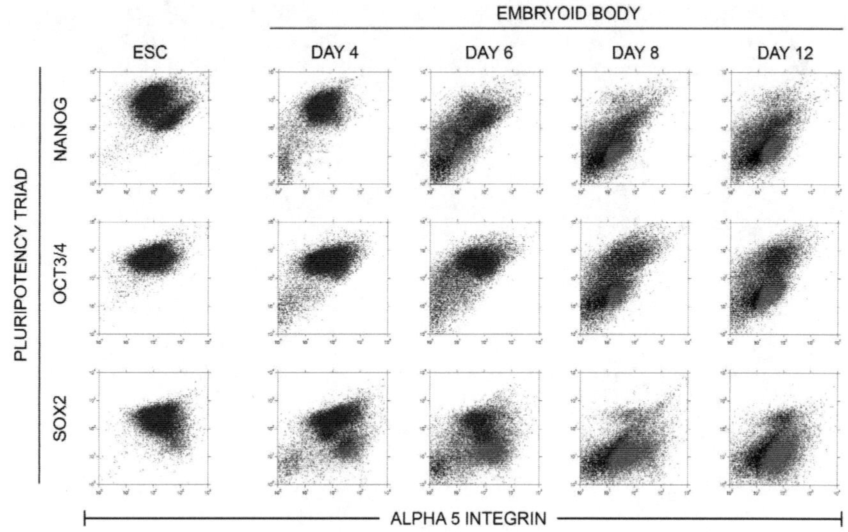

Whole Population Expression Kinetics

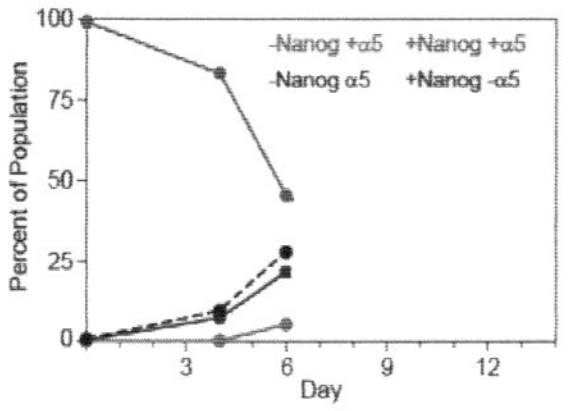

Days 0 – 6

- $\alpha 5$ Integrin is expressed in pluripotent cells
- $\alpha 5$ integrin is lost faster than Nanog with differentiation

Sub-population Kinetics

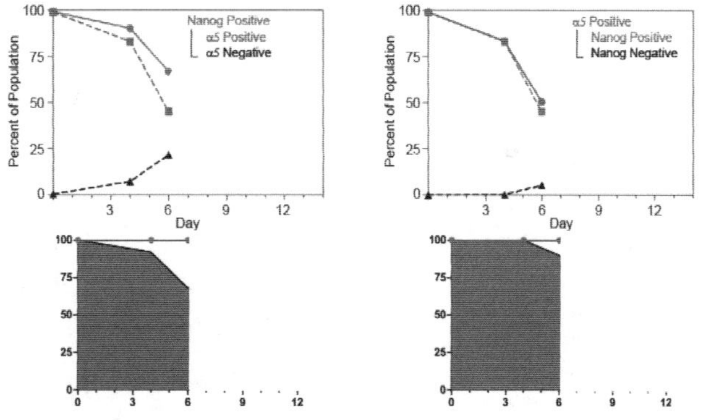

Whole Population Expression Kinetics

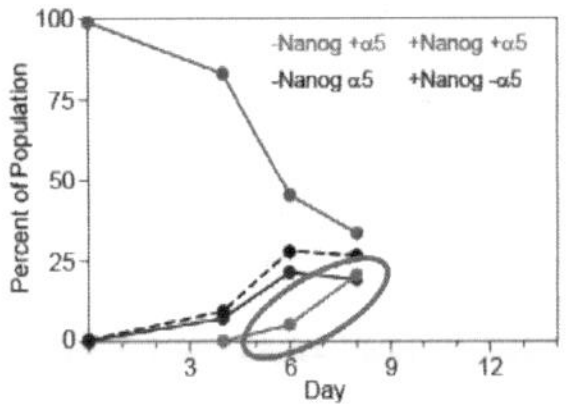

Days 6 - 8

- Emergence of α5 Integrin in differentiated cells

Sub-population Kinetics

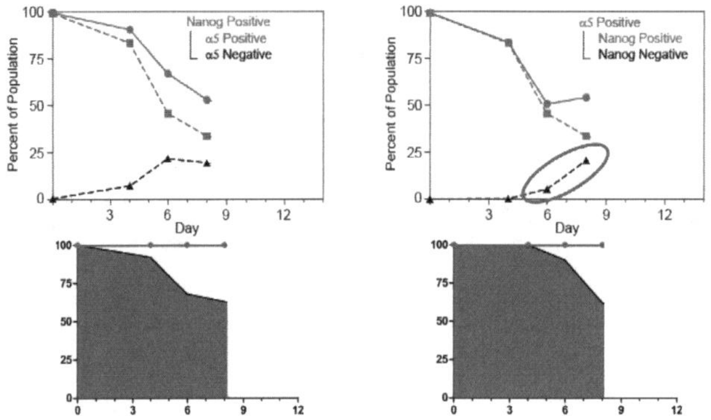

Whole Population Expression Kinetics

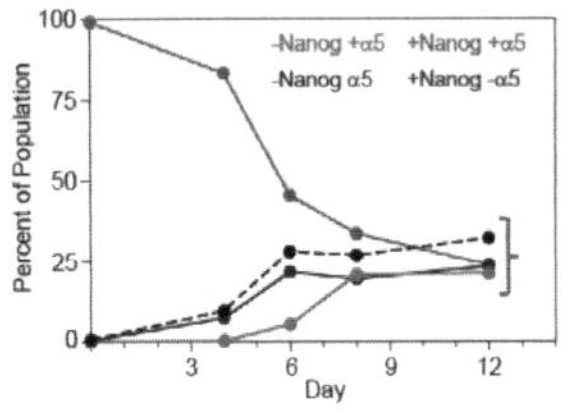

Days 8 – 12

- α5 integrin expressed equally in pluripotent and differentiated cells

Sub-population Kinetics

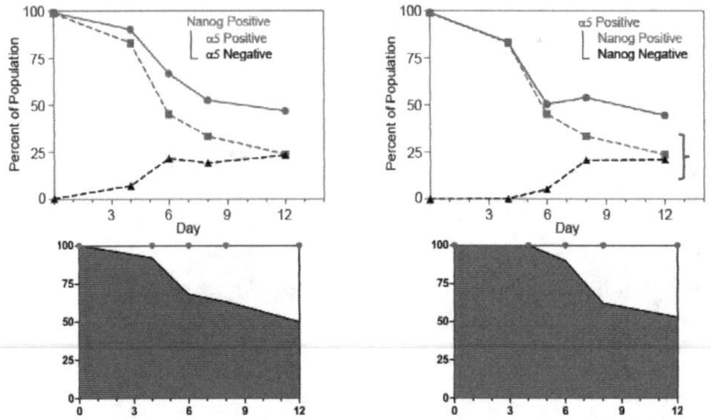

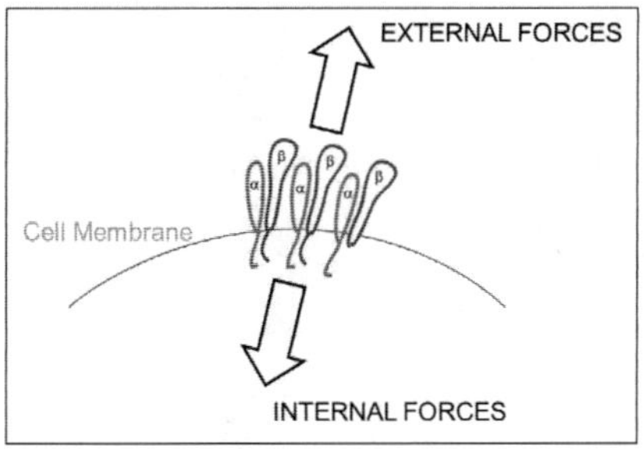

We saw changes in mechano-sensitivity with differentiation. Is this due to observed changes in the cytoskeleton, force transmitters, or mechanosensors?

How do changes in forces via actin-myosin motors affect differentiation?

Inhibiting Actin-Myosin II in Differentiating ESCs

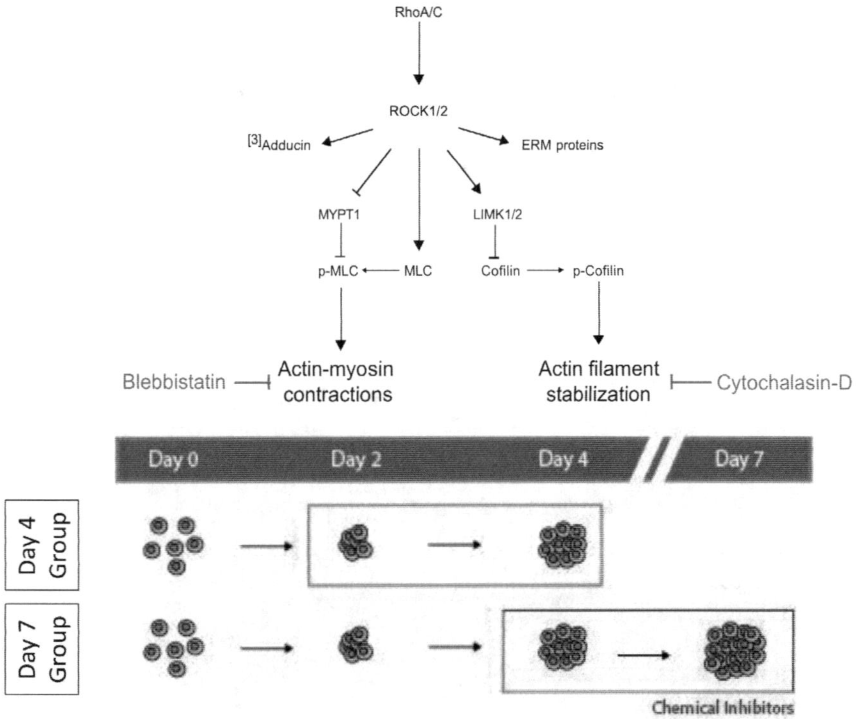

Changes in EB size

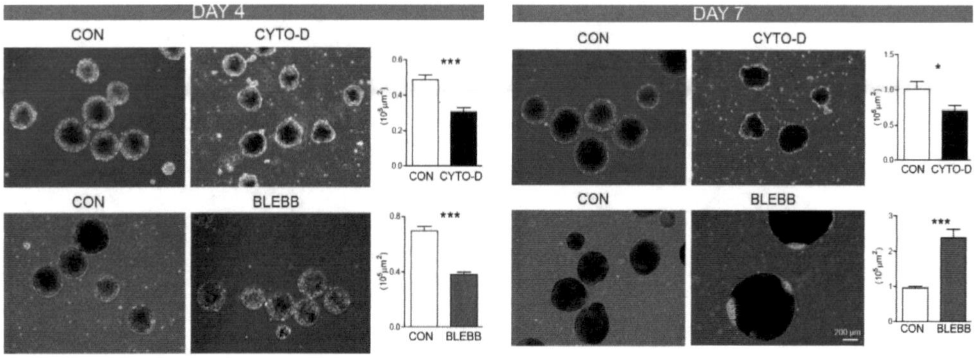

- CYTO-D decreased the size of EBs when applied early and later in differentiation.
- BLEBB had different effects early (decreased size) and late (increased size).

Morphological Changes

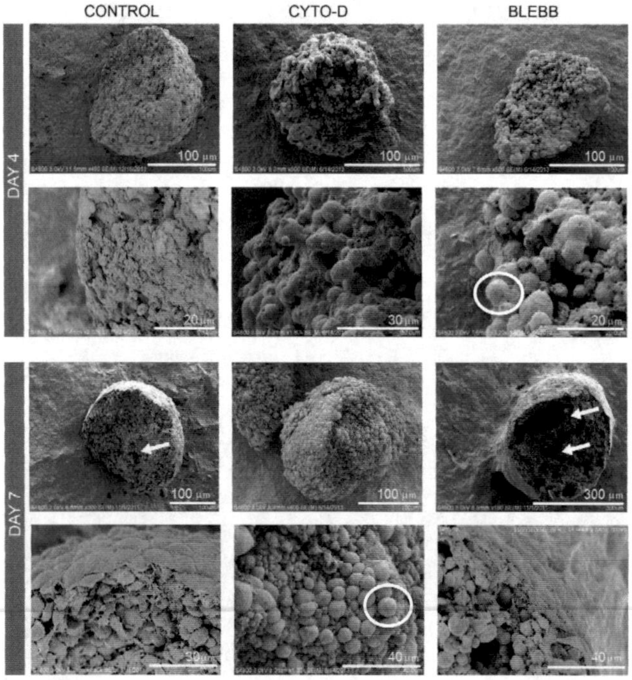

Changes in Pluripotency

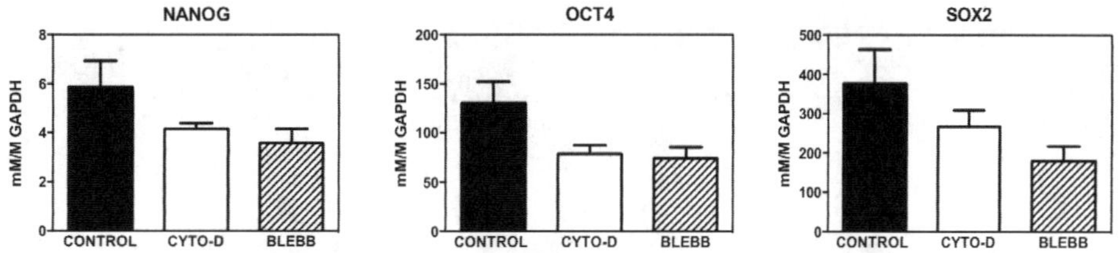

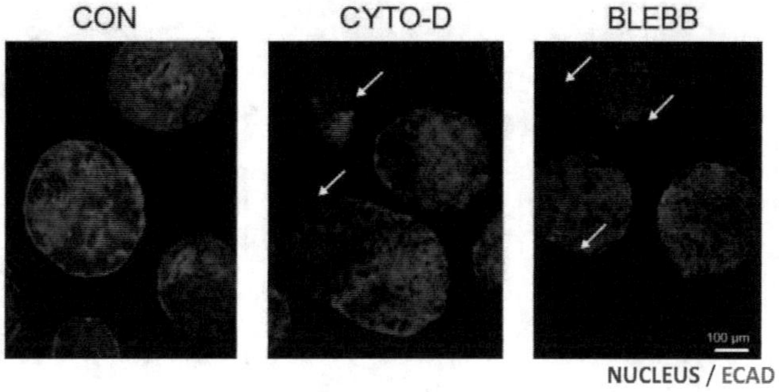

Actin and Myosin II inhibitors decrease pluripotency under differentiating conditions, while ROCK and Myosin II inhibitors maintain pluripotency under maintenance conditions [Harb, *PloS one*, 2008; Walker, *Nature Communications*, 2010].

Inhibitors Lowered Mesendodermal Commitment

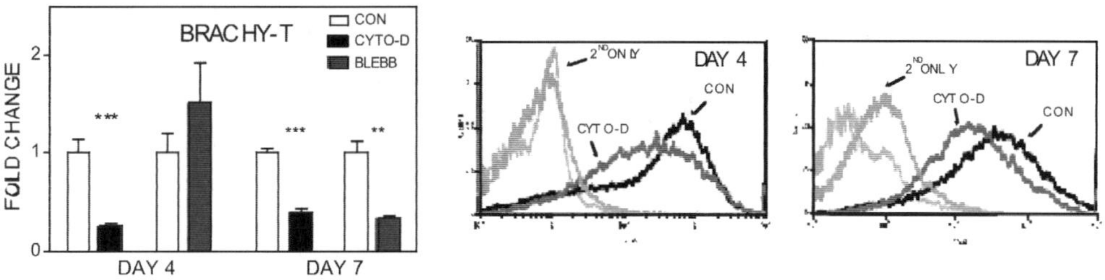

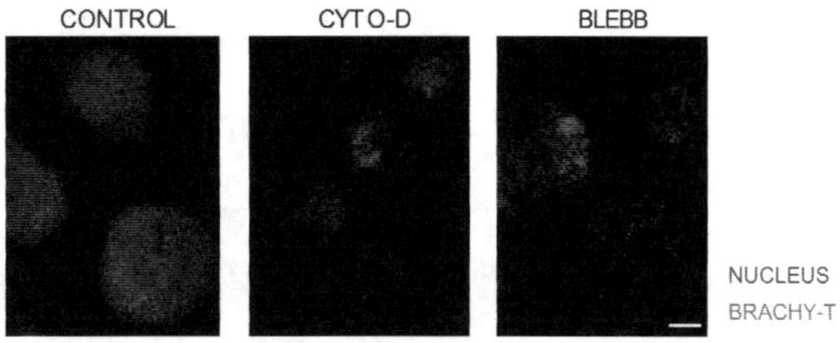

- BRACHY-T, a marker of mesendodermal commitment was decreased by both CYTO-D and BLEBB at Day 7.
- Protein distribution was altered in treated samples at Day 7.

Inhibitors Modulate Early Mesodermal Differentiation

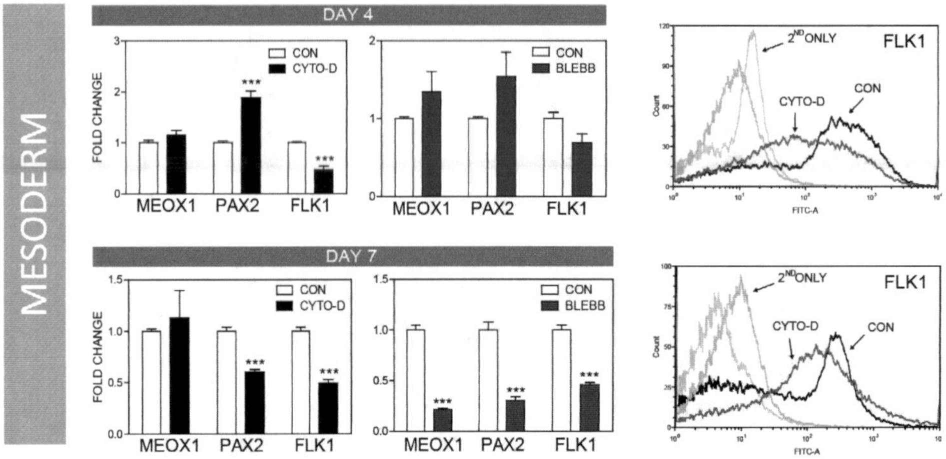

- At earlier time points, CYTO-D had higher expression of the intermediate mesoderm marker PAX2 and lower expression of the lateral plate mesoderm marker FLK1.
- No changes were detected with BLEBB in any of the markers assessed when applied from Day 2-4.
- CYTO-D and BLEBB decreased early mesodermal marker expression at Day 7.

EBs as a Model of Development

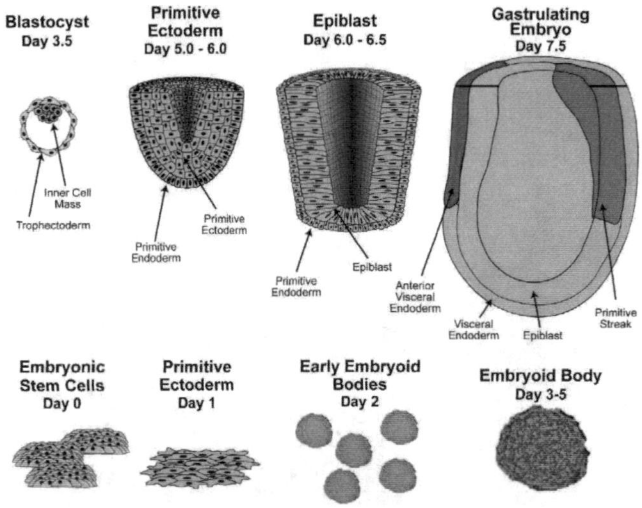

[Keller *Gene Dev* '05]

What is the role of actin-myosin II motors during gastrulation events?

Cell-Matrix vs Cell-Cell Interactions

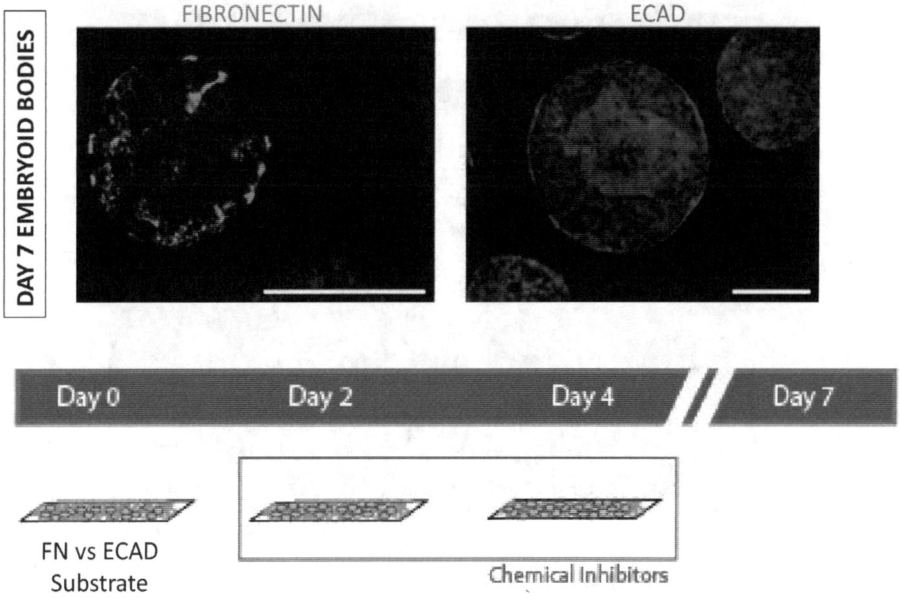

Isolating Integrin Binding

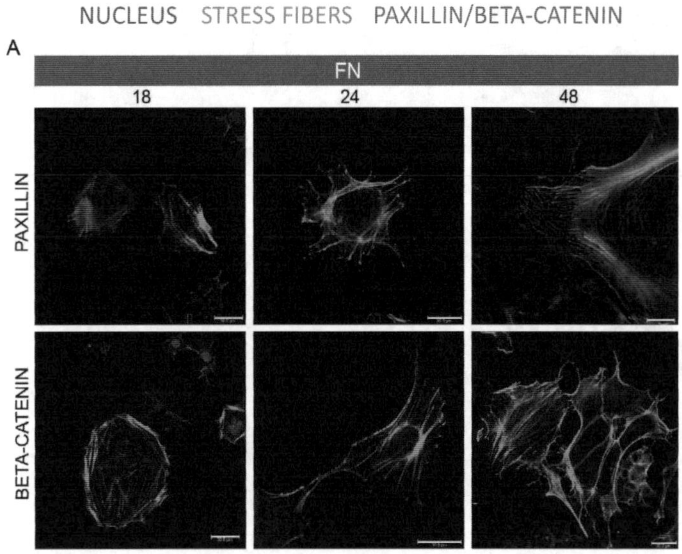

Cells on fibronectin recruited paxillin to their focal adhesions, not beta-catenin.

Isolating Cadherin Binding

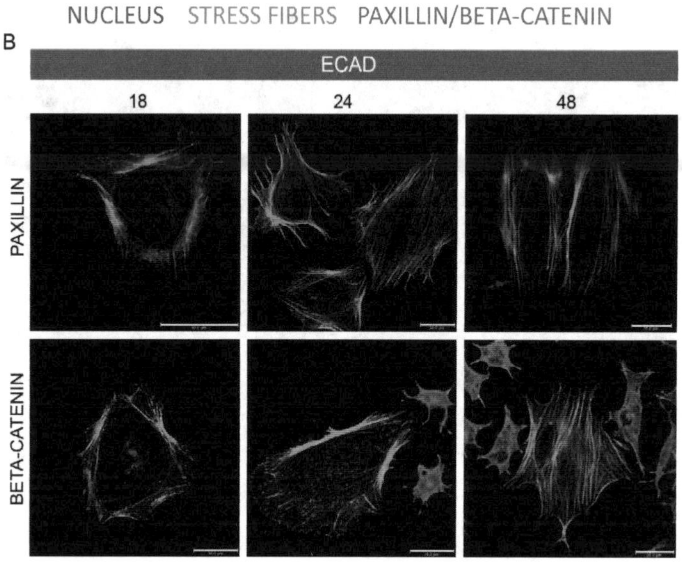

Cells on ECAD recruited beta-catenin to their cell adhesions, not paxillin.

Isolating Integrin vs Cadherin Interactions

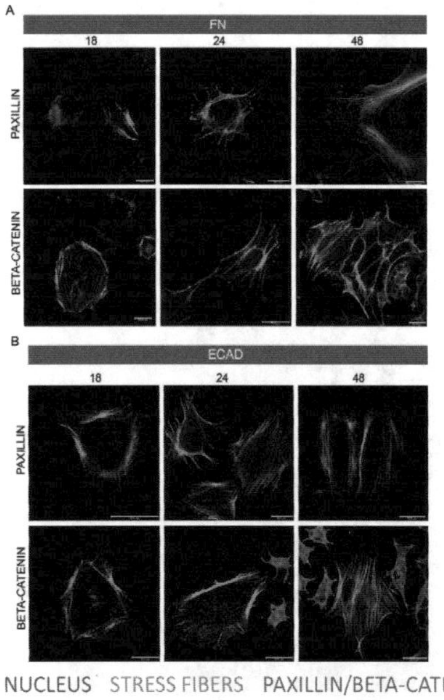

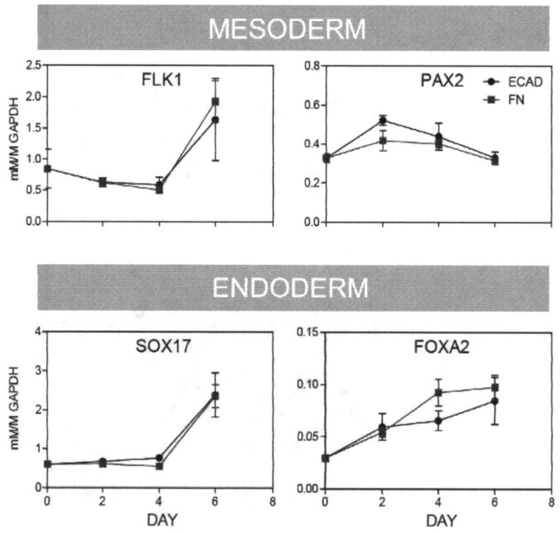

- Cells on fibronectin recruited paxillin to their focal adhesions, not beta-catenin.
- Cells on ECAD recruited beta-catenin to their cell adhesions, not paxillin.
- Culture on FN and ECAD resulted in similar early mesodermal and endodermal differentiation.

Inhibition of Actin Polymerization in 2D

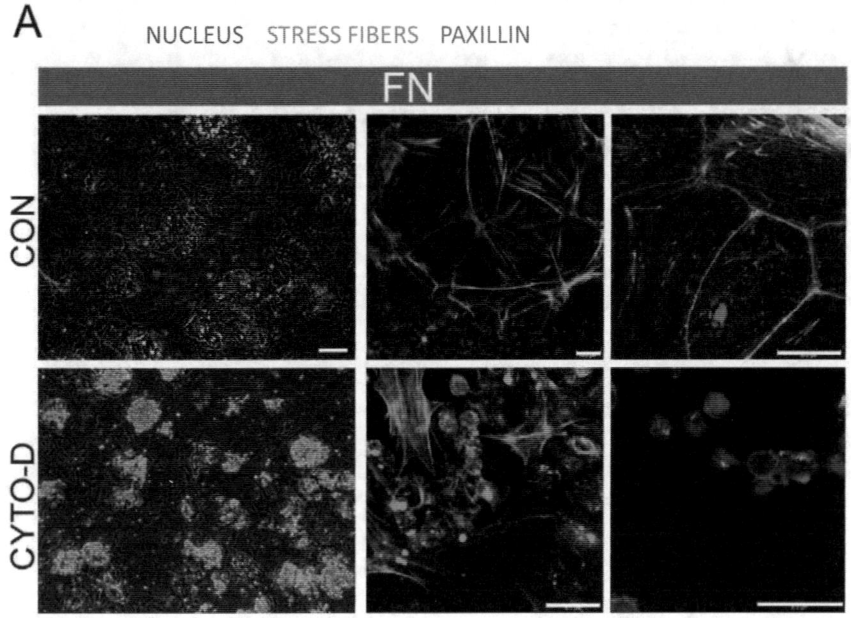

CYTO-D caused cell rounding and disassembly of stress fibers on FN.

Inhibition of Actin Polymerization in 2D

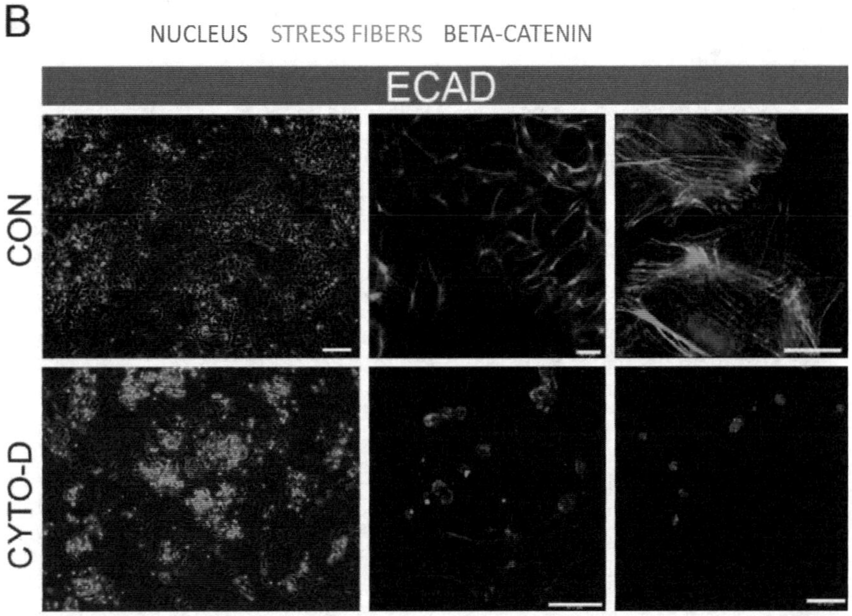

CYTO-D caused cell rounding and disassembly of stress fibers on ECAD.

Inhibition of Actin Polymerization in 2D Increases Early Mesoderm and Endoderm

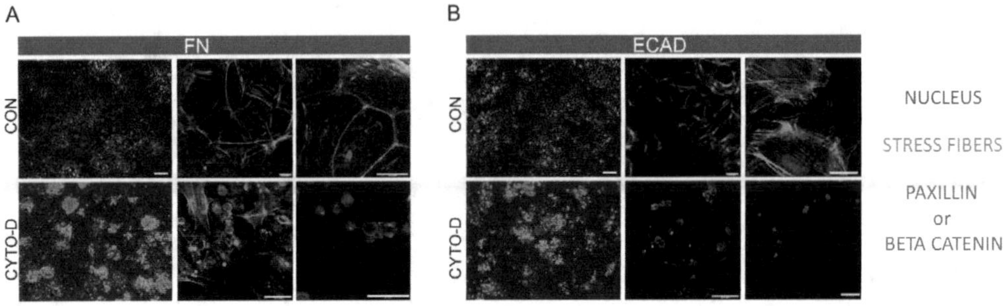

Force & Mesodermal Differentiation

Applied Shear Stress

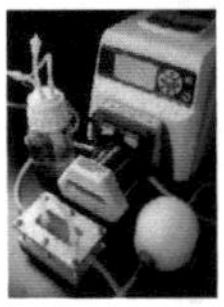

↑ External Force → ↑ Mesoderm

Actin-Myosin II Inhibition in EBs

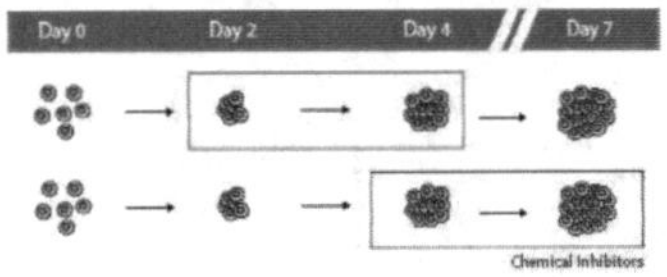

↓ Internal Force
External Force → ↓ Mesoderm

Actin-Myosin II Inhibition in 2D

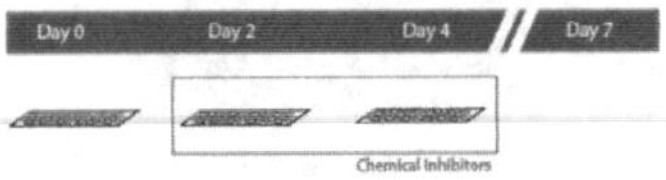

↓ Internal Force → ↑ Mesoderm

Questions

- Does force regulate differentiation through biochemical signals or force transmission?
- Is force or stress part of a multi-variate control with other co-factors?
- Does applied force or actin-myosin motor modulation alter processes that in turn regulate differentiation?
- How does the dynamic system in ESCs with respect to the mechnosensing and mechanotransduction processes modulate the effects of force or stress?

Acknowledgements

Laboratory Members

Liana Boraas
Erika Broadnax
Michelle Janaszak
Lina Maria Quijano

Julia Guidry
Stephanie Messina
Douglas James

Kristen Lynch, PhD
Emma Pineda, PhD
Russell Wolfe, PhD

David Dao
Austin Dobbins
Amaris Genemaras
Todd Johnson
Jardin Leleux
Elliot Neal, MS
Kate Peuker
Lindsey Shepard
Elizabeth Zipf

Collaborators

Bruce Bunnell, PhD
Patrice Delafontaine, MD
S. Michal Jazwinski, PhD
Ken Muneoka, PhD

EMD MILLIPORE
NATIONAL INSTITUTES OF HEALTH
STATE OF LOUISIANA BOARD of REGENTS
National Institute of General Medical Sciences

Mechanotransduction

- Shear stress may activate several biochemical signaling pathways and we need to determine how those pathways affect the transcription factors that regulate differentiation.
- Is “force” information also transmitted to the nucleus mechanically?

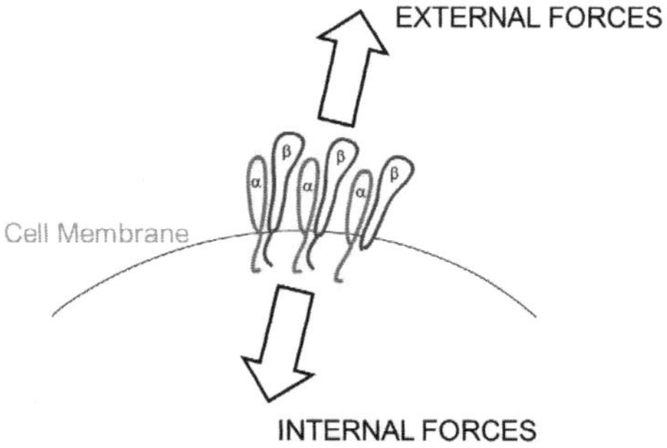

Force & Mesodermal Differentiation

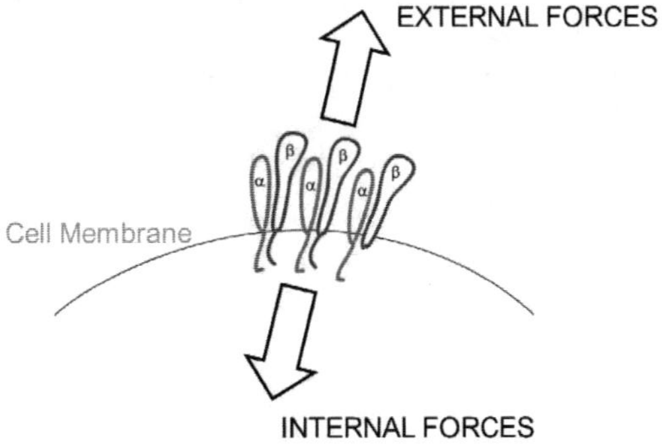

41

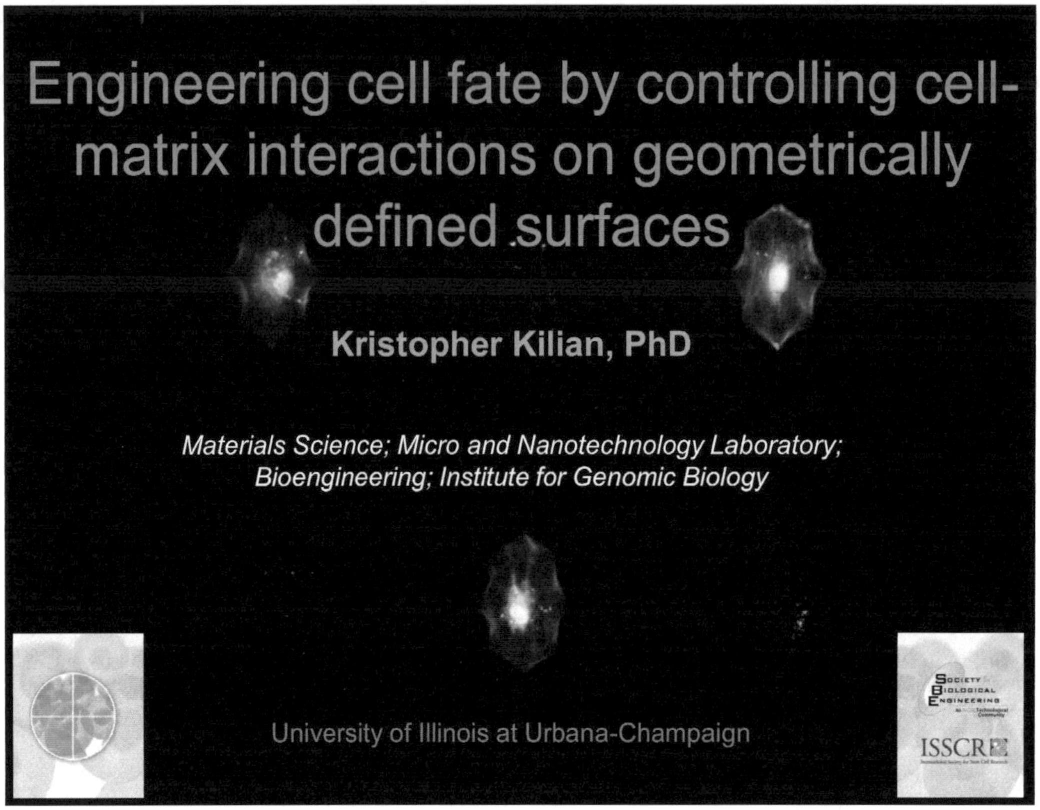
Engineering cell fate by controlling cell-matrix interactions on geometrically defined surfaces
Kristopher Kilian, PhD
Materials Science; Micro and Nanotechnology Laboratory; Bioengineering; Institute for Genomic Biology
University of Illinois at Urbana-Champaign
ISSCR

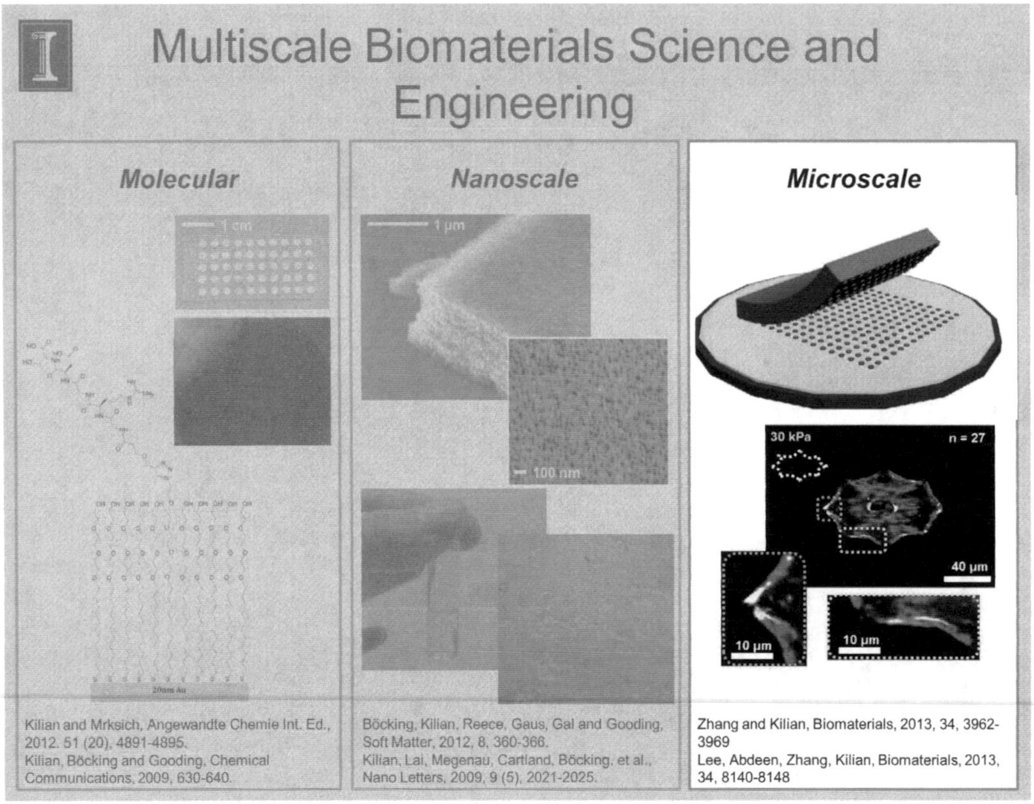
Multiscale Biomaterials Science and Engineering
Molecular
1 cm
20nm Au
Kilian and Mrksich, Angewandte Chemie Int. Ed., 2012, 51 (20), 4891-4895.
Kilian, Böcking and Gooding, Chemical Communications, 2009, 630-640.
Nanoscale
1 µm
100 nm
Böcking, Kilian, Reece, Gaus, Gal and Gooding, Soft Matter, 2012, 8, 360-366.
Kilian, Lai, Megenau, Cartland, Böcking, et al., Nano Letters, 2009, 9 (5), 2021-2025.
Microscale
30 kPa
n = 27
40 µm
10 µm
10 µm
Zhang and Kilian, Biomaterials, 2013, 34, 3962-3969
Lee, Abdeen, Zhang, Kilian, Biomaterials, 2013, 34, 8140-8148

The influence of matrix stiffness and ligand composition on cell fate determination

Matrix stiffness…

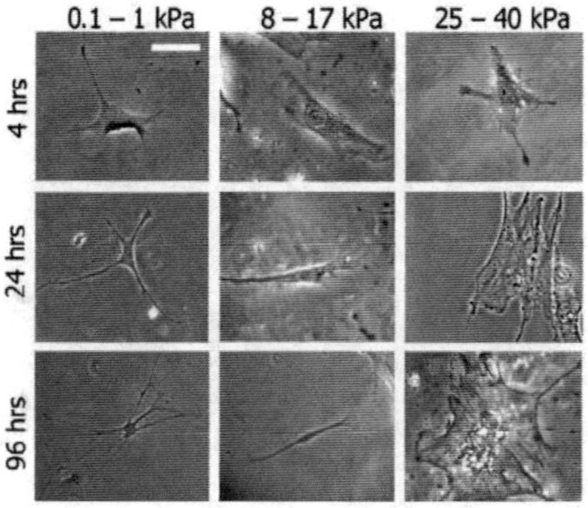

Engler AJ, Sen S, Sweeney HL, Discher DE., Cell (2006), 126, 677-689

…and composition

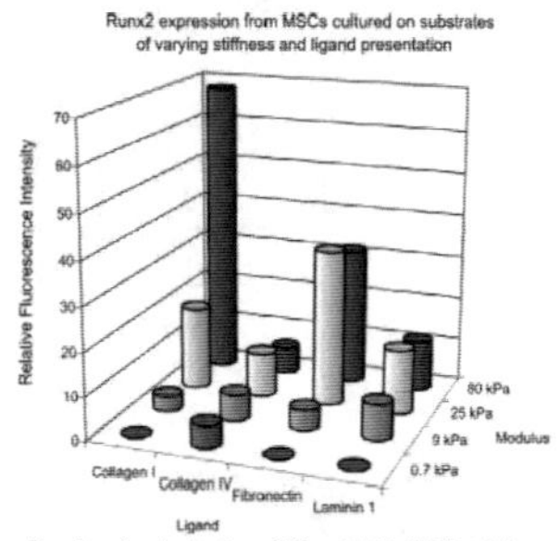

Rowlands et al., Am J Physiol Cell Physiol. 2008 295(4), 1037-1044

polyacrylamide < 1 kPa

Neuronal markers Adipogenic markers

Round (spread) cells express higher adipogenic (neurogenic) markers…

Junmin Lee

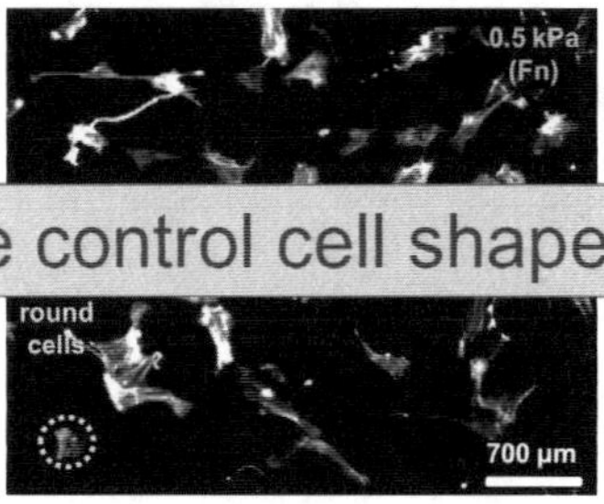

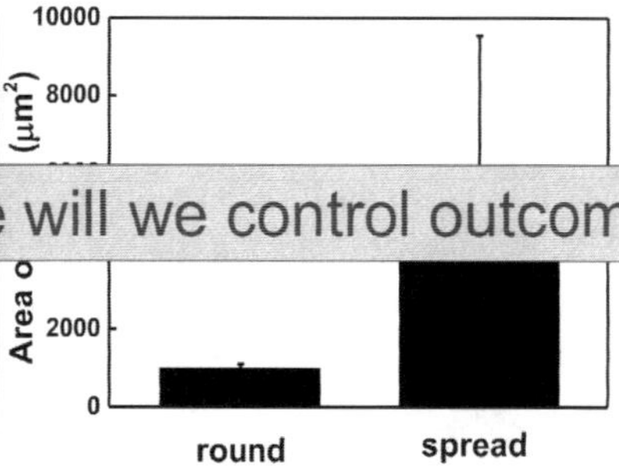

If we control cell shape will we control outcome?

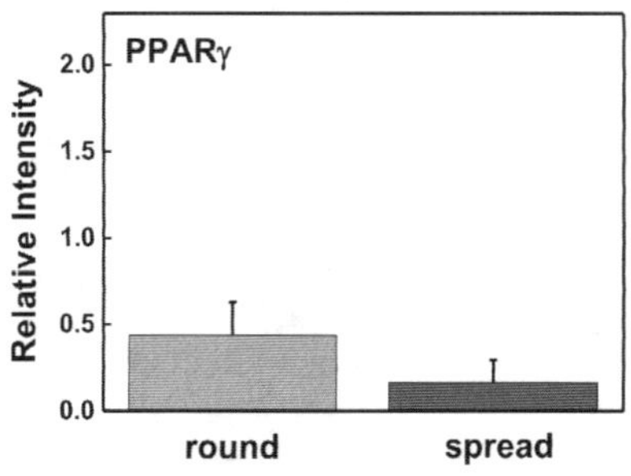

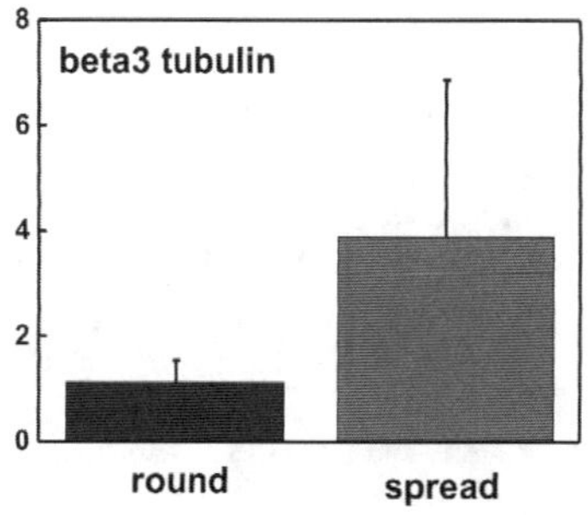

Micropatterning MSCs on soft hydrogels

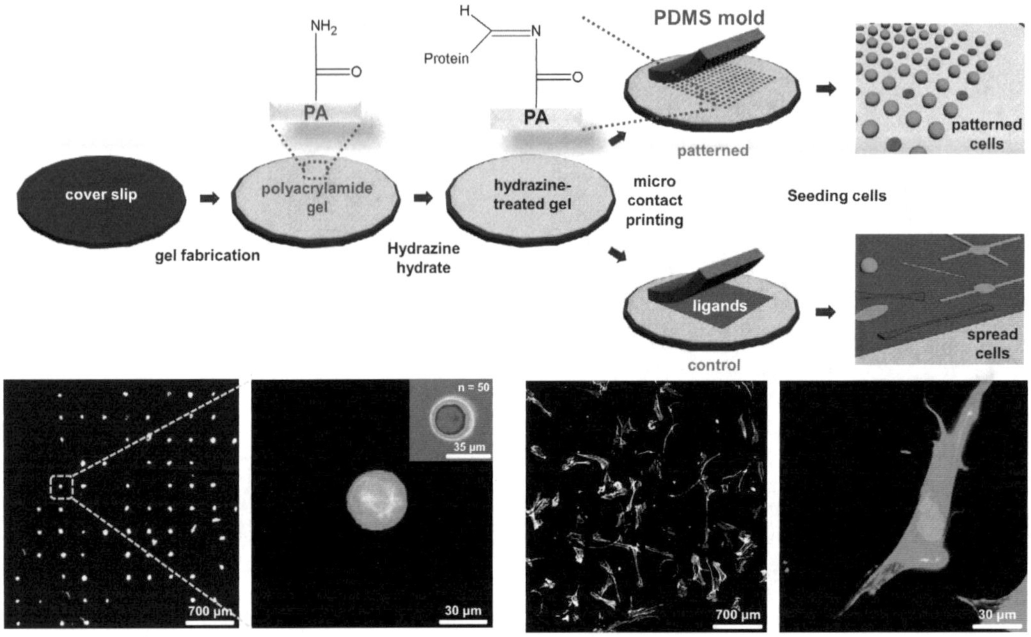

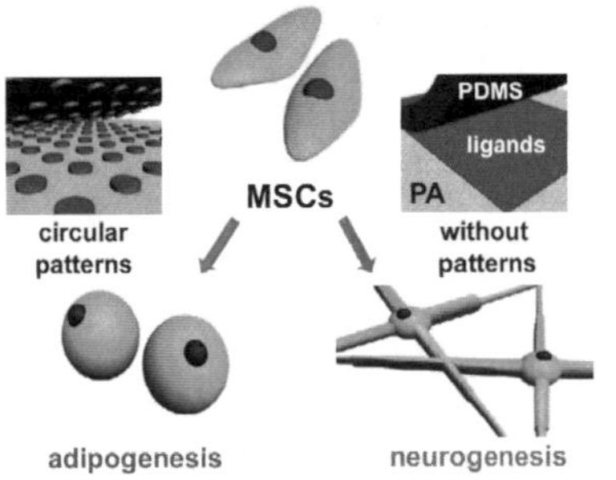

Geometry guides expression of adipogenesis and neurogenesis markers

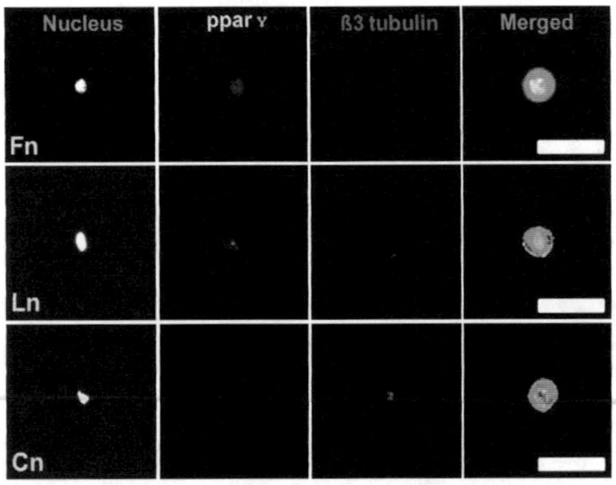

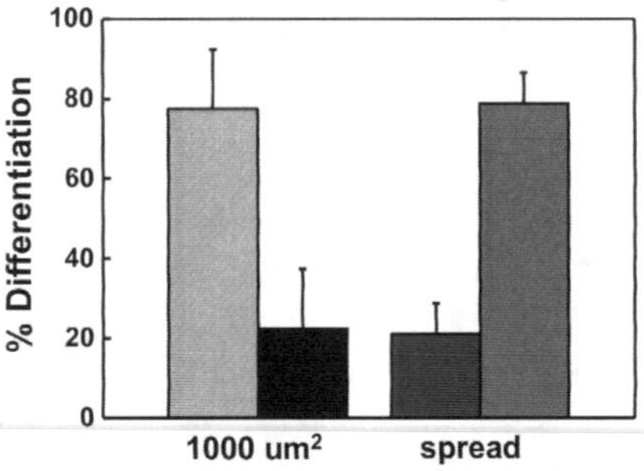

Lee, Abdeen, Zhang, Kilian, Biomaterials, 2013, 34, 8140-8148

Matrix mechanics AND ligand composition

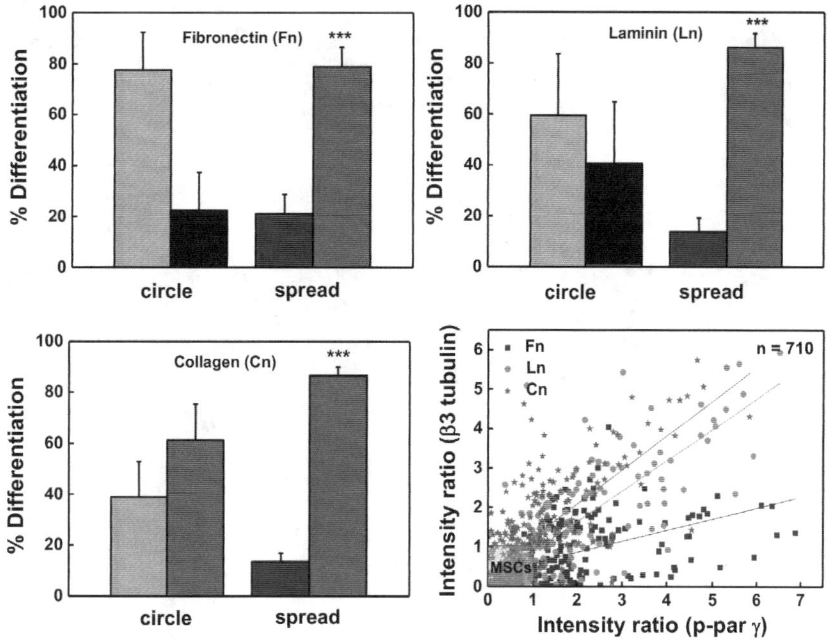

Lee, Abdeen, Zhang, Kilian, Biomaterials, 2013, 34, 8140-8148

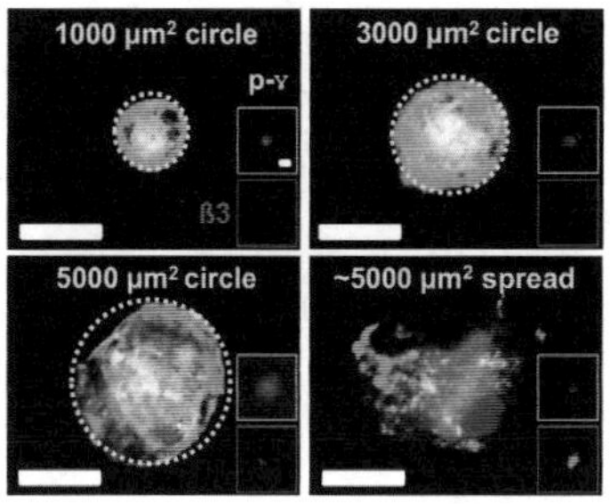

Geometric cues guide expression of neurogenesis markers

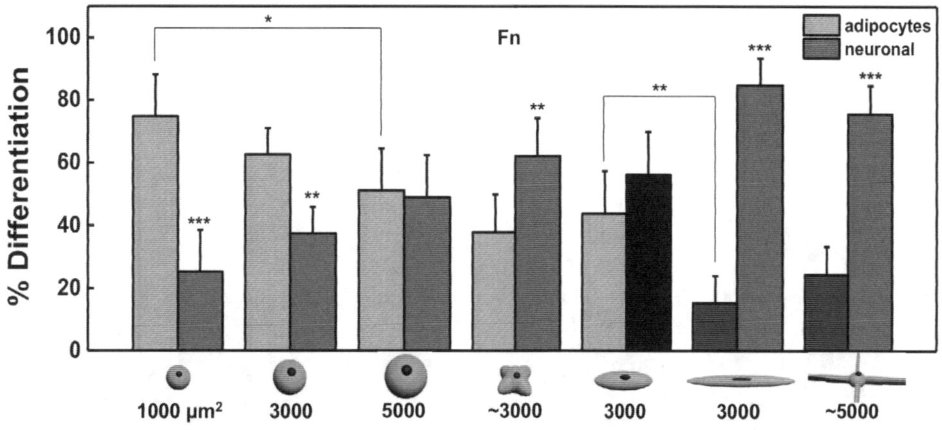

Lee, Abdeen, Zhang, Kilian, Biomaterials, 2013, 34, 8140-8148

Matrix composition directs neurogenesis

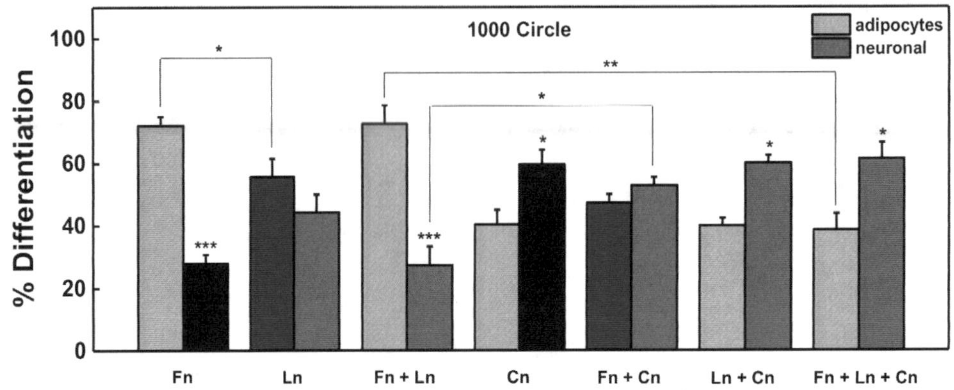

ligands / shapes	Fn	Ln	Fn + Ln	Cn	Fn + Cn	Ln + Cn	Fn + Ln + Cn
1000 circle	adipogenic	adipogenic	adipogenic	neurogenic	neurogenic	neurogenic	neurogenic
spread	neurogenic	neurogenic	neurogenic	neurogenic	neurogenic	neurogenic	neurogenic

Lee, Abdeen, Zhang, Kilian, Biomaterials, 2013, 34, 8140-8148

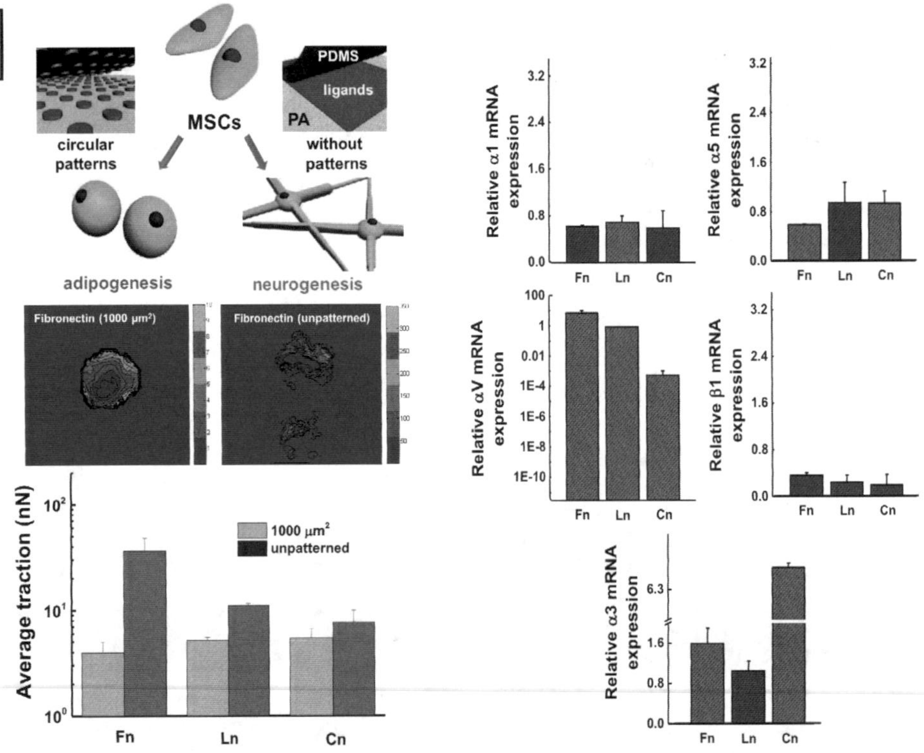

Multiscale Biomaterials Science and Engineering

Molecular

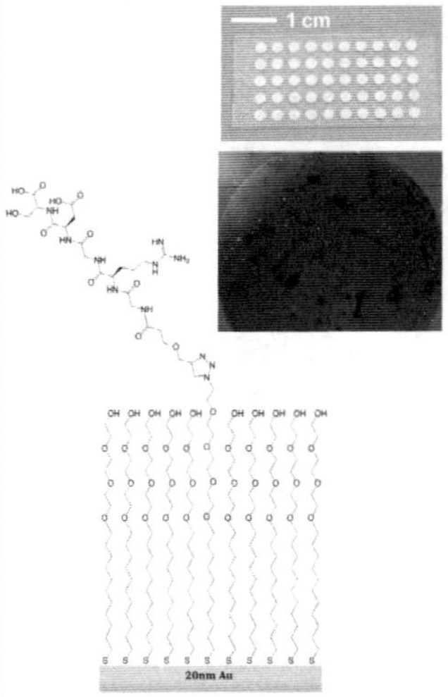

Kilian and Mrksich, Angewandte Chemie Int. Ed., 2012. 51 (20), 4891-4895.
Kilian, Böcking and Gooding, Chemical Communications, 2009, 630-640.

Nanoscale

Böcking, Kilian, Reece, Gaus, Gal and Gooding, Soft Matter, 2012, 8, 360-366.
Kilian, Lai, Megenau, Cartland, Böcking, et al., Nano Letters, 2009, 9 (5), 2021-2025.

Microscale

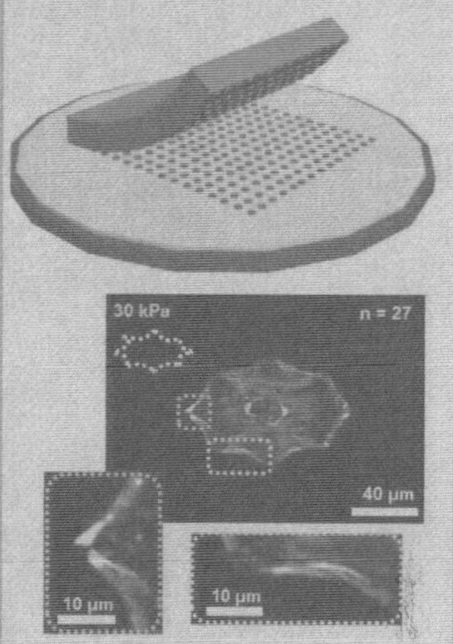

Zhang and Kilian, Biomaterials, 2013, 34, 3962-3969
Lee, Abdeen, Zhang, Kilian, Biomaterials, 2013, 34, 8140-8148

…short peptides as model matrices

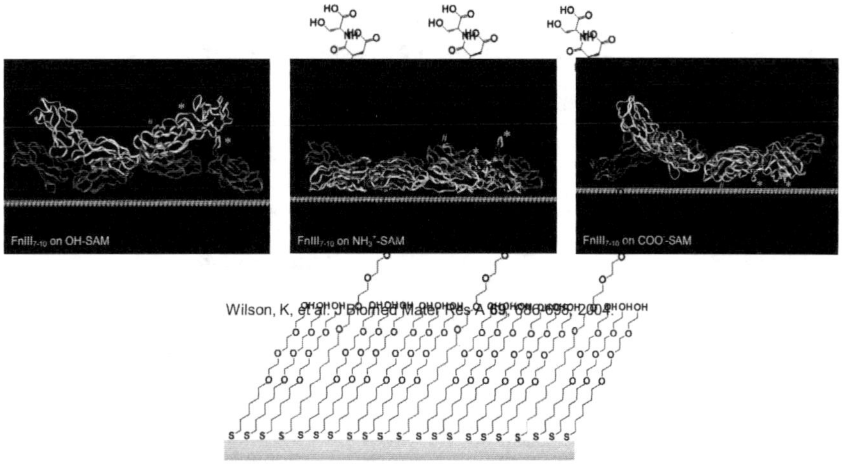

Wilson, K, et al. J Biomed Mater Res A 69, 686-698, 2004.

Ligand affinity can influence the expression of smooth muscle markers

Doug Zhang

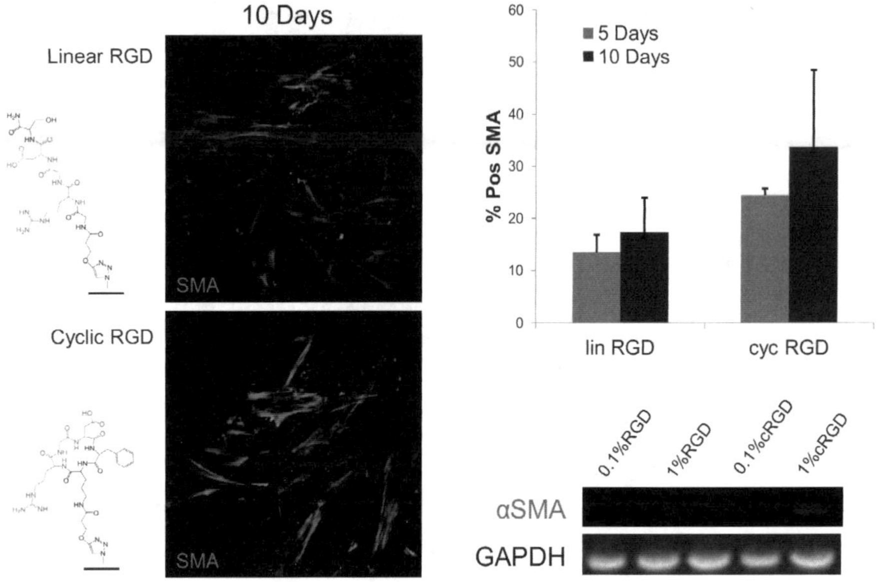

Patterning peptides to study the influence of cell geometry on specific ECM interactions

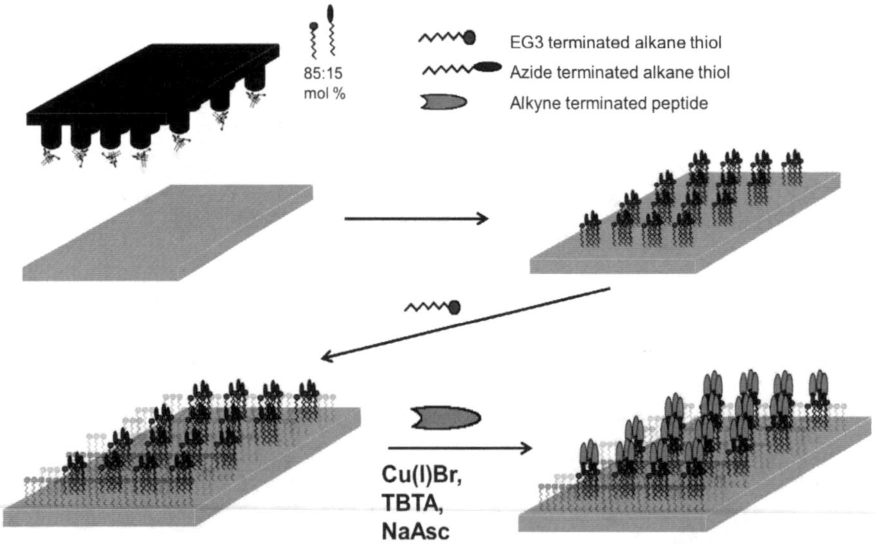

Doug Zhang

Controlling cell geometry and adhesion ligand

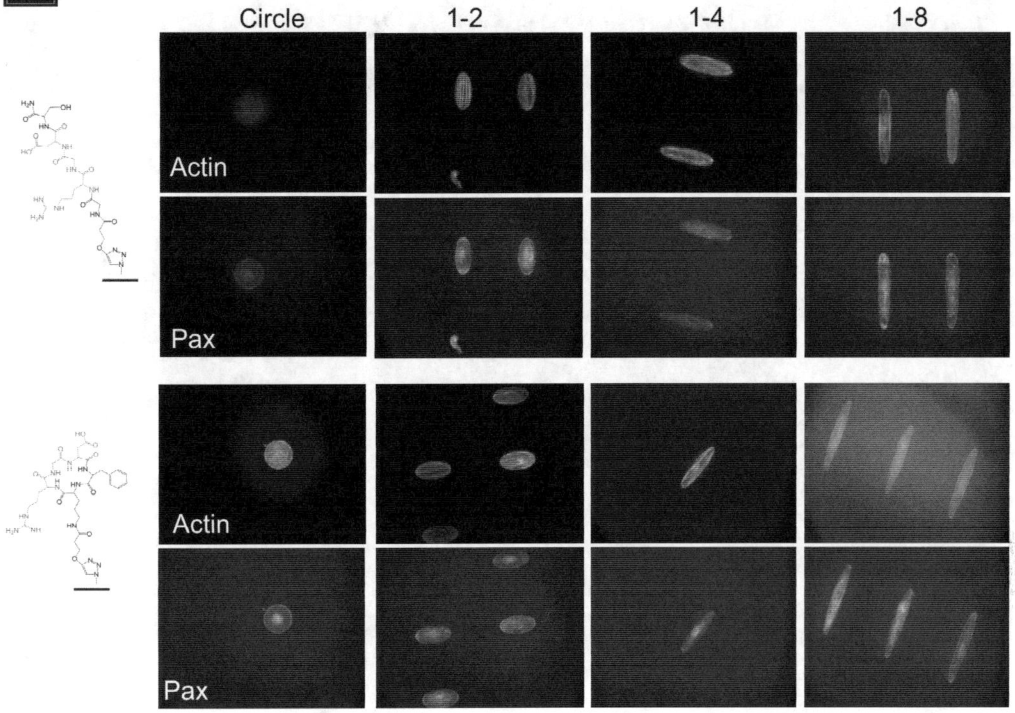

Quantitation of focal adhesion

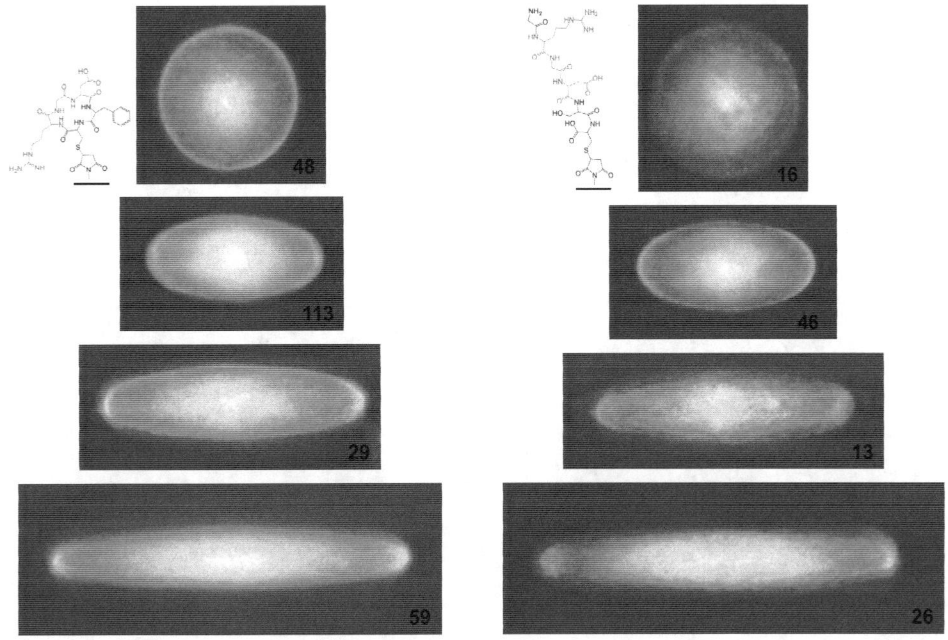

Cell elongation and ligand affinity influence smooth muscle myogenesis

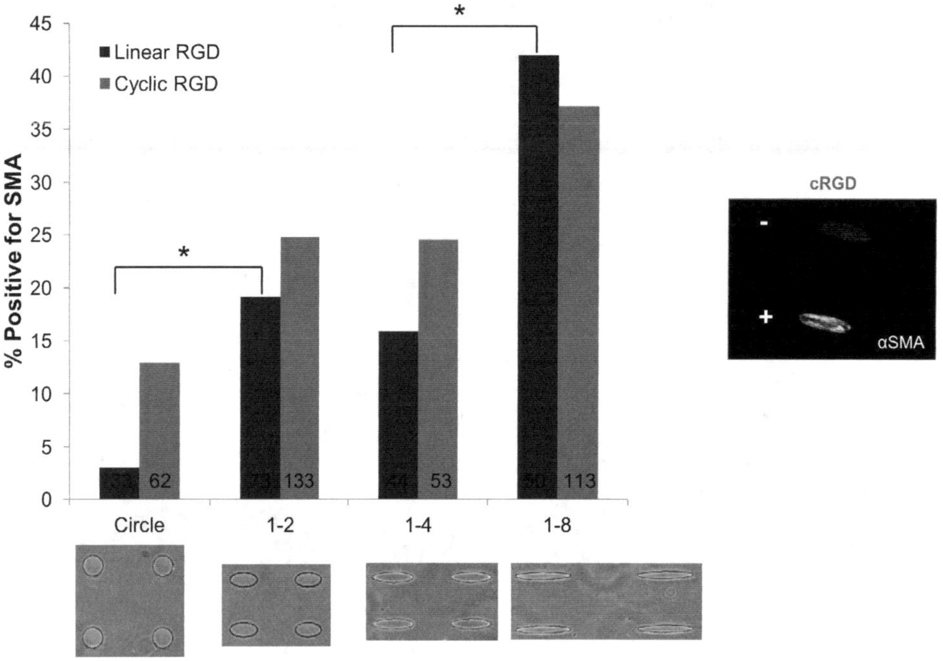

Pharmacological inhibitors…

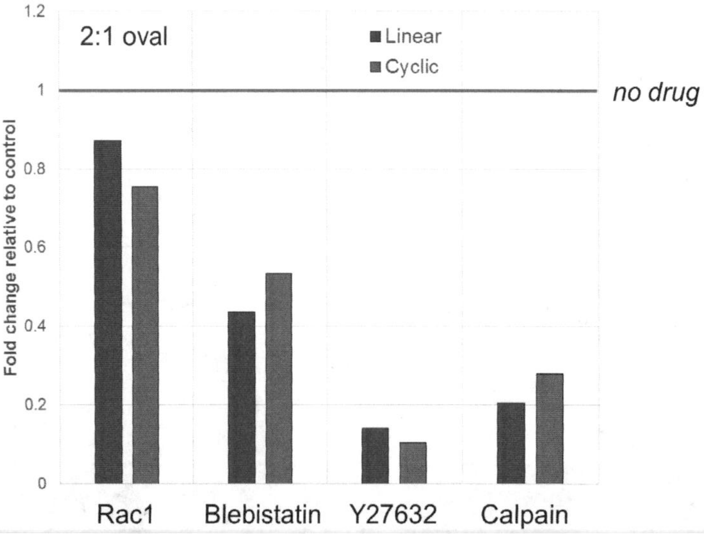

What about patterning other bioactive peptides?

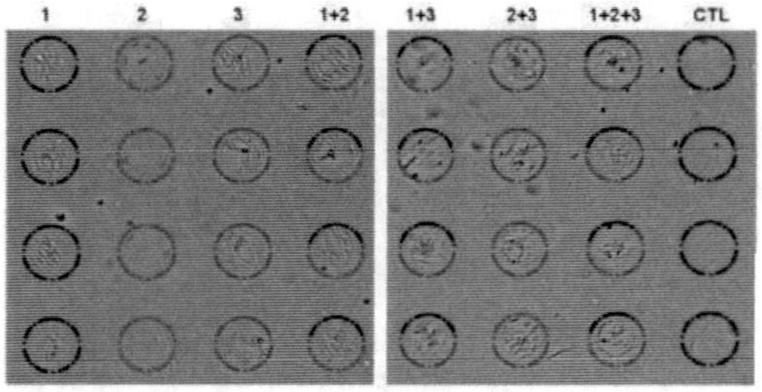

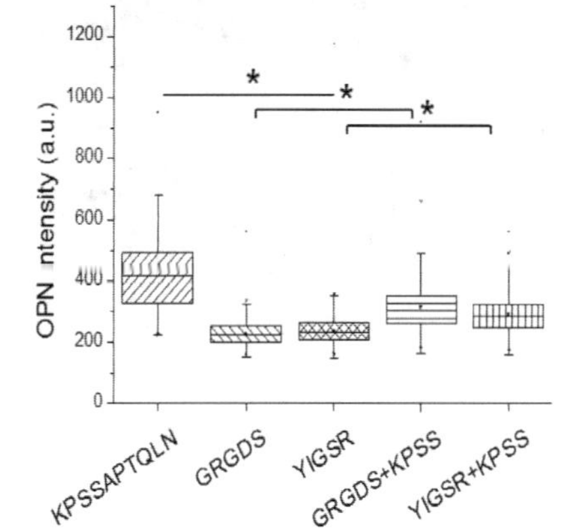

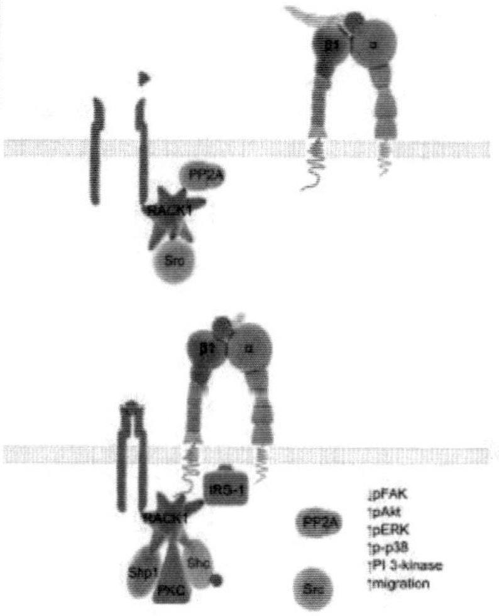

Legate, K.R. *et al.*, 2009

Patterning combinations of bioactive peptides

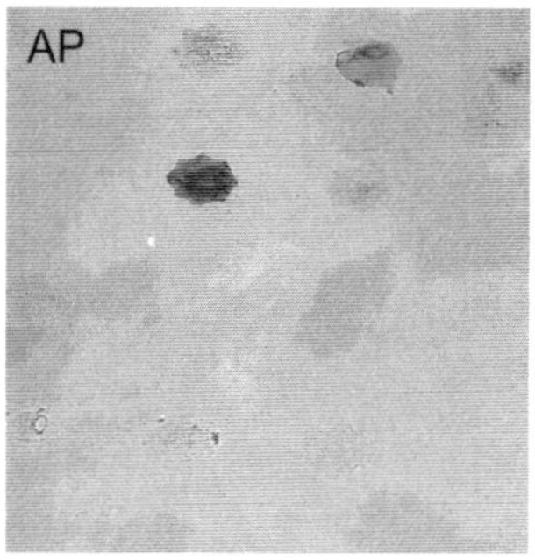

Extracellular Matrix	RGD
Growth Factor	KPSSAPTQLN (KPSS)

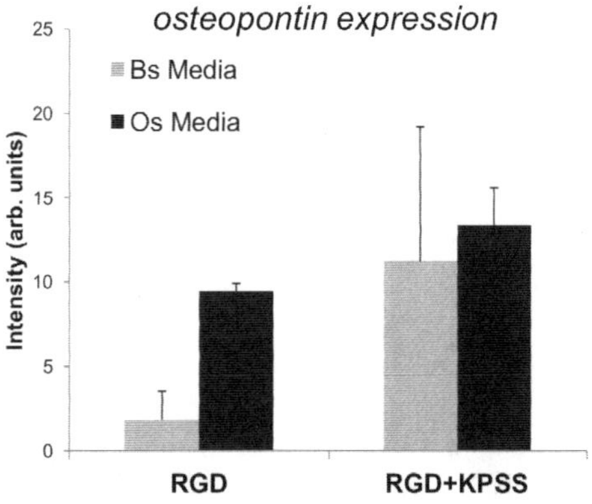

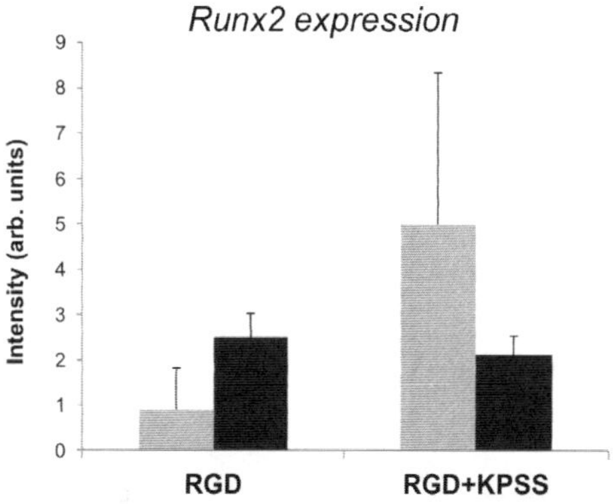

Michael Sun

Acknowledgements

The Kilian Lab

Amr Abdeen
Tiffany Huang
Yanfen Li
Junmin Lee
Douglas Zhang
Michael Sun
Samuel Mo
Alex Kim
Claire Tang
Margaret Barbero

Collaborators

Gabi Popescu, UIUC
Andrew Smith, UIUC
Martha Gillette, UIUC
Bruce Lahn, UC
J. Justin Gooding, UNSW
Partha Ray, Carle Hospital

American Cancer Society

Campus Research Board Programs
University of Illinois at Urbana–Champaign

Peptide surfaces for biomaterials development

Extracellular Matrix

Proteoglycan

Growth Factor

ECM/PG

ECM/GF

PG/GF

ECM/PG/GF

SPPS

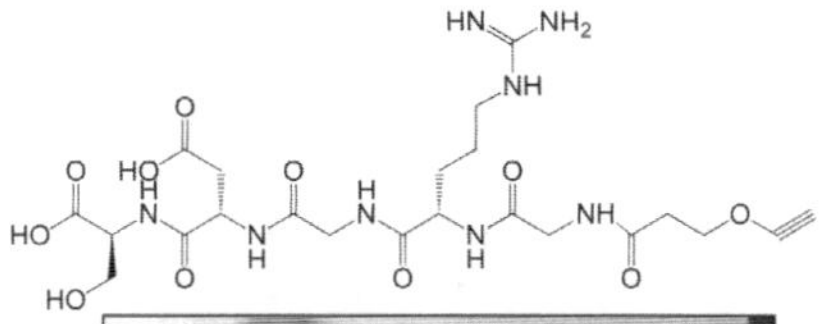

high throughput arraying
high content imaging

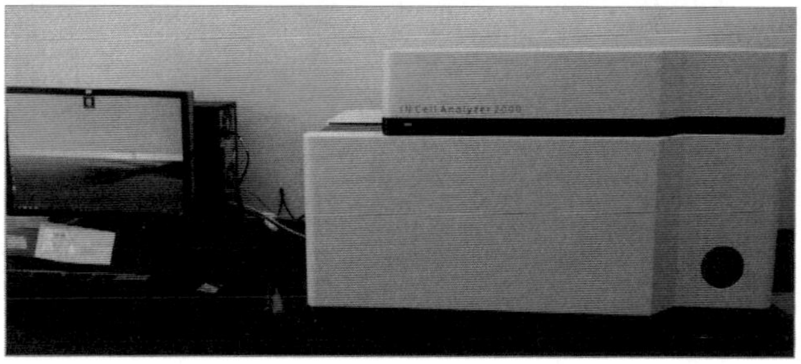

Sequence	Origin/Mimetic	Binding
GRGDS	Fibronectin, Vitronectin	Integrin αvβ3, α5β1, α3Bβ
YIGSR	Laminin β1	Integrin α4β1
IKVAV	Laminin α chain	Integrin α3β1
DGEA	Collagen I	Integrin α2β1
REDV	Fibronectin	Integrin α4β1
KPHSRN	Fibronectin	Integrin α5β1, αIIbβ3
SVVYGLR	Osteopontin	Integrin α4β1
KRSR	Heparan sulfate	Proteoglycan
KGR	Laminin α3 chain	Proteoglycan
SHWSPWSS	hThrombospondin	Proteoglycan
FHRRIKA	Bone Sialoprotein	Proteoglycan
DWIVA	BMP-2	Growth Factor
KPSSAPTQLN	BMP-7	Growth Factor
YSDKSLPHP	JAGGED-2	Growth Factor
HYQASVSPEPP	DELTA-1	Growth Factor
IPKVELVPAG	ACTIVIN-1	Growth Factor
ECPNPVGEEFH	NODAL-1	Growth Factor

Doug Zhang; Michael Sun

Mimicking the niche – the potential of cell-derived extracellular matrices to support human bone marrow stem cells

Marina Prewitz
Leibniz Institute for Polymer Research
Max Bergmann Center for Biomaterials
Dresden, Germany

SBE – International Conference on Stem Cell Engineering 17.03.2014

Leibniz Association

Leibniz-Institut für Polymerforschung Dresden e. V.

MAX BERGMANN center of biomaterials dresden

Bone marrow stem and progenitor cells

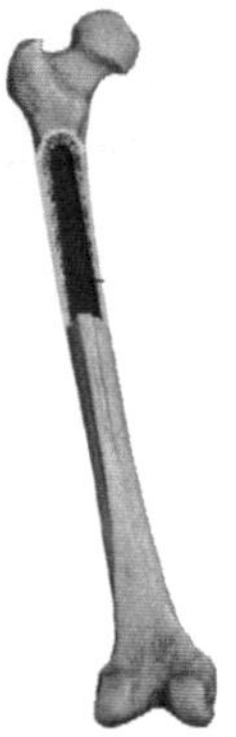

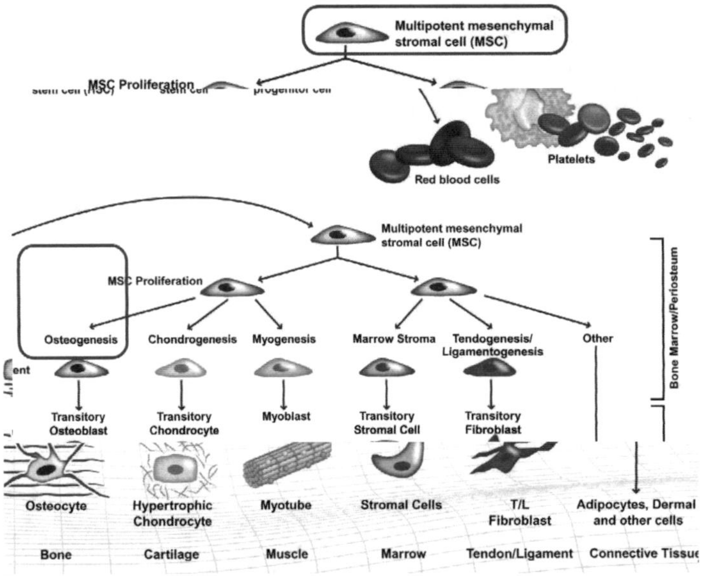

Clinical applications

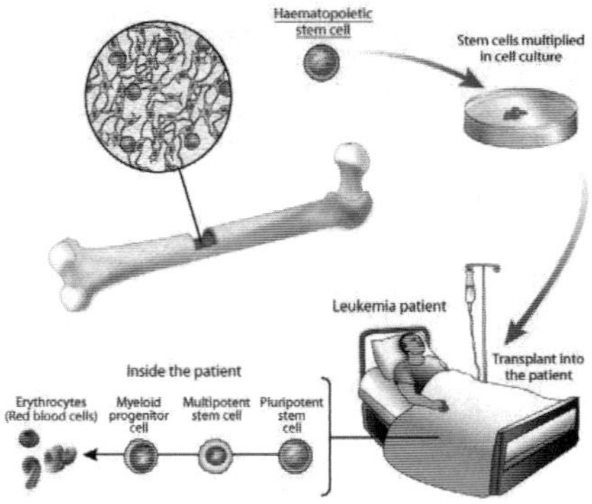

www.kumc.edu/

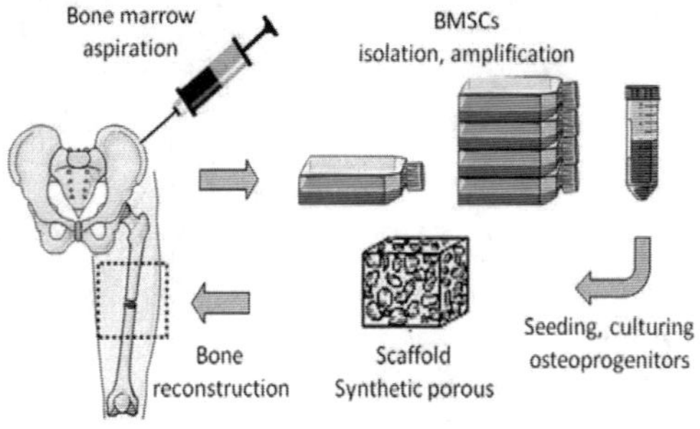

www.reborne.org/

Expansion of stem and progenitor cells *in vitro*.

Creating cellular microenvironments in vitro from extracellular matrix

Standard *in vitro* culture conditions

Plasma treated polystyrene (PTP)

Reconstituted ECM molecules

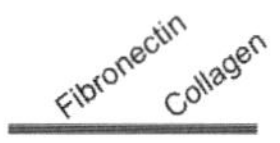

Matrigel

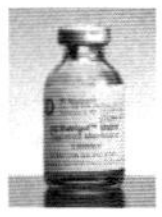

Organ & Cell - derived ECM scaffolds

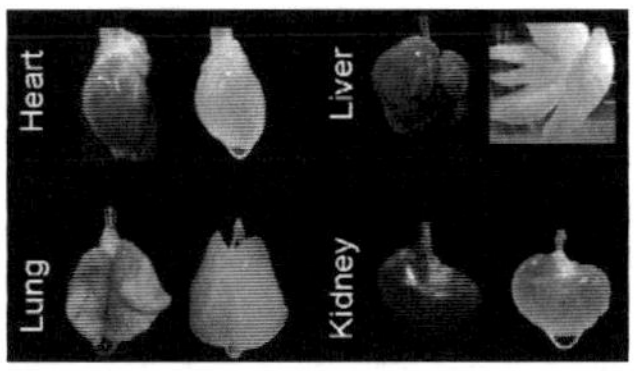

http://midwestresearchswine.com

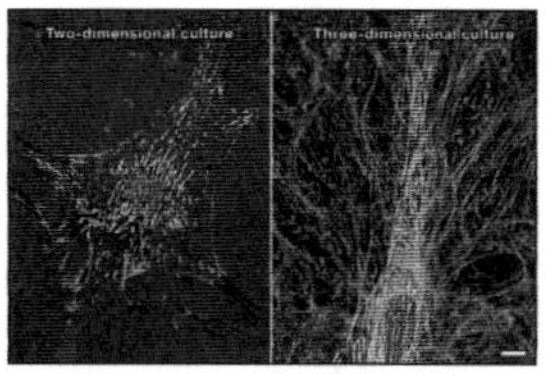

Yamada, K. M., & Cukierman, E. (2007). *Cell*, *130*(4), 601–610.

Tissue-specific cell types for tissue-mimetic ECM

Cell culture → ECM secretion → Decellularised matrix

Bone marrow MSC-derived tissue mimetic ECM

confluent MSC layer

PTP

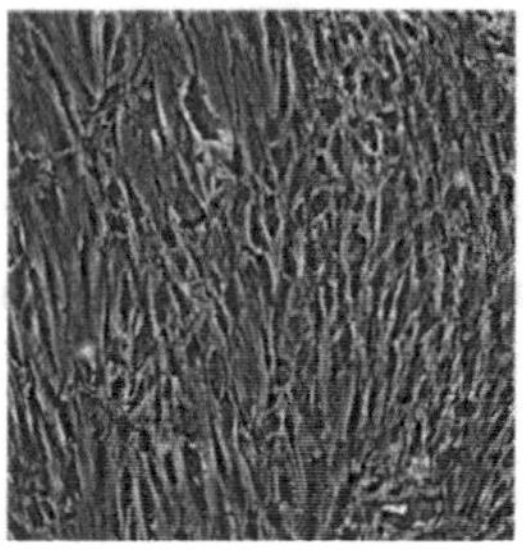

Fibronectin

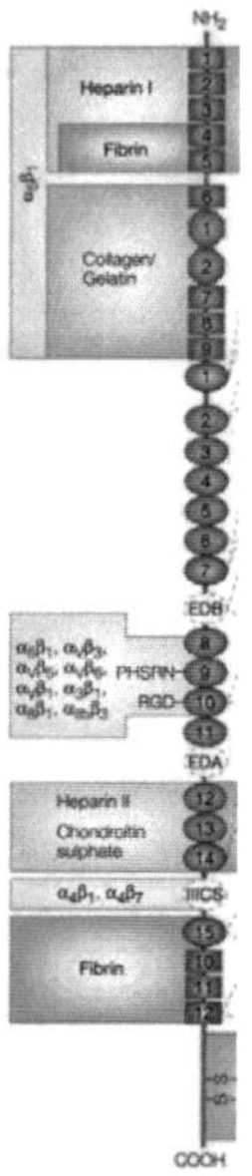

Nature Reviews Molecular Cell Biology 2, 793-805 (2001)

Immobilization of cell-secreted ECM to a culture carrier

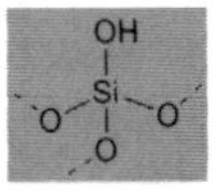

Immobilization of cell-secreted ECM to a culture carrier

APDS = 3-aminopropyl-dimethyl-ethoxy silane

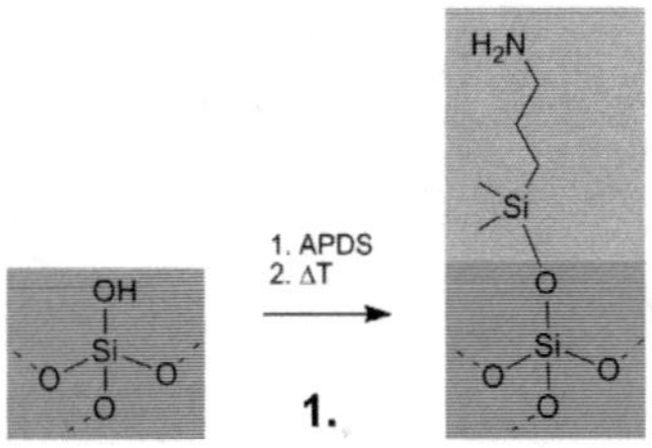

Immobilization of cell-secreted ECM to a culture carrier

APDS = 3-aminopropyl-dimethyl-ethoxy silane

POMA = poly (octadecene *alt* maleic anhydride)

$R^1 = (CH_2)_{15}CH_3$

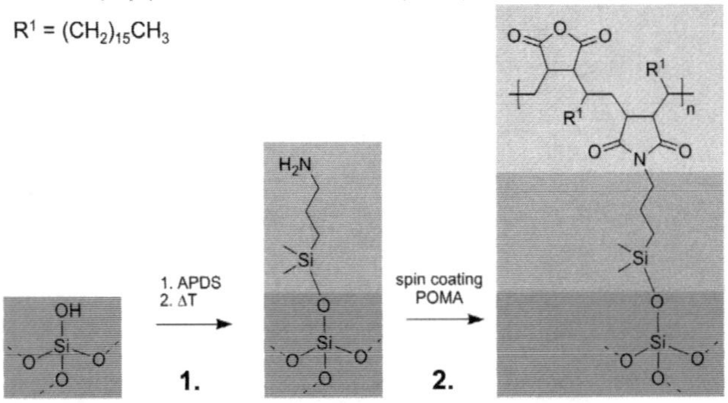

Immobilization of cell-secreted ECM to a culture carrier

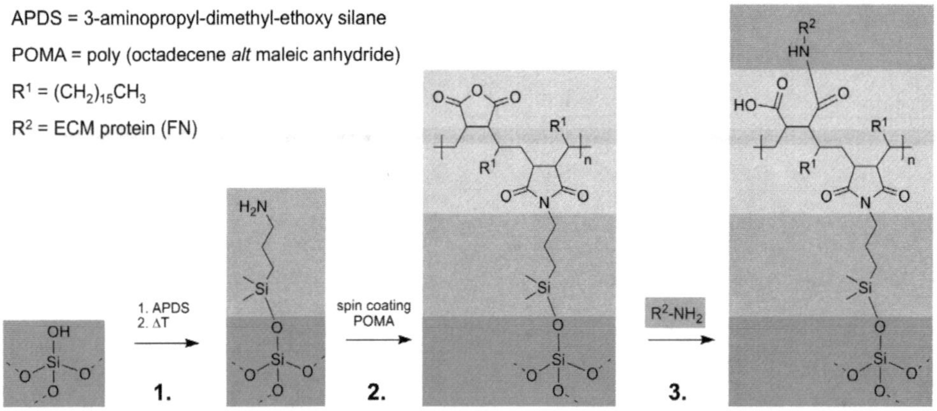

Immobilization of cell-secreted ECM to a culture carrier

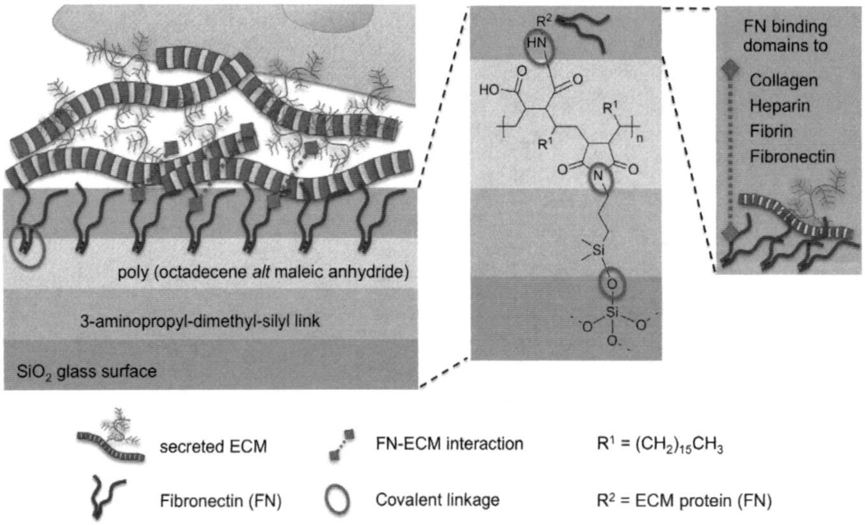

Immobilization of cell-secreted ECM to a culture carrier

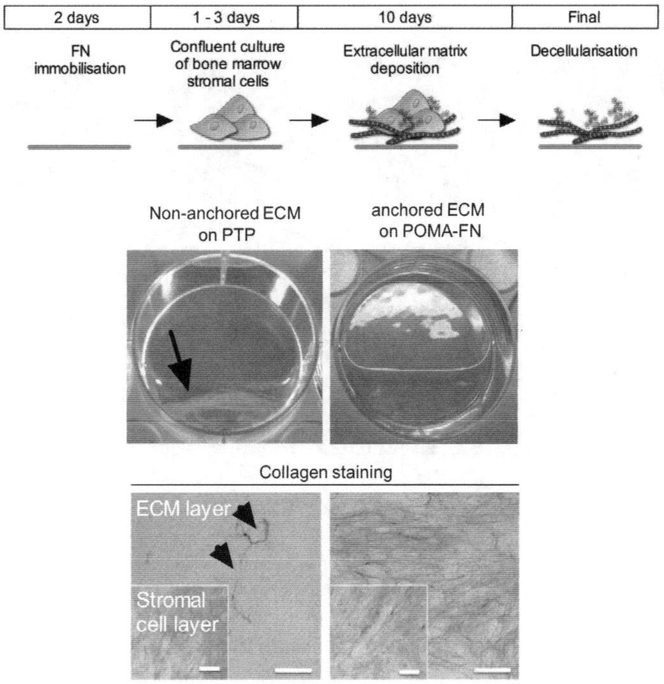

Varying cell types for matrix secretion

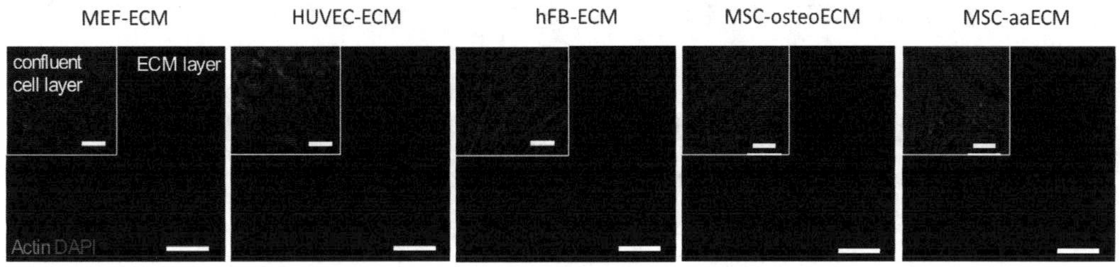

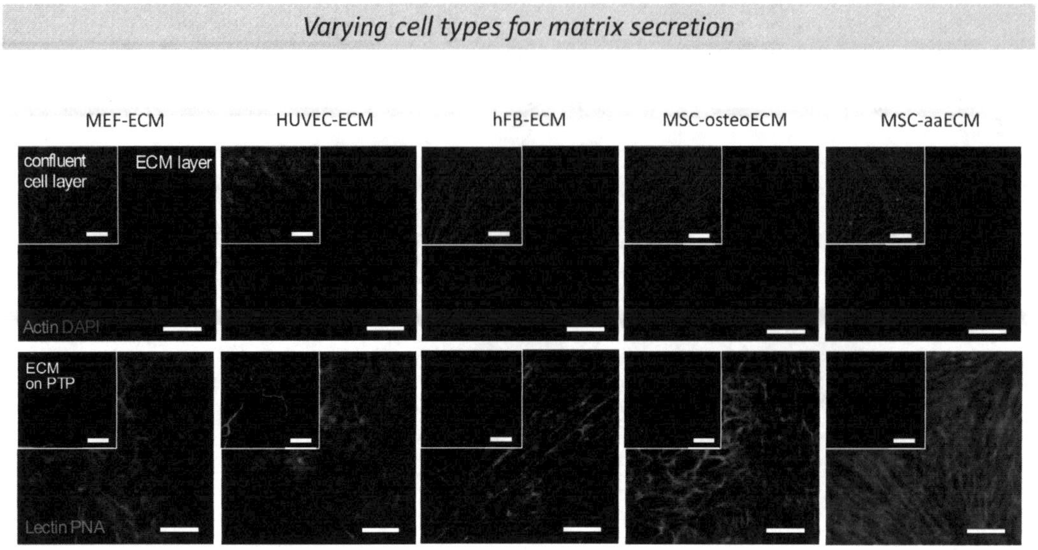
Varying cell types for matrix secretion
MEF-ECM
HUVEC-ECM
hFB-ECM
MSC-osteoECM
MSC-aaECM
confluent cell layer
ECM layer
Actin DAPI
ECM on PTP
Lectin PNA

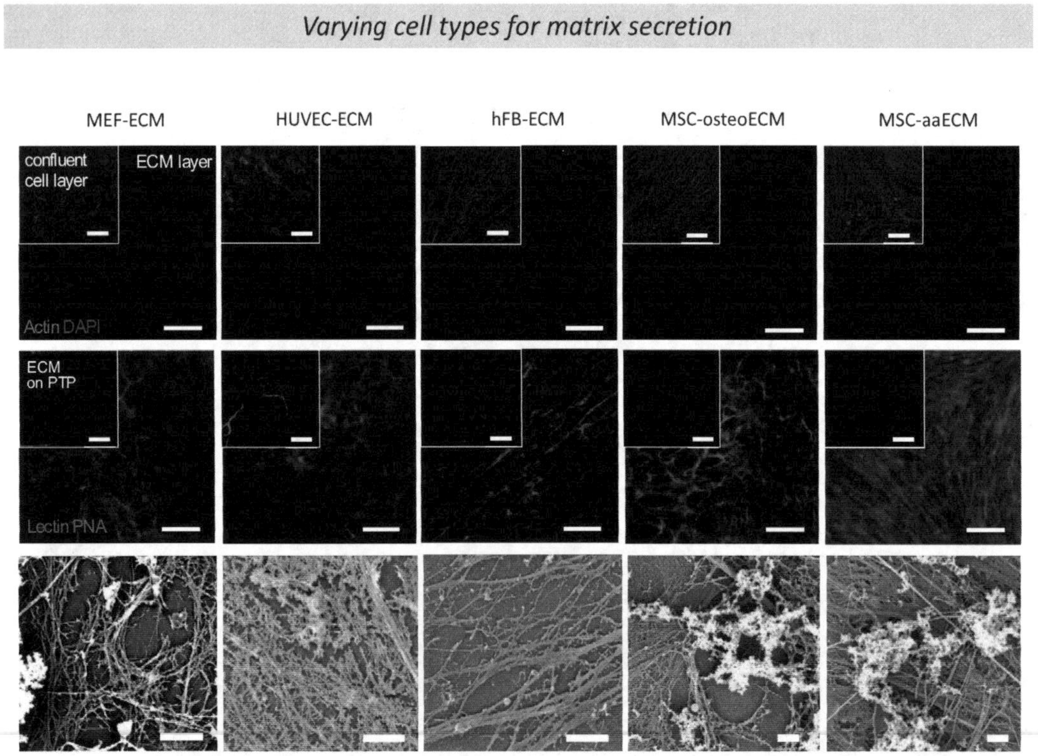
Varying cell types for matrix secretion
MEF-ECM
HUVEC-ECM
hFB-ECM
MSC-osteoECM
MSC-aaECM
confluent cell layer
ECM layer
Actin DAPI
ECM on PTP
Lectin PNA

Physical properties of decellularized ECM

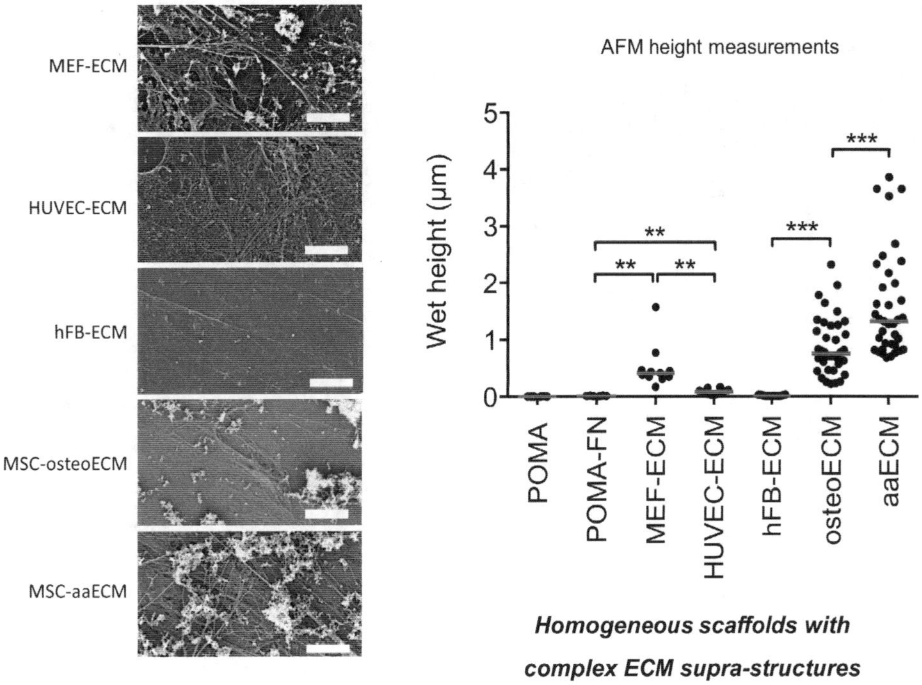

Mechanical properties of decellularized ECM

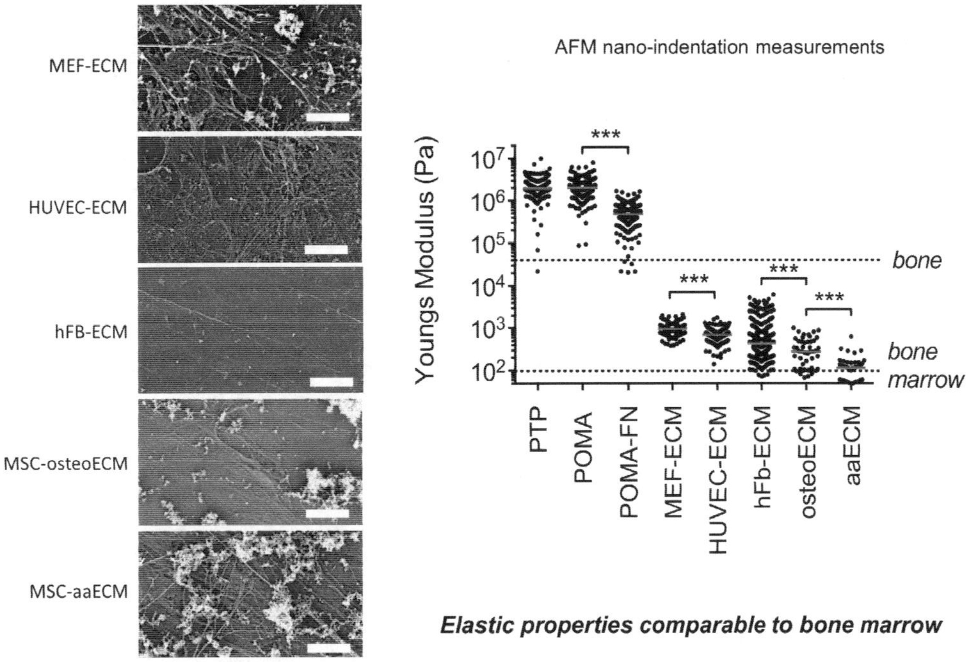

Identified molecules within decellularized MSC matrices

Proteome analysis via mass spectrometry

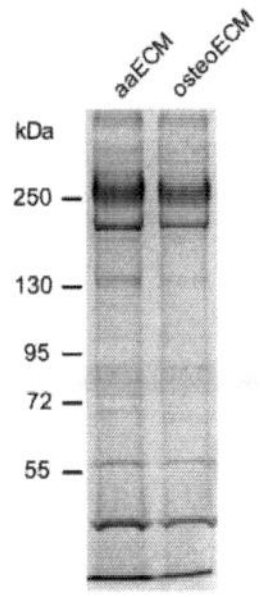

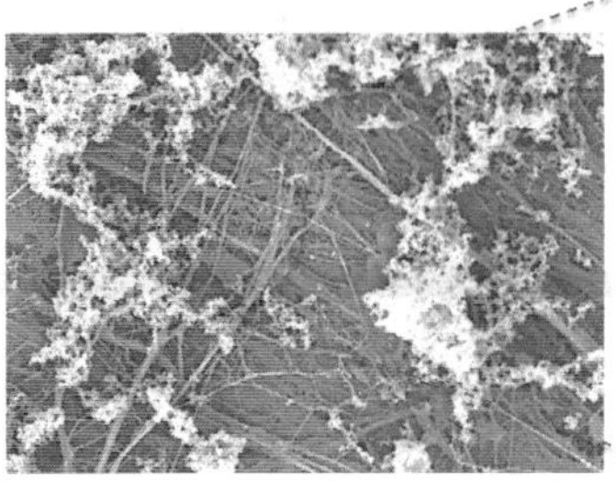

ECM proteins:

- Collagens I, IV, V, VIII, XI, ...
- Fibulins
- Nidogens
- Elastin
- Laminin
- Hemicentin
- Osteonectin
- Thrombospondins
- Perlecan, Podocan, Tenascin, ...

Growth factors:

- bFGF
- SDF-1
- Wnt-5b
- von Willebrand factor, ...

MultiPlex quantification of numerous cytokines released from decellularized ECM

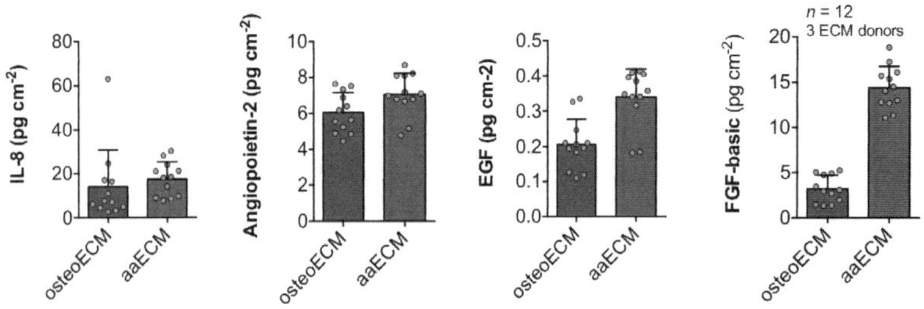

MSC growth on tissue culture plastic

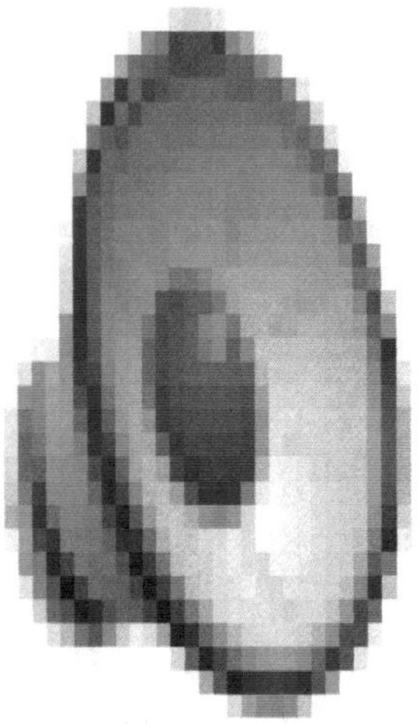

MSC growth on MSC-derived osteoECM

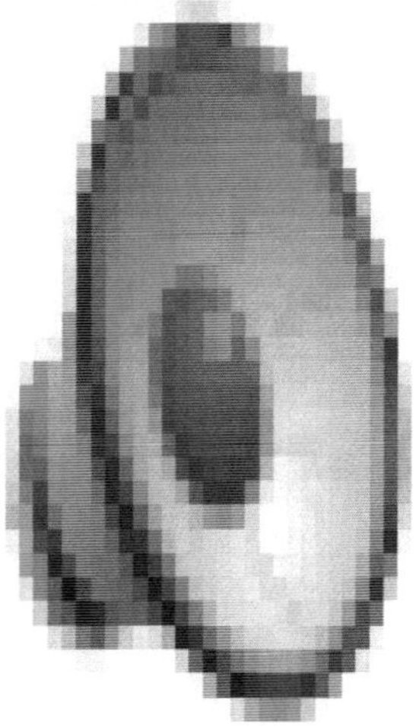

Haematopoietic stem cells on MSC-derived ECM

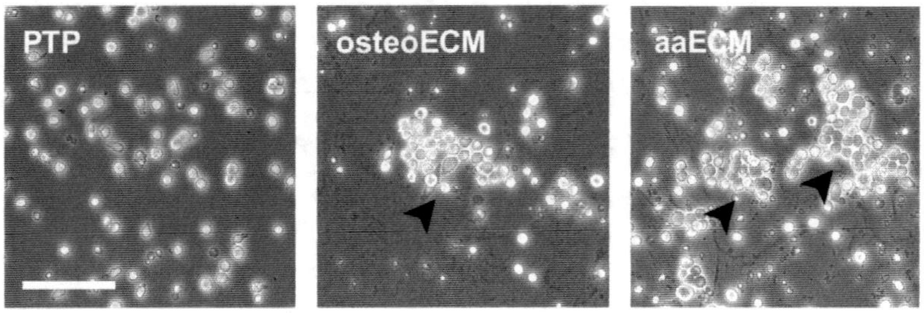

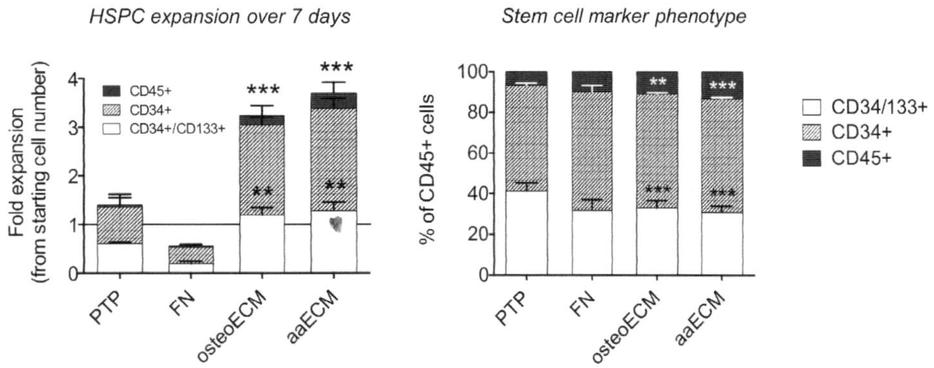

Haematopoietic stem cells on MSC-derived ECM

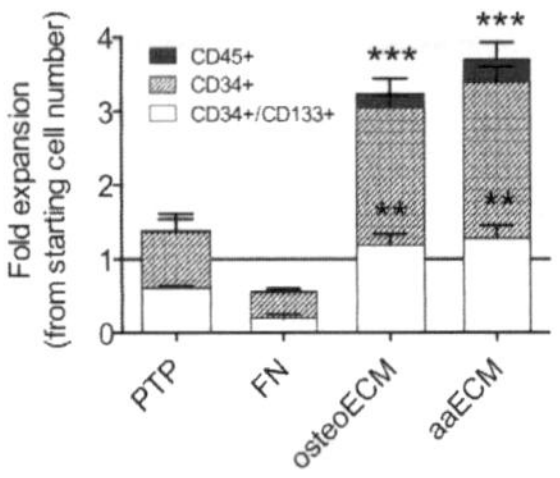

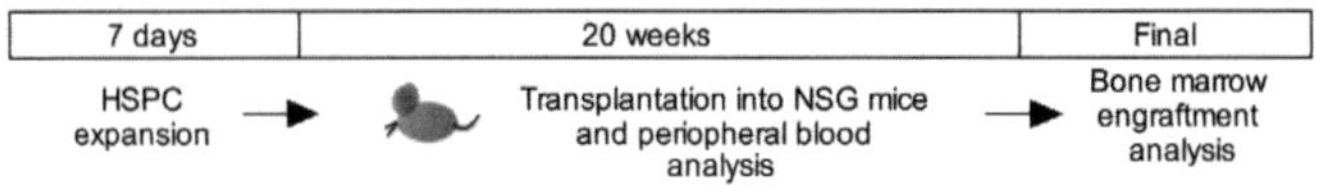

Peripheral blood chimerism

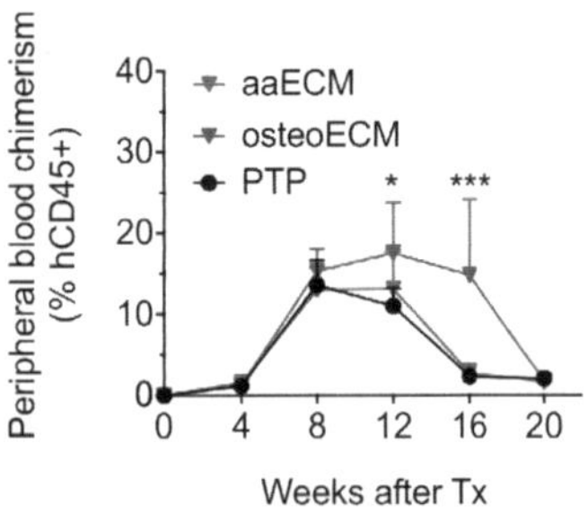

Long-term engraftment
Bone marrow chimerism at week 20

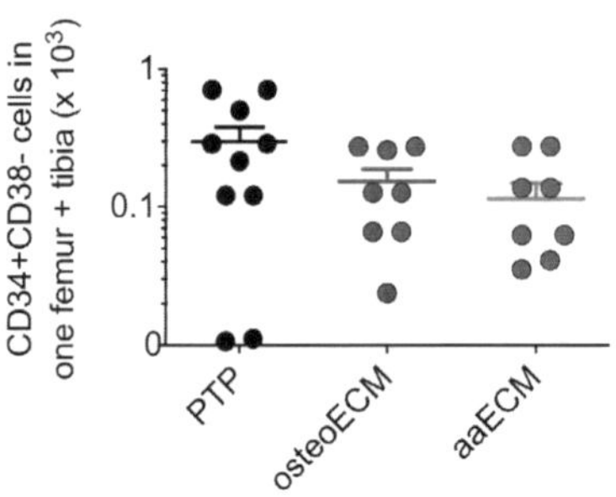

Expansion of HSPC without loss in long-term engraftment capacity

Outlook

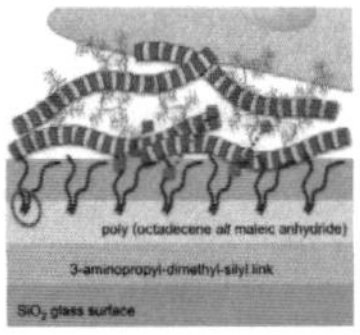

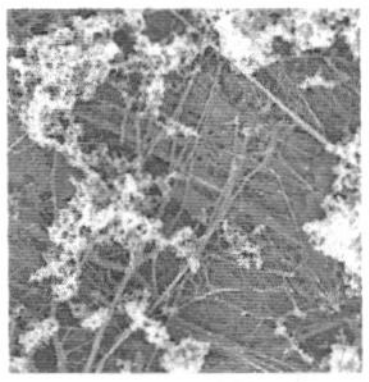

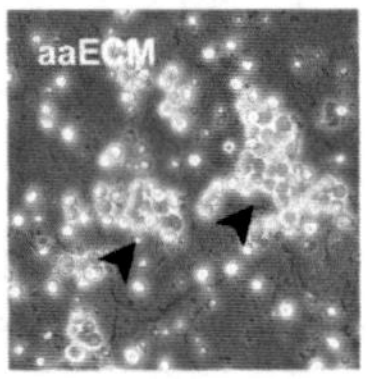

- *In vitro* immobilization of cell-secreted ECM from epithelial or mesenchymal cell types to create basement membrane or connective tissue mimetic ECM scaffolds
- Additional functionalization of decellularized ECM by uptake and release of further cytokines and growth factors
- Co-culture of haematopoietic stem cells with other niche supporting cells within the ECM scaffold (MSC, osteoblasts, reticular CAR cells)
- Further characterization of cell-derived ECM will aid the design of artificial scaffolds and microenvironments

Leibniz-Institut für Polymerforschung Dresden e. V.
MAX BERGMANN center of biomaterials dresden
TECHNISCHE UNIVERSITÄT DRESDEN
CRTD
Center for Regenerative Therapies Dresden
Medizinische Akademie Carl Gustav Carus
Dresden
DFG SFB 655
Dresden
Cells into tissues
Leibniz Institute for Polymer Research, Dresden
Prof. Carsten Werner
Dr. Philipp Seib
Dr. Jens Friedrichs
Aline Stissel
University Hospital Dresden, Carl Gustav Carus
Prof. Martin Bornhäuser
Dr. Malte von Bonin
Katrin Müller
Kristin Heidel
University of Technology, Dresden
Dr. Konstantinos Anastassiadis
Dr. Bernard Hoflack

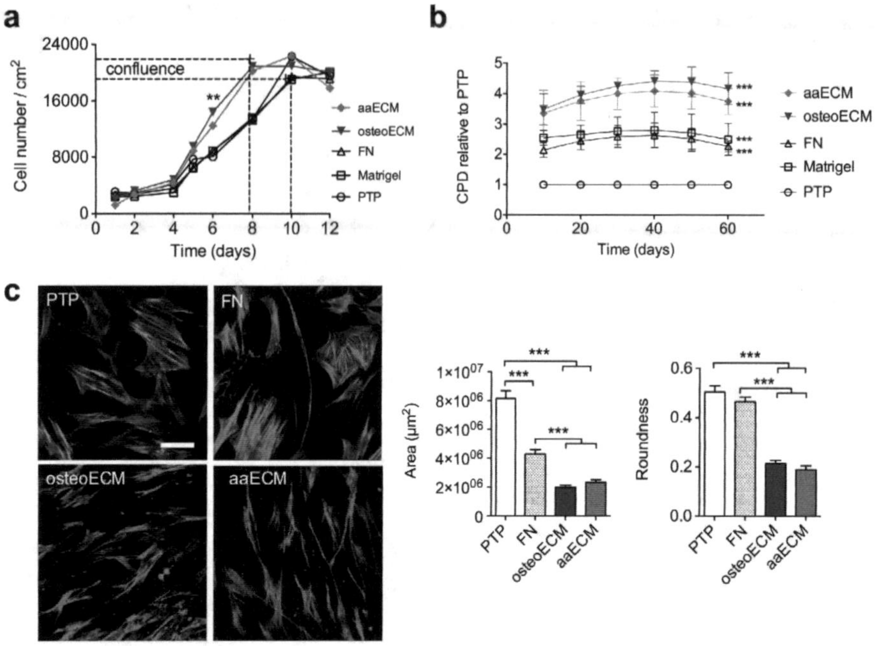
a
24000
confluence
16000
8000
0
0 2 4 6 8 10 12
Time (days)
Cell number / cm2
**
aaECM
osteoECM
FN
Matrigel
PTP
b
CPD relative to PTP
5 4 3 2 1 0
0 20 40 60
Time (days)
***
aaECM
osteoECM
FN
Matrigel
PTP
c
PTP
FN
osteoECM
aaECM
Area (µm2)
1×10^07
8×10^06
6×10^06
4×10^06
2×10^06
0
***
PTP FN osteoECM aaECM
Roundness
0.6
0.4
0.2
0.0
***
PTP FN osteoECM aaECM

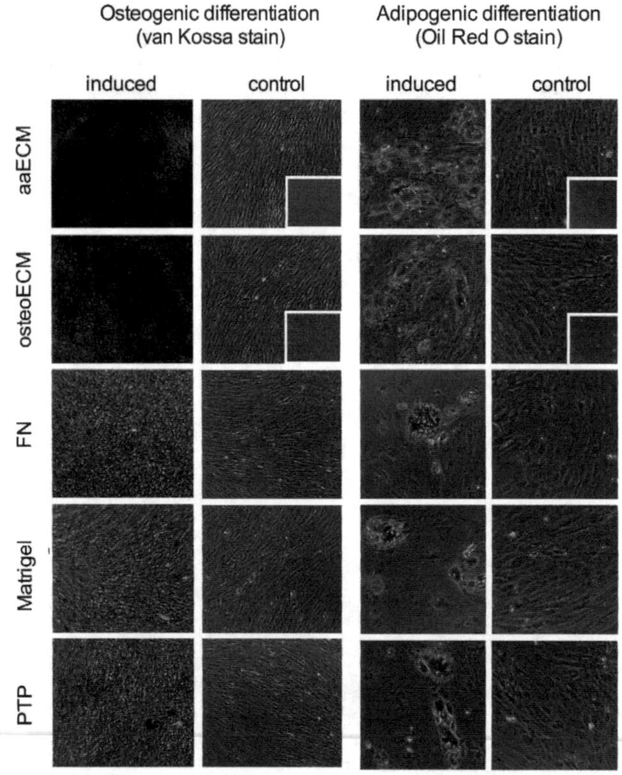
Osteogenic differentiation
(van Kossa stain)
Adipogenic differentiation
(Oil Red O stain)
induced
control
induced
control
aaECM
osteoECM
FN
Matrigel
PTP

p53 Inhibits the Direct Conversion of Fibroblasts into Motor Neurons

Kimberley Babos
University of Southern California
Department of Stem Cell Biology and Regenerative Medicine
March 17, 2014

Background

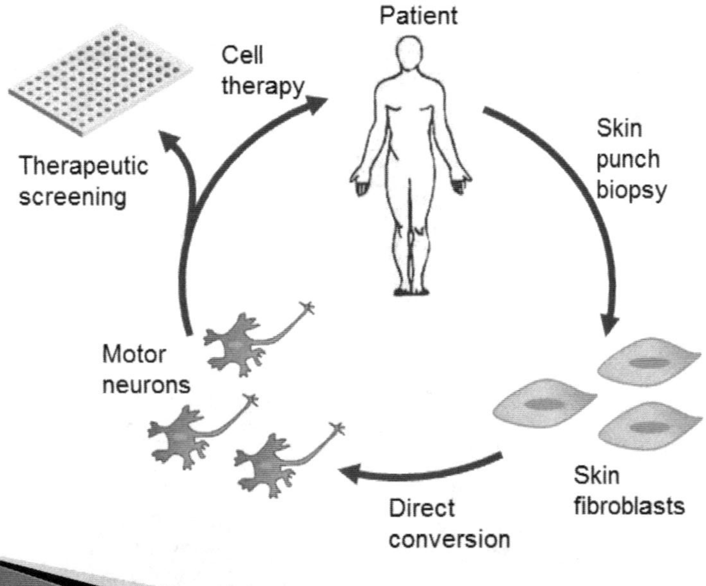

Figure courtesy of K. Galloway

Background

- Cellular reprogramming allows for conversion from one cell type to another

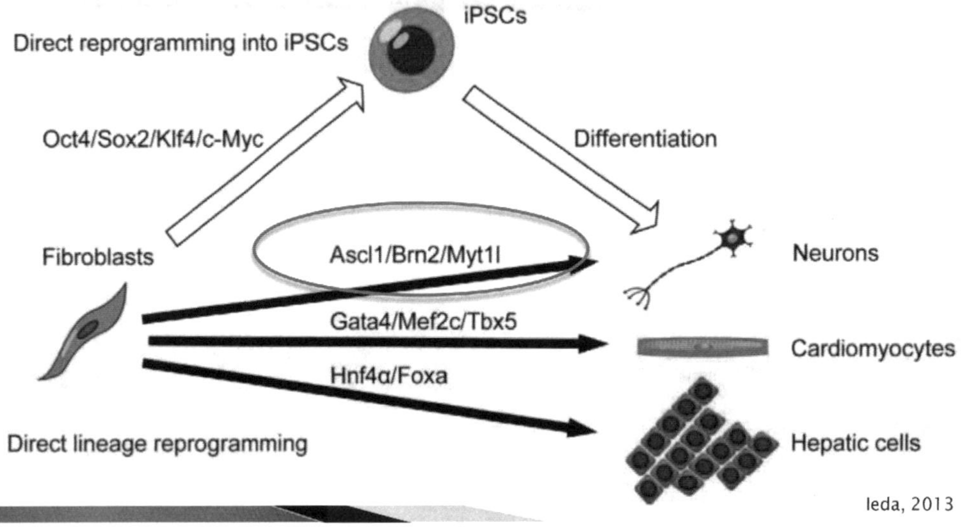

Ieda, 2013

Background

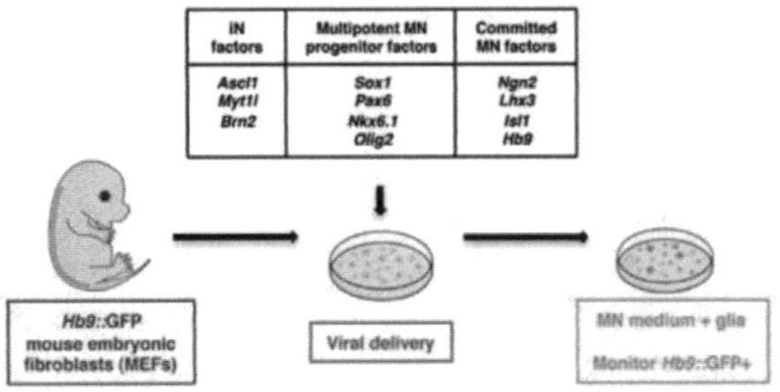

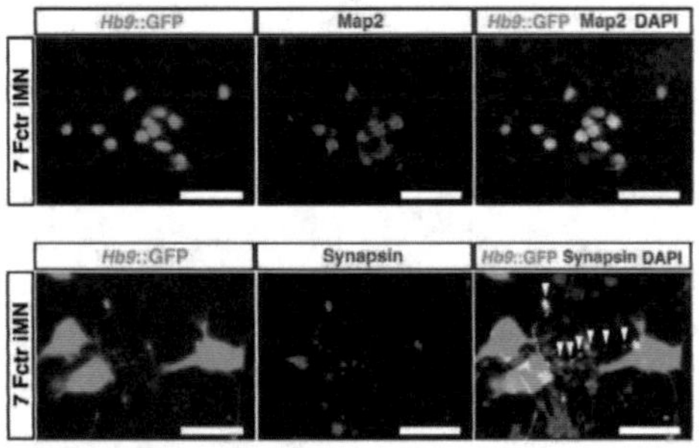

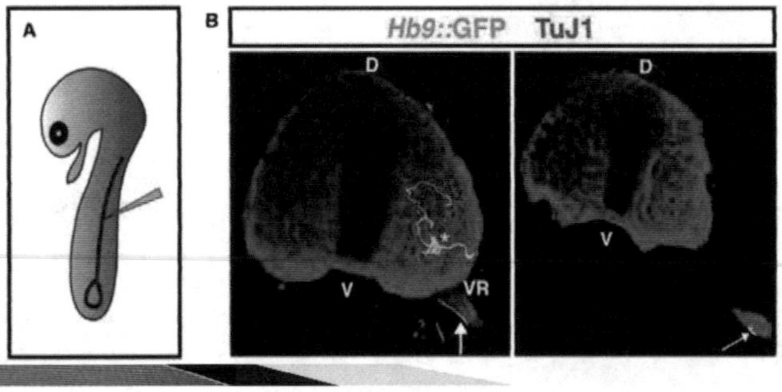

Son, and Ichida *et al*, 2011

Background

Suppression of induced pluripotent stem cell generatio

Hyenjong Hong[1,2], K
Masato Nakagawa[1,2]

Does p53 play a role in direct lineage reprogramming?

Linking the
somatic ce

Teruhisa Kawamura[1,2]*, Jotaro Suzuki[1,3]*, Yunyuan V. Wang[1], Sergio Menendez[4], Laura Batlle Morera[4], Angel Raya[4,5,6], Geoffrey M. Wahl[1] & Juan Carlos Izpisúa Belmonte[1,4]

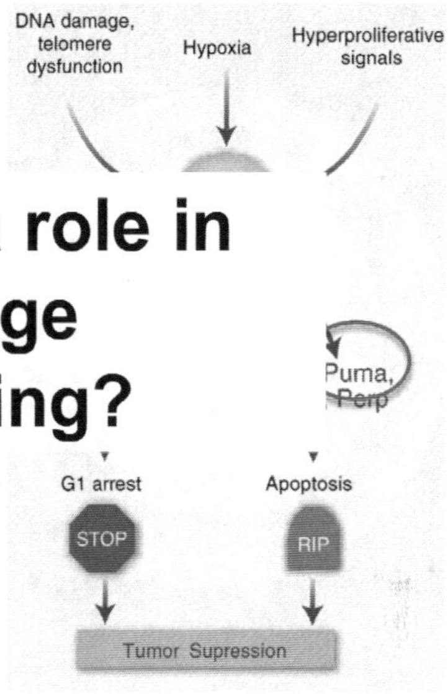

Attardi and DePinho, 2004

Approach

iMN Factors		p53 Inhibitors
Ascl1	Brn2	p53DD OR
mIsl-1	Lhx3	p53 shRNA OR
Mytl-1	Ngn2	p53 siRNA OR
Hb9		p53 SM Inhibitor

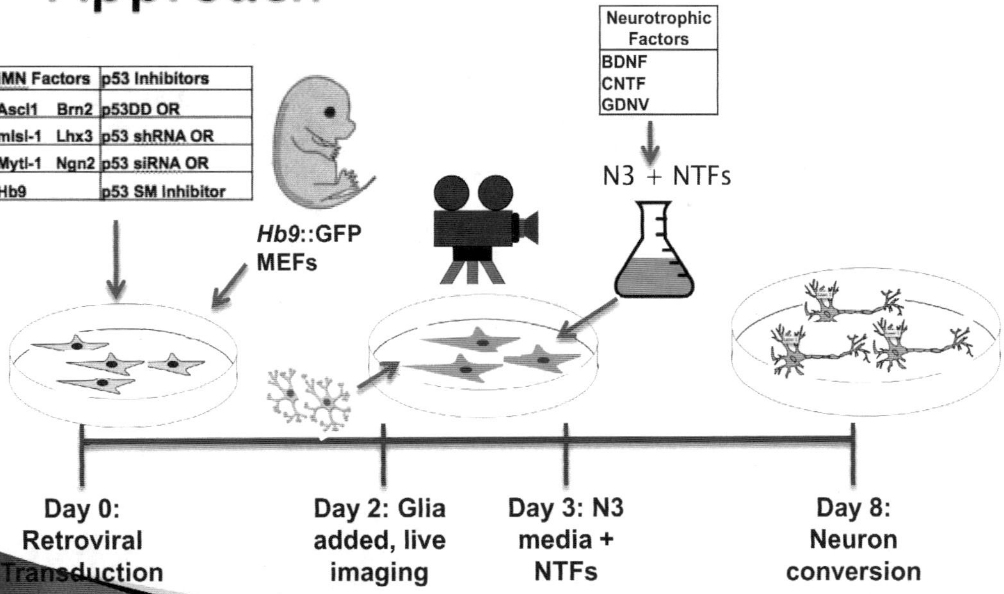

p53 Inhibition

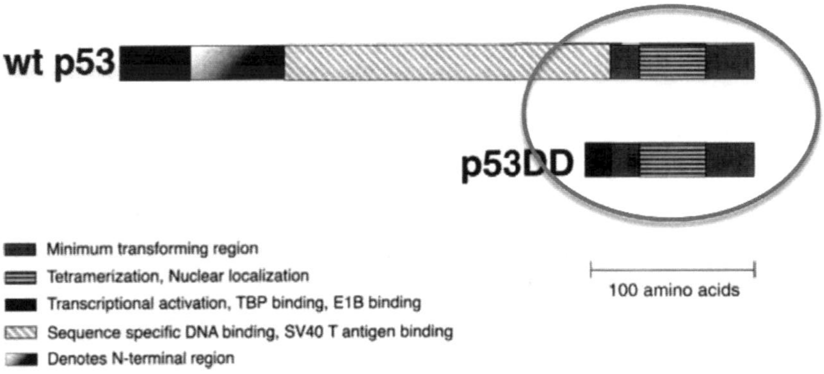

Bowman, *et al*, 1996

p53 Inhibition Increases Efficiency of Mouse iMN Generation

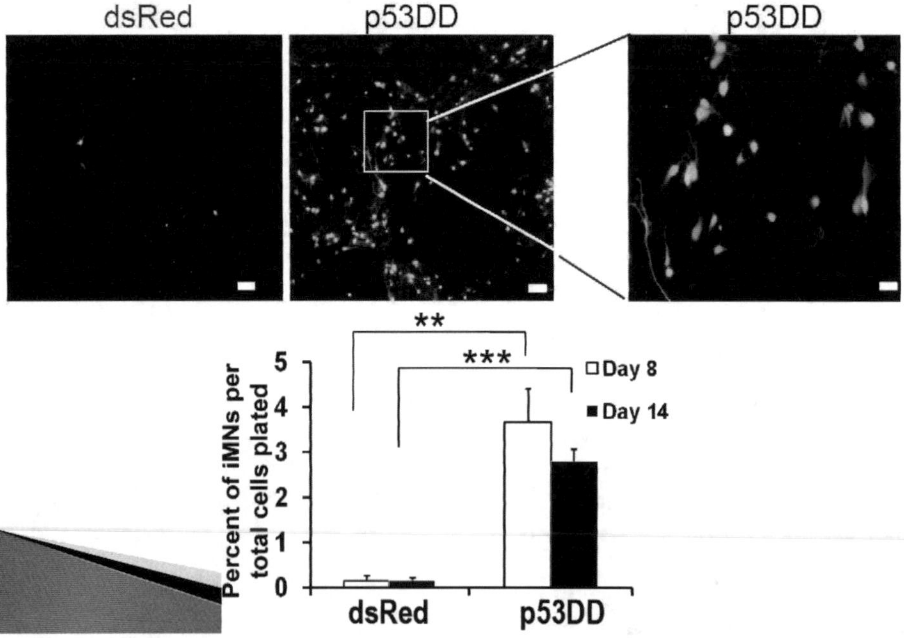

p53 Inhibition Increases Efficiency of Human iMN Generation

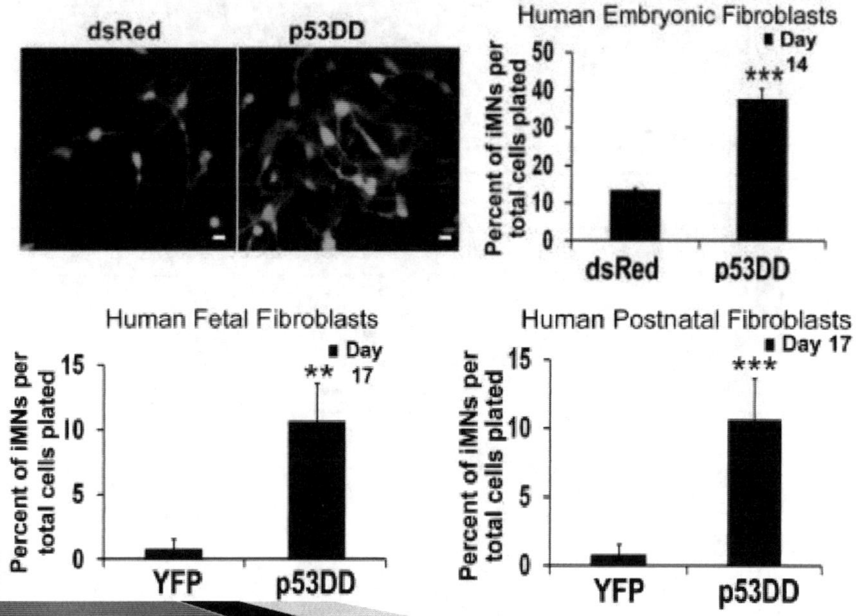

p53 Inhibition Increases Efficiency of Human iMN Generation

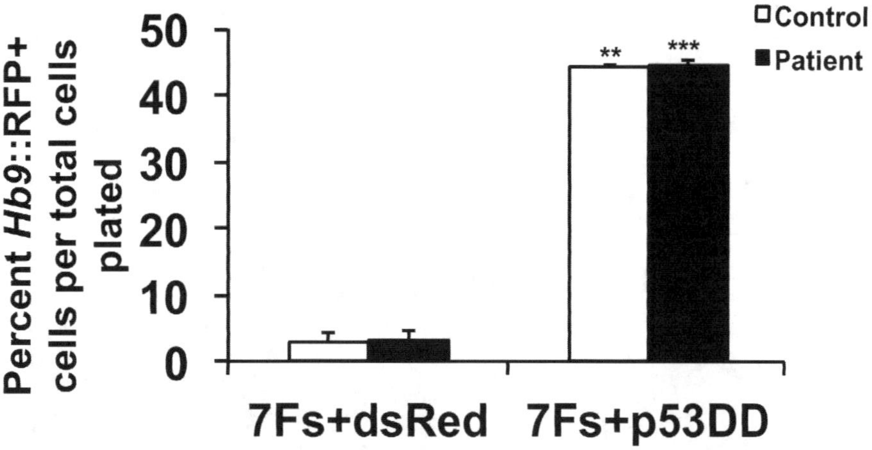

Molecular Characterization of p53DD-iMNs

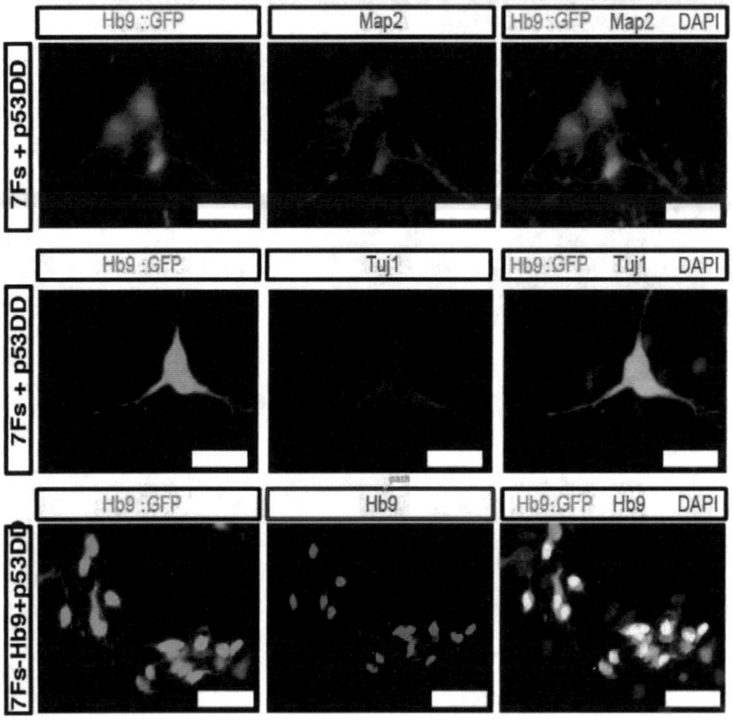

Molecular Characterization of p53DD-iMNs

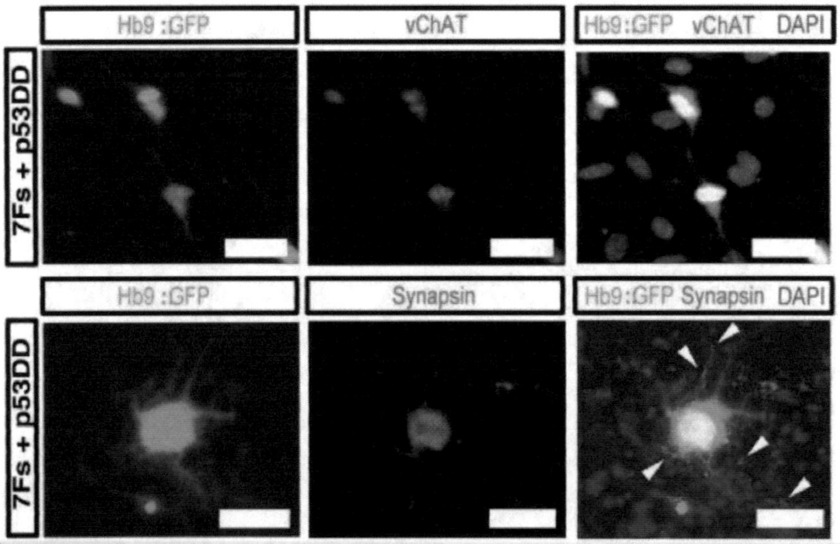

p53DD-iMNs Generate Functional NMJs

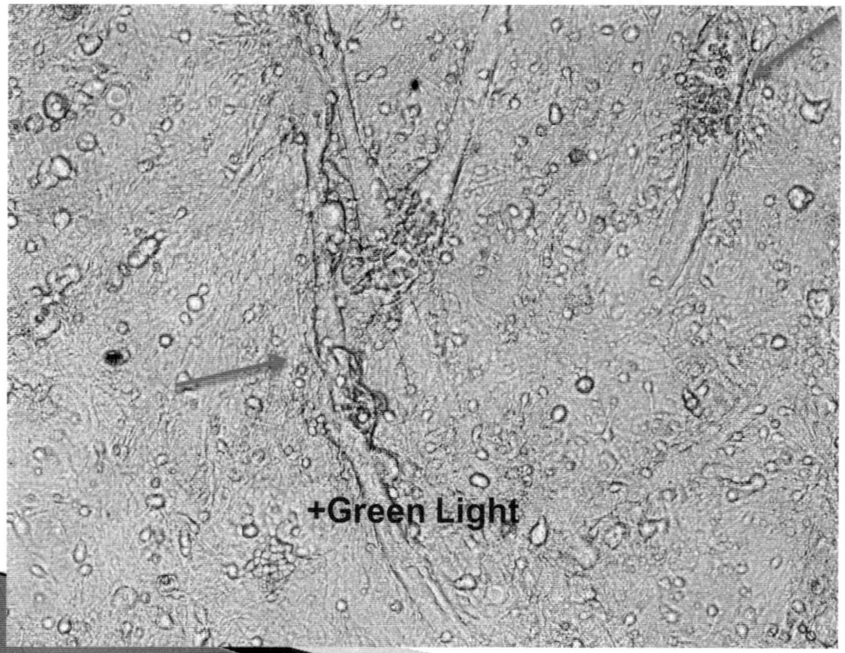

p53 Inhibition Promotes Enhanced Conversion

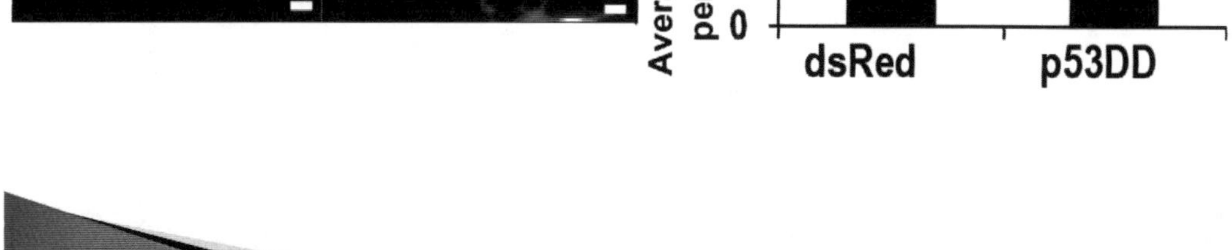

p53 Inhibition Promotes Enhanced Conversion

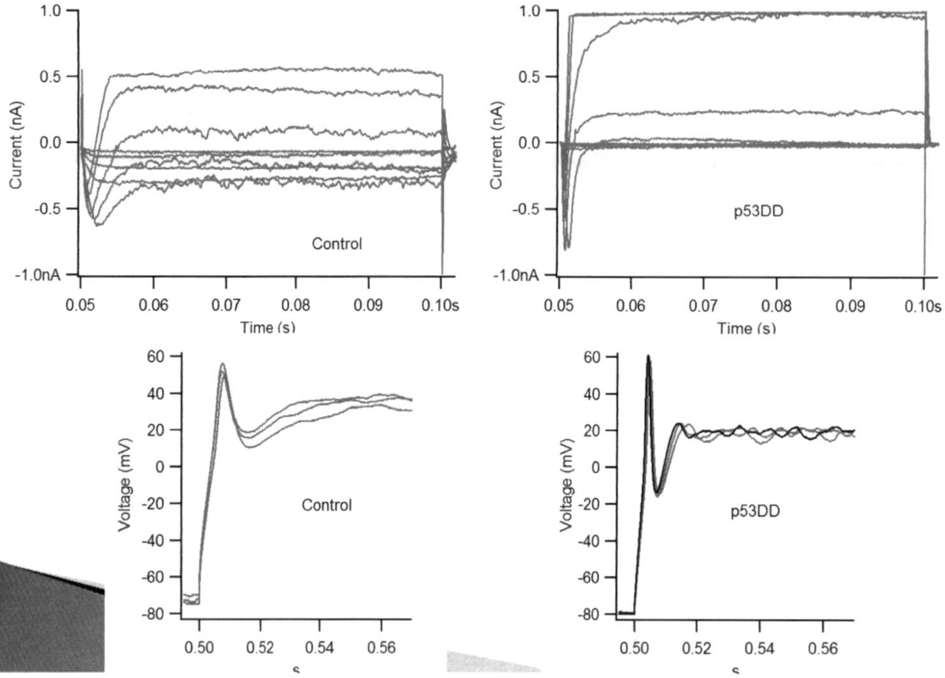

Improved Conversion Not Limited to p53DD

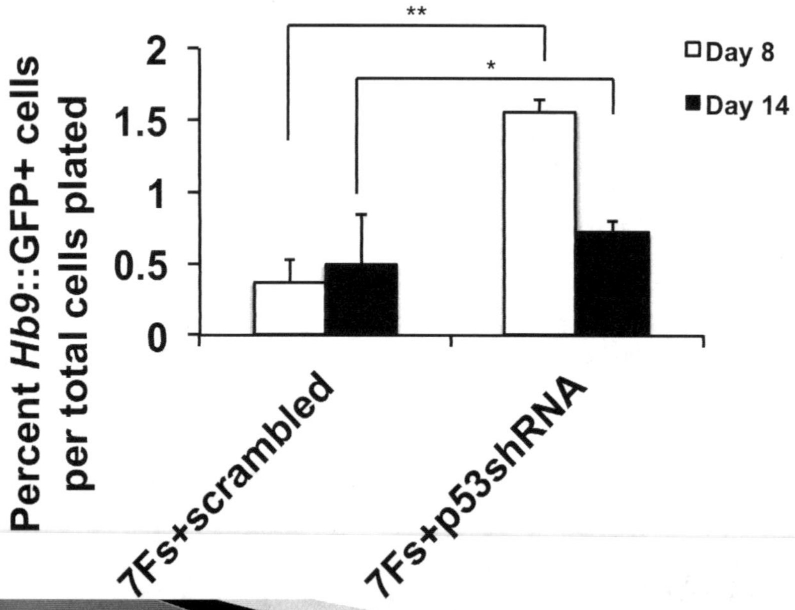

p53 Inhibition Increases Conversion Efficiency into General Neuronal Classes

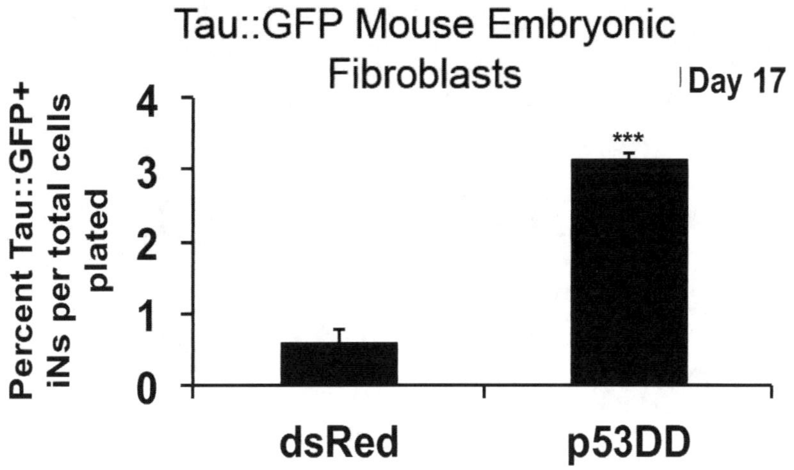

p53 Inhibition Does Not Reduce Programmed Cell Death

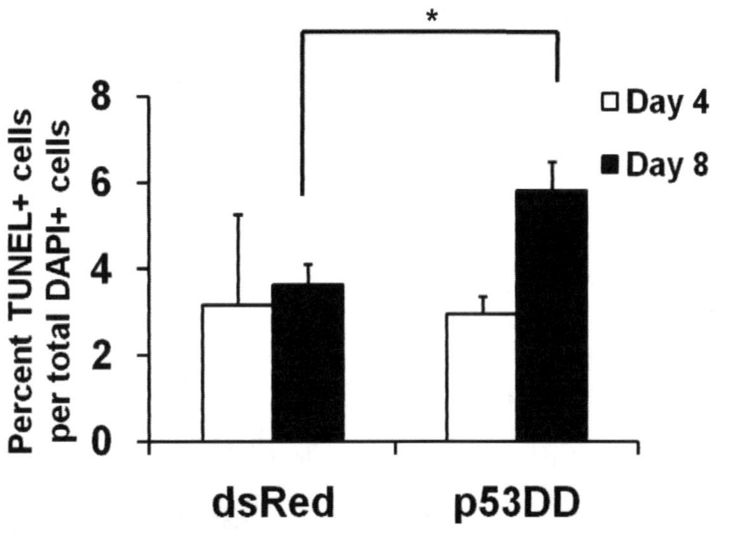

p53 Inhibition Does Not Enhance iMN Survival

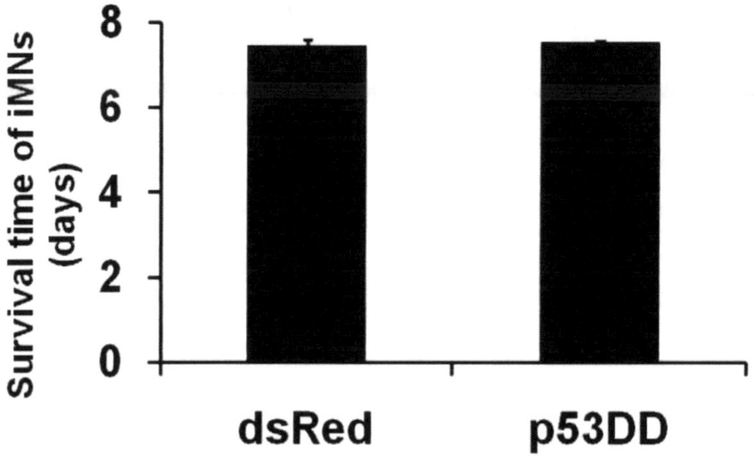

p53 Inhibition Enhances Early-Phase iMN Reprogramming

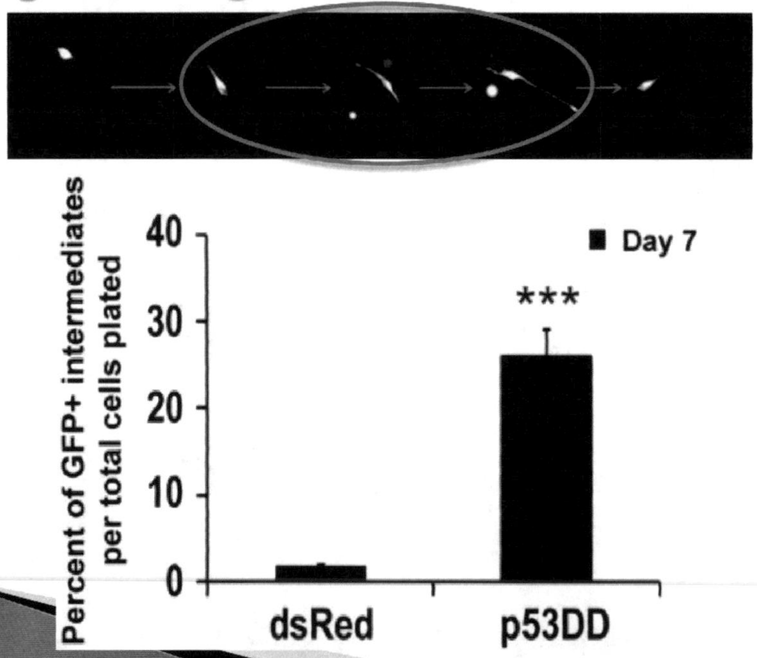

p53 Inhibition Enhances Early-Phase iMN Reprogramming

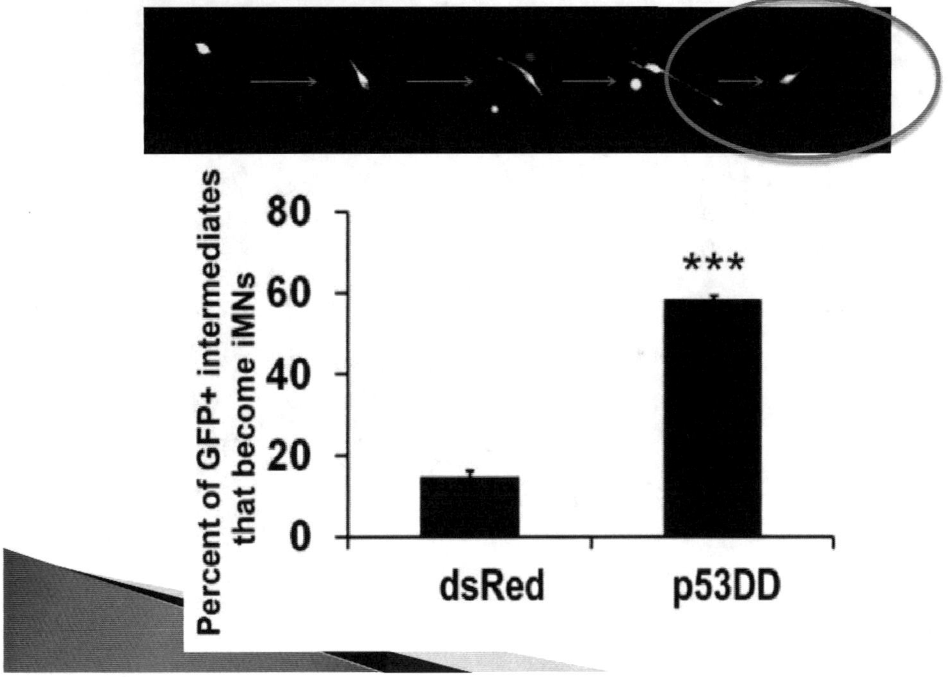

p53 Inhibition Increases Expression of Neural-Related Genes

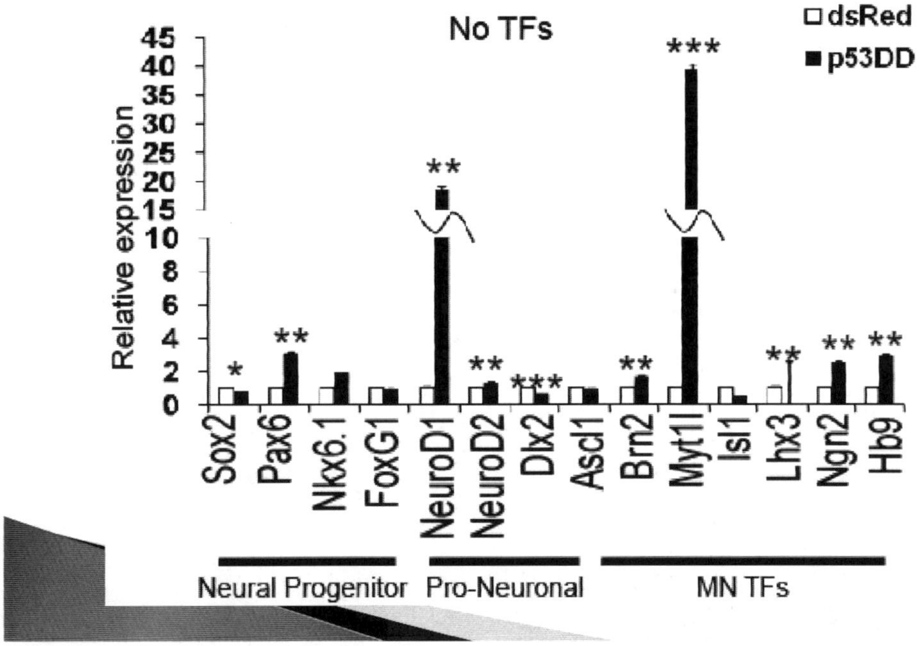

p53 Inhibition Increases Expression of Neural-Related Genes

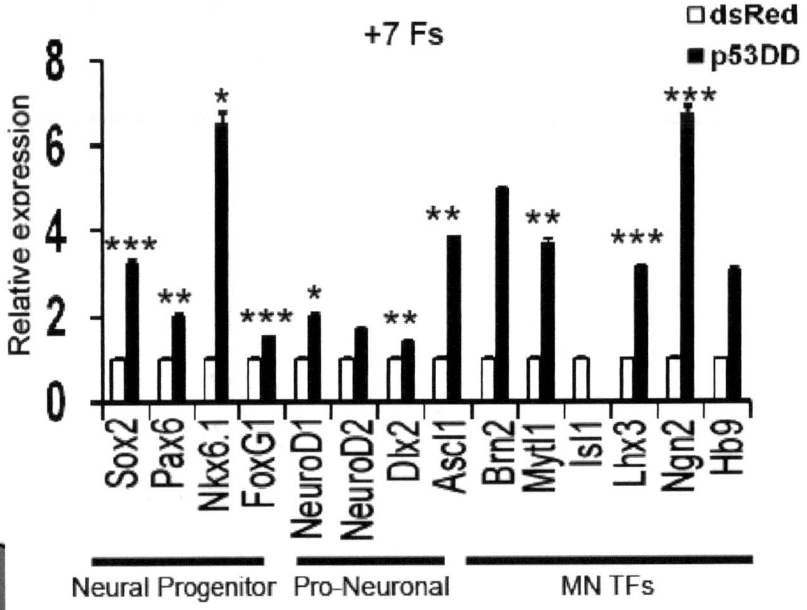

Summary

- Inhibiting p53 improves the direct conversion of mouse and human somatic cells into iMNs
- p53 inhibits reprogramming to post-mitotic cell types
- p53 acts early in reprogramming
- p53 plays an apoptosis-independent inhibitory role in transdifferentiation
- p53 can inhibit neuronal lineage commitment

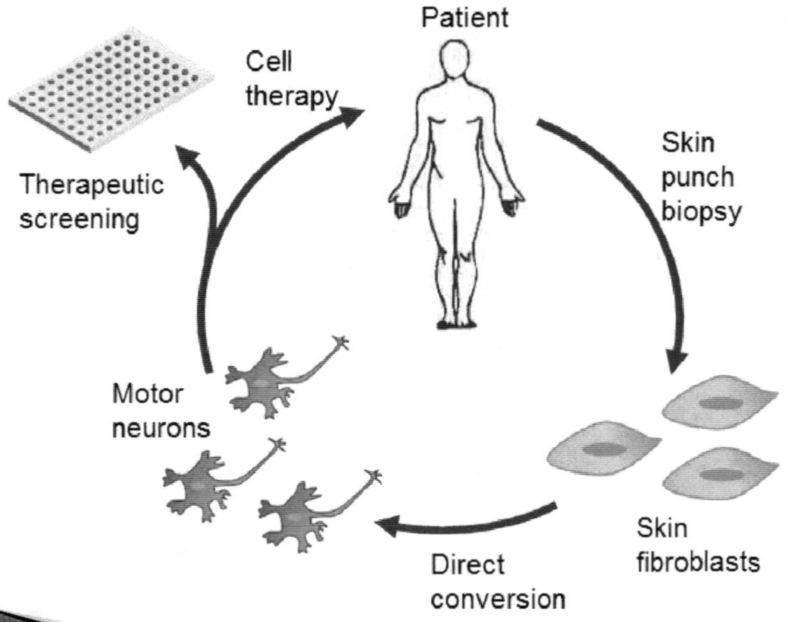

Figure courtesy of K. Galloway

Acknowledgments

- Dr. Justin Ichida and Lab Members
  - Yingxiao "TK" Shi
  - Yichen Li
  - Sebastian Lin
  - Katie Galloway
  - Suhasni Gopalakrishnan
  - Pooja Hor
  - Valerie Hennes
  - Qing Liu-Michael
- Dr. Berislav Zlokovic and Lab Members
  - Kassandra Kisler Elliot
- Dr. Wange Lu and Lab Members
  - Wen-Hsuan Chang

USC University of Southern California

ThermoFisher
SCIENTIFIC

Creation of iPSC Based Model Systems to Study Parkinson's Disease

Bob Harrington, PhD
Global Leader, Strategic Alliance Management
Discovery Biology and Stem Cell Sciences
Cell Biology
Life Sciences Solutions Group

The world leader in serving science

Thermo Fisher Scientific's stem cell workflow: Products and services to drive program success

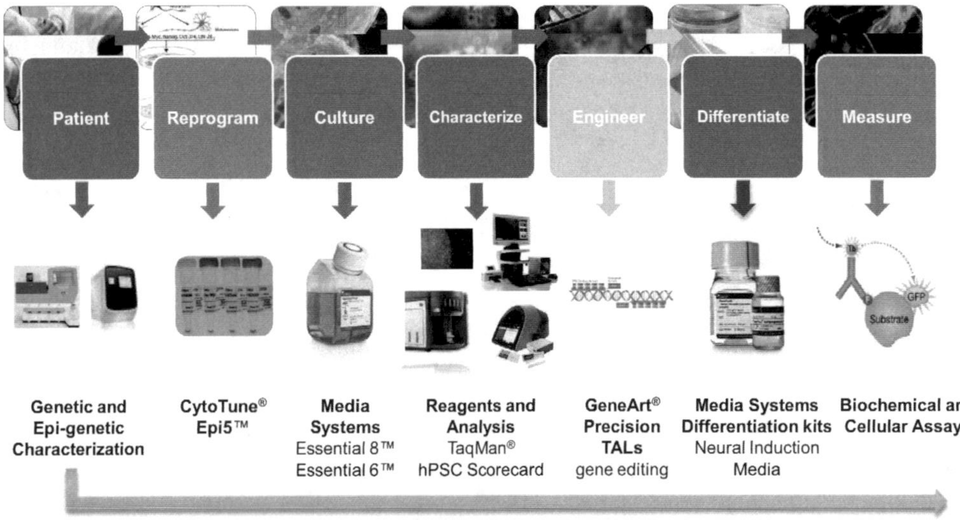

Tools for iPSCs and ESCs
Services available to expand your capabilities

2

ThermoFisher
SCIENTIFIC

CellModel™ Services workflow

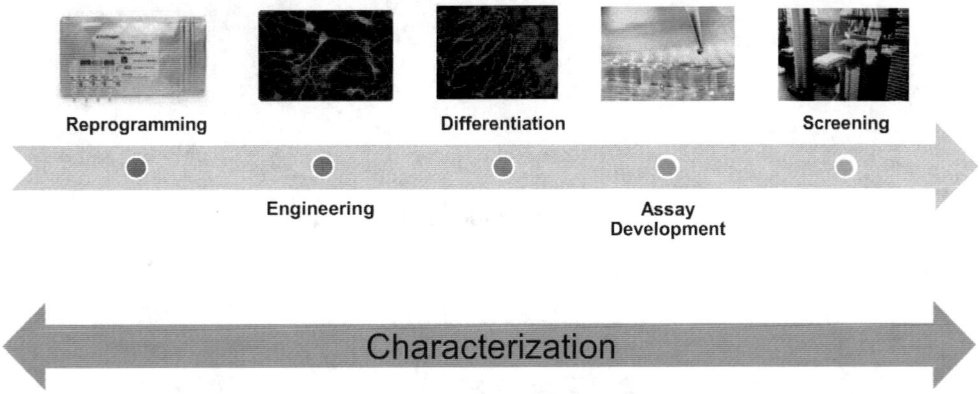

Choose a single step or a complete workflow

Information: www.lifetechnologies.com/cellmodel

Outline

1. Background:
   Parkinson's Disease, The Parkinson's Institute, and Thermo Fisher Scientific

1. Generation and Characterization of iPSCs from PD-affected donors using the CytoTune®-iPS Sendai Reprogramming Kit

2. Generation and Characterization of NSCs and neurons using the Gibco® PSC Neural Induction Medium

3. Gene Editing using GeneArt® Precision TALs

1. Development of disease-relevant phenotypic assays

All products and applications in this presentation are for research only. Not for use in diagnostic procedures.

Partnership Between the Parkinson's Institute and Thermo Fisher Scientific

The Parkinson's Institute and Clinical Center

causecarecure

Birgitt Schuele

J. William Langston

Parkinson's Institute, Sunnyvale CA

Dedicated to finding the cause and a cure for Parkinson's disease

life technologies™

ThermoFisher SCIENTIFIC

The world leader in serving science

Harnessing the power of science and technology to shape biological discovery and improve life

Background: Parkinson's Disease

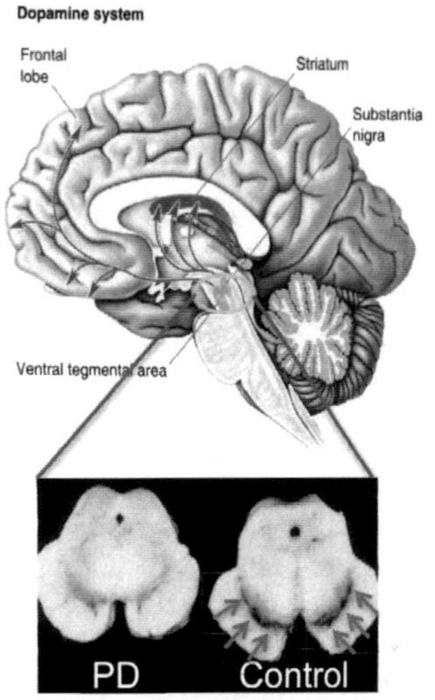

Section of SNPC from autopsy sample. Dark area in control sample shows healthy neurons.

- Neurodegenerative disorder affecting 1-2 % of the general population over age 65.
- 2nd most common ND disorder (after Alzheimer's)
- Cardinal symptoms include resting tremors, bradykinesia (slowness of movements) and muscle rigidity. Progression can include cognitive impairment, depression, and dementia.
- Marked by a loss of midbrain dopaminergic neurons (mDA) in the substantia nigra pars compacta (SNPC)

Collaboration Overview

Fibroblasts

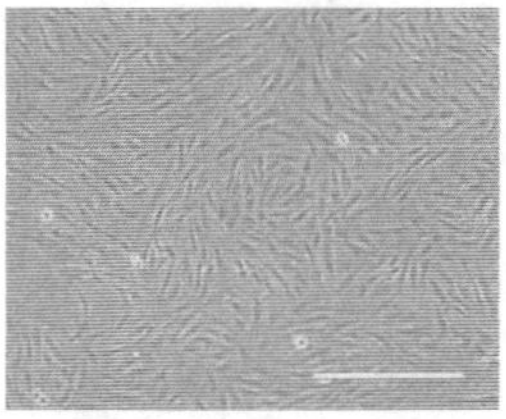

iPSCs

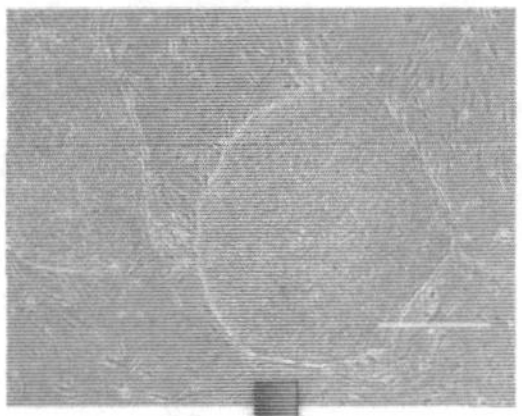

Neural Stem Cells

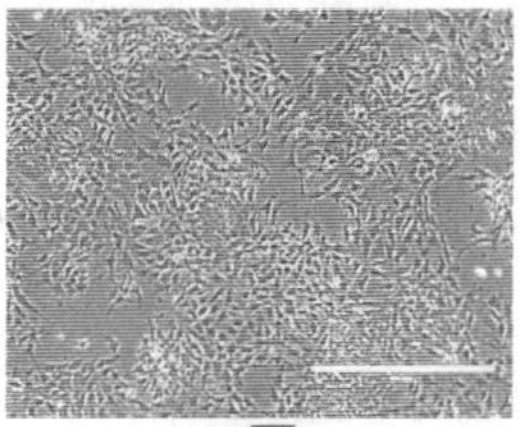

ID	Age at biopsy	Disease	Mutation
PD-1	19	YOPD	PARKIN
Ctrl-1	25	control	
PD-2	74	PD	LRRK2
PD-3	61	PD	LRRK2/GBA
MSA	62	MSA	
MSA-2	79	MSA	
Ctrl-2	69	control	
Ctrl-3*	46	control	

Use TAL-based gene editing to specifically modify disease-relevant genes.

YOPD: Young onset Parkinson's disease
PD: Parkinson's disease
MSA: Multiple system atrophy
Ctrl-3: Son of MSA-2 (unaffected)

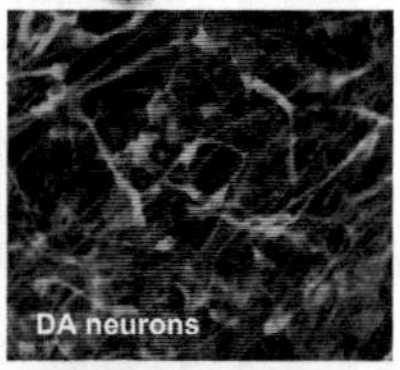

Functional Assays

Goals:

- Understand the molecular basis for Parkinson's Disease
- Identify compounds that can impact disease phenotypes

Generation and Characterization of iPSCs from PD Affected Donors using the CytoTune®-iPS Sendai Reprogramming Kit

CytoTune®-iPS Sendai Reprogramming Kit addresses major challenges in reprogramming

Reprogramming Challenges:

- **Integrating technologies** can affect cell function
- **Efficiency** in hard to reprogram cell types, or with limited samples
- **Ease-of-use**: certain technologies require multiple transfections

Sendai Virus:

- RNA virus
- Non-pathogenic
- Broad tropism

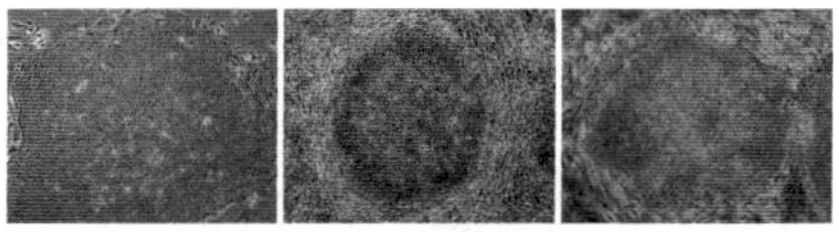

Simple Workflow:

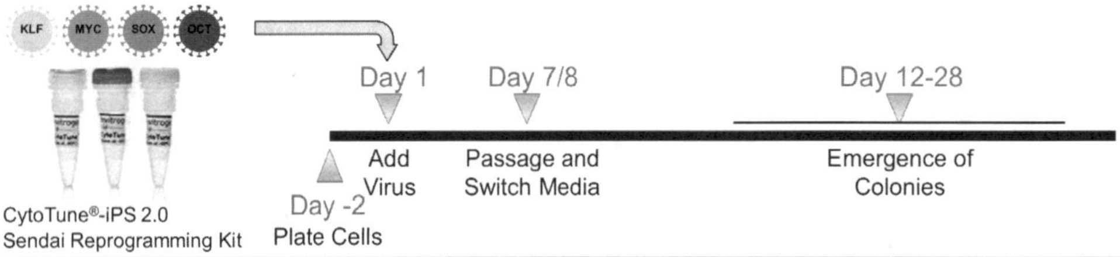

iPSC generation from PD Donors in Essential 8™ media

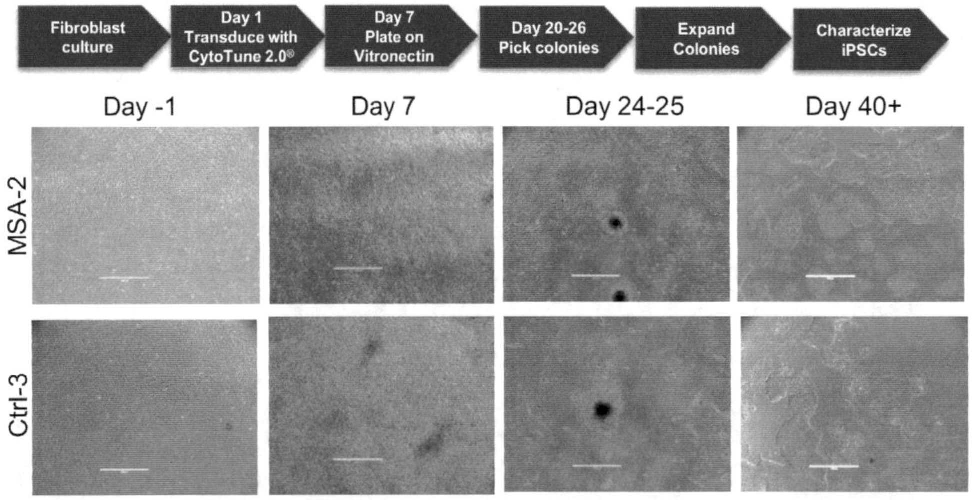

Fibroblasts cultured in DMEM / FBS

7 days post-transduction, passage to 10 cm Vitronectin plates in Essential 8 medium

Pick colonies on day 24 (bottom) and day 25 (top)

Expand individual colonies for cryopreservation and characterization

Reprogramming efficiency varied widely across lines

Donor	Mutation	Day 7 plating density (# cells plated/10-cm MEF dish) 50,000 cells SSEA4⁺ colonies (Efficiency)	100,000 cells SSEA4⁺ colonies (Efficiency)	200,000 cells SSEA4⁺ colonies (Efficiency)
PD-1	Parkin	55 (0.110%)	132 (0.13%)	583 (0.300%)
Ctrl-1	-	35 (0.070%)	81 (0.08%)	97 (0.050%)
PD-2	LRRK2	55 (0.110%)	63 (0.06%)	193 (0.100%)
PD-3	LRRK2/GBA	3 (0.006%)	7 (0.007%)	17 (0.009%)
MSA		3 (0.006%)	20 (0.02%)	40 (0.020%)
Ctrl-2		22 (0.044%)	18 (0.02%)	63 (0.031%)

Reprogramming efficiency varied from <0.01% to 0.3%
Original CytoTune®-iPS Reprogramming technology was used for this study.
CytoTune®–iPS 2.0 (launched July 2013) doubles reprogramming efficiencies

Thermo Fisher SCIENTIFIC

Characterization of iPSCs

Karyotype (all normal)

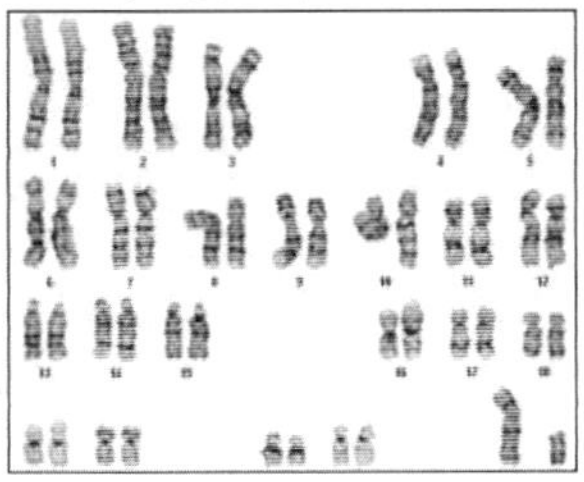

Markers by ICC

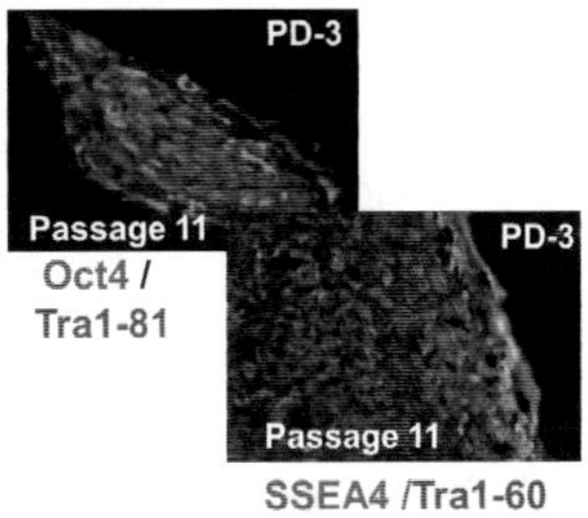

EB Formation and Staining

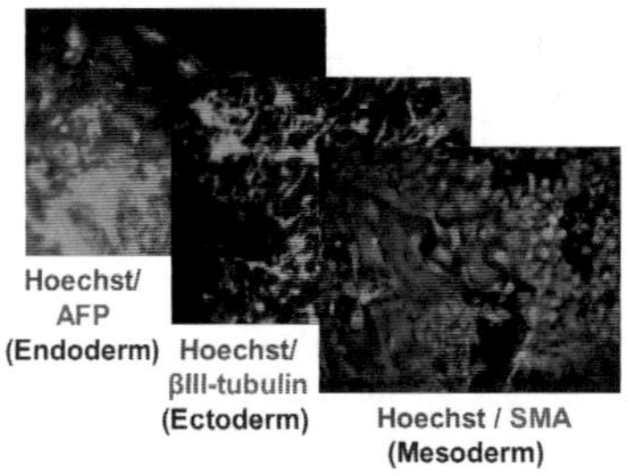

Markers by TaqMan® qPCR

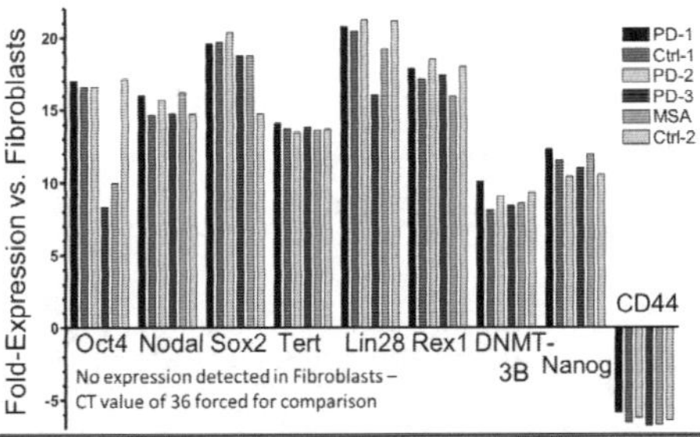

Markers by Flow Cytometry

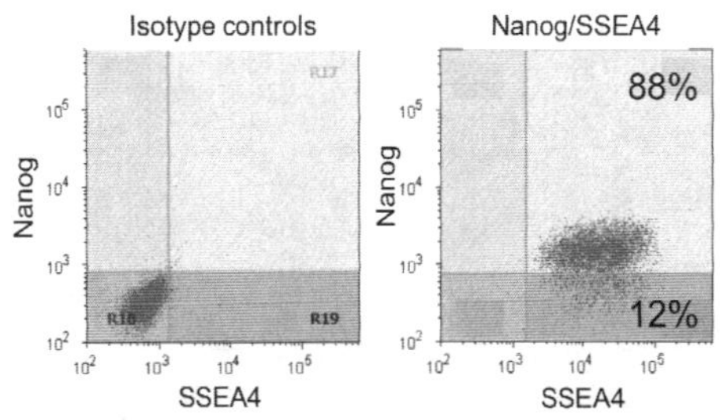

Thermo Fisher SCIENTIFIC

TaqMan® hPSC Scorecard™ Panel

- Developed under research collaboration with Dr. Alex Meissner
- Based on his landmark publication (Bock et al. *Cell* 144, 439–452 2011)
- Reduced content includes only the key predictive markers and controls

SCORE!

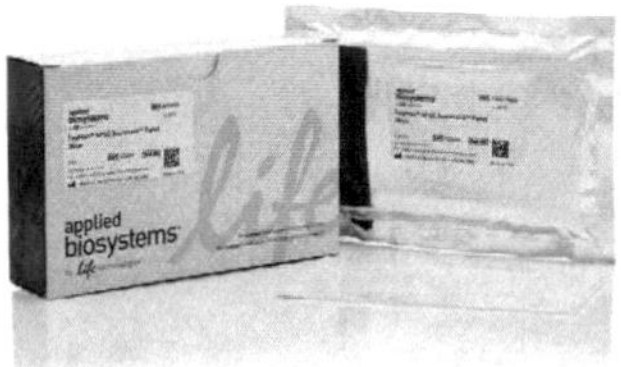

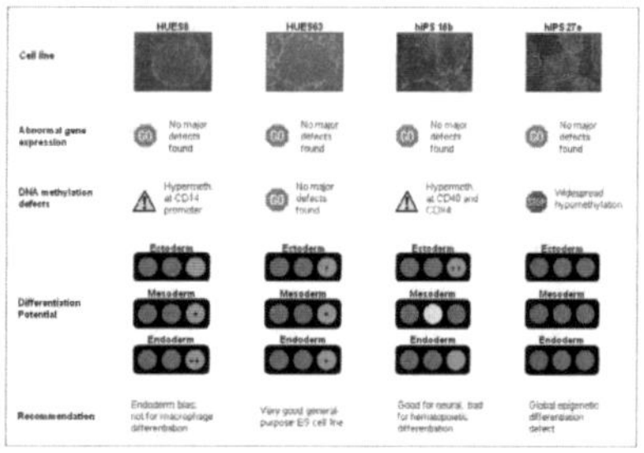

Resource

Cell

Reference Maps of Human ES and iPS Cell Variation Enable High-Throughput Characterization of Pluripotent Cell Lines

Product information and how-to videos:
lifetechnologies.com/scorecard

iPSC lines were analyzed using the TaqMan® hPSC Scorecard Panel

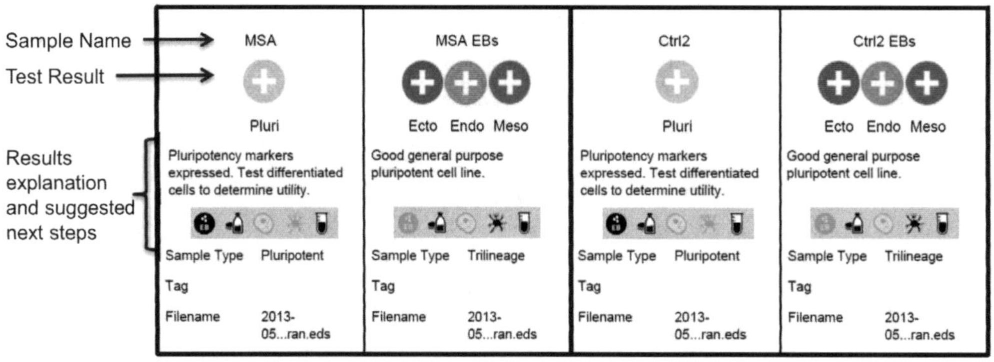

- iPSC cells from donor lines tested positive for pluripotency
- Late passage iPSC cells from MSA and Ctrl2 tested positive for pluripotency and EBs show potential for all three lineages.

Generation and Characterization of NSCs using the Gibco® PSC Neural Induction Medium

Gibco® PSC Neural Induction Medium

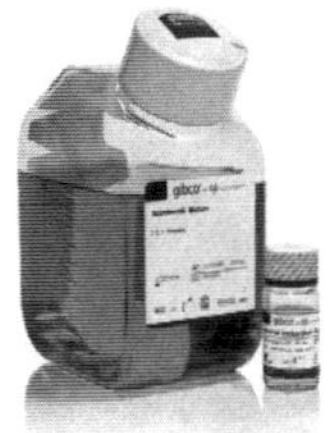

- Simplified workflow
  - ✓ High efficiency neural induction in only 7 days
  - ✓ Eliminates the need to form EB or Rosettes
  - ✓ Adherent culture conditions
- Scalable production
  - ✓ Generate >20M NSCs starting from 1M PSCs
  - ✓ Less variability (No EBs)
- Enables modeling of the brain
  - ✓ High expression of NSC markers
  - ✓ Validated for further differentiation into other neural cell types

More information at lifetechnologies.com/nim

NSCs were generated from PD iPSCs in 7 days

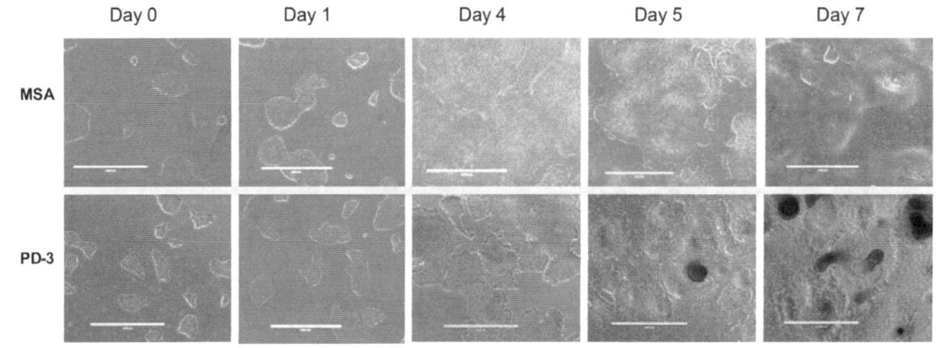

- Novel method allows facile NSC generation, bypassing EB formation / rosette isolation.
- Substantial differences in morphology were observed between lines during the 7-day induction, all NSCs had proper morphology.

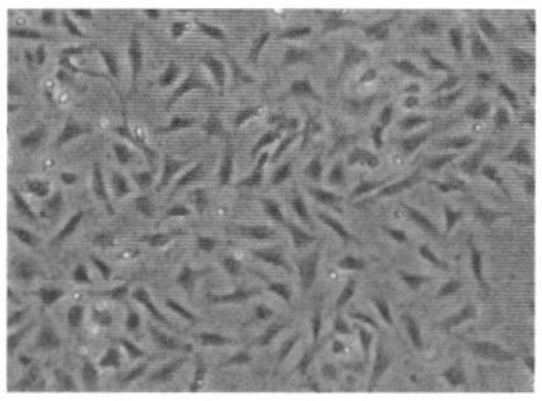

Representative NSC morphology

Karyotypically normal

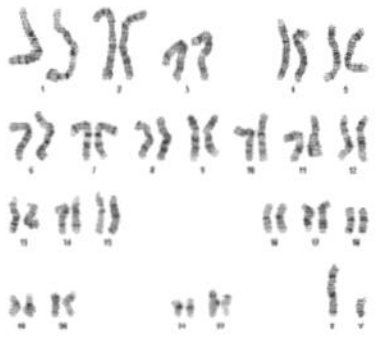

Thermo Fisher SCIENTIFIC

NSCs were characterized for appropriate markers

Markers by ICC

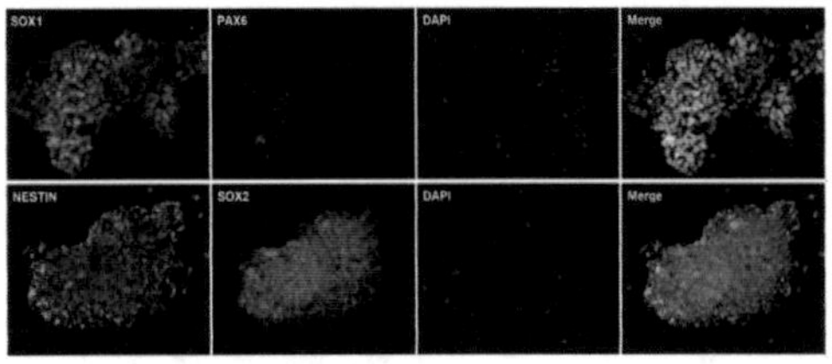

Images obtained using Evos® FLoid® imaging station

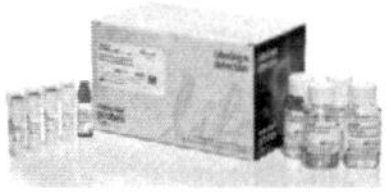

Human Neural Stem Cell Immunocytochemistry Kit

Markers by TaqMan® qPCR

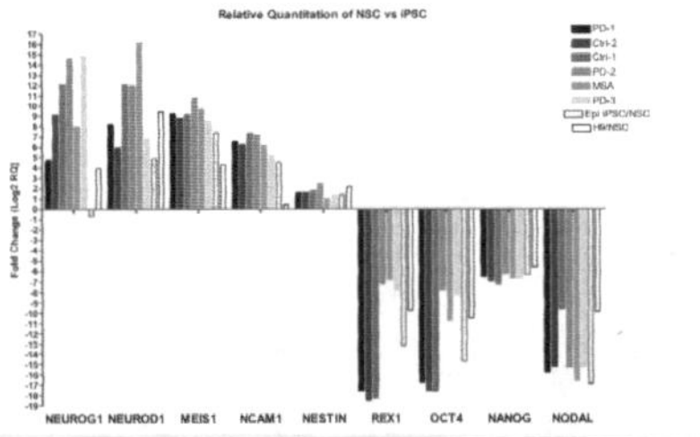

Markers by Flow Cytometry

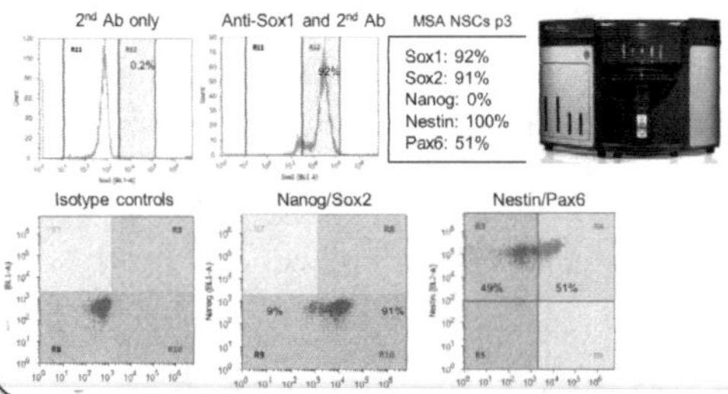

Thermo Fisher SCIENTIFIC

Differentiation of NSCs to midbrain DA Neurons

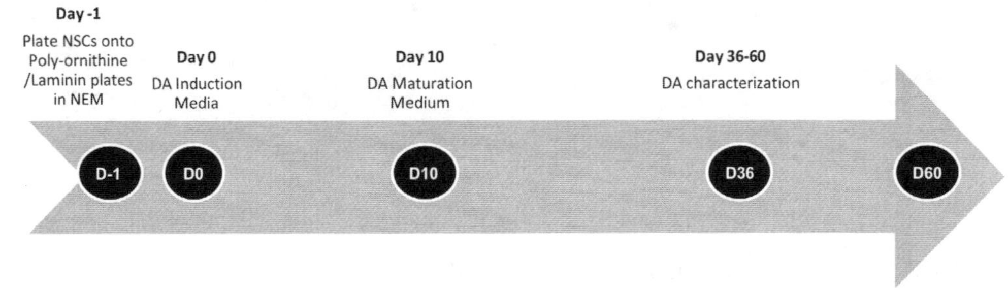

Day -1: Start with **high quality** NSCs (Sox1/Sox2/Nestin positive, <7 passages in NEM).
Day 0: DA Induction Media (Neurobasal / B27 + SHH / FGF8) – 10 Days
Day 10: DA Maturation Media (Neurobasal / B27 + BDNF / GDNF) – 26-50 Days

Protocol based on:
Xeno-free defined conditions for culture of human embryonic stem cells, neural stem cells and dopaminergic neurons derived from them.
Swistowski A, Peng J, Han Y, Swistowska AM, Rao MS, Zeng X.
PLoS One. **2009** Jul 14;4(7)

19 ThermoFisher SCIENTIFIC

NSCs can be differentiated into DA neurons

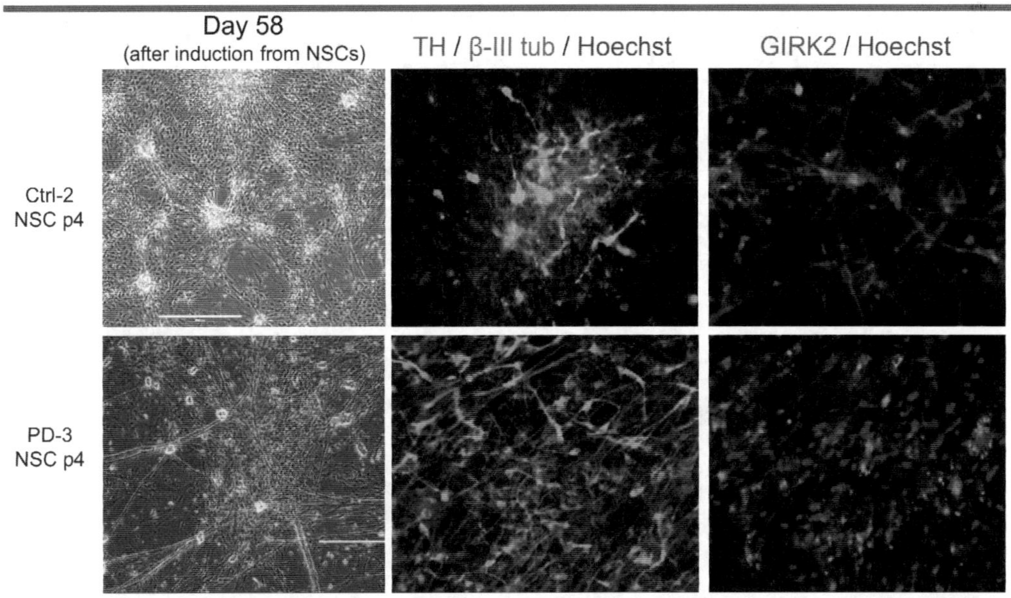

Cells are positive for DA marker TH and A9 subtype DA neuron marker GIRK2.

20 ThermoFisher SCIENTIFIC

Genomic Editing of PD iPSC lines

21

Genome editing of iPSCs to generate disease models

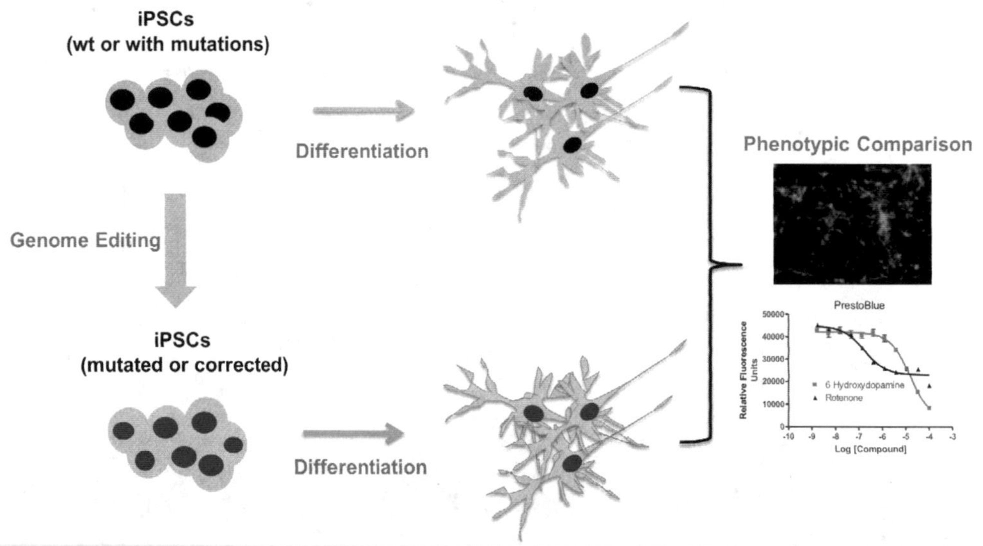

22

Genome editing using GeneArt® Precision TALs

- Overview of TAL technology
- Overview of Editing Process
- Screening and Isolation of Edited Colonies

- Three examples:
  - Disruption of SNCA / α-synuclein
  - Correction of LRRK2 G2019S
  - Correction of GBA N370S

An in-depth webinar that goes through the methods used in greater technical detail is available:
https://event.webcasts.com/starthere.jsp?ei=1022254

Transcription Activator Like (TAL) Nuclease

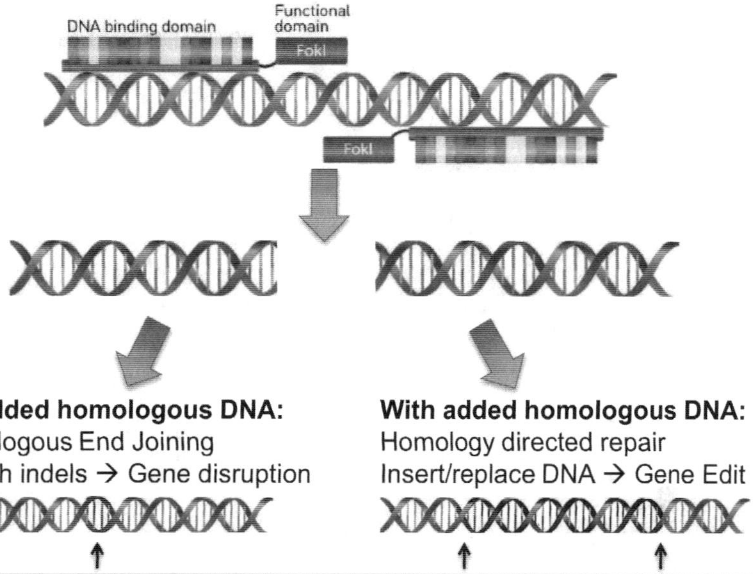

Without added homologous DNA:
Non-Homologous End Joining
Repairs with indels → Gene disruption

With added homologous DNA:
Homology directed repair
Insert/replace DNA → Gene Edit

Background – Multiple System Atrophy (MSA)

MSA:

- Neurodegenerative disorder with Parkinsonian features (tremors at rest, rigidity of muscles, slowness in moving)
- No known genetic basis or risk-factors.
- Poor response to standard dopamine-replacement therapy (levadopa) used in treating PD
- Characterized by glial α-synuclein deposits (Lewy Bodies)

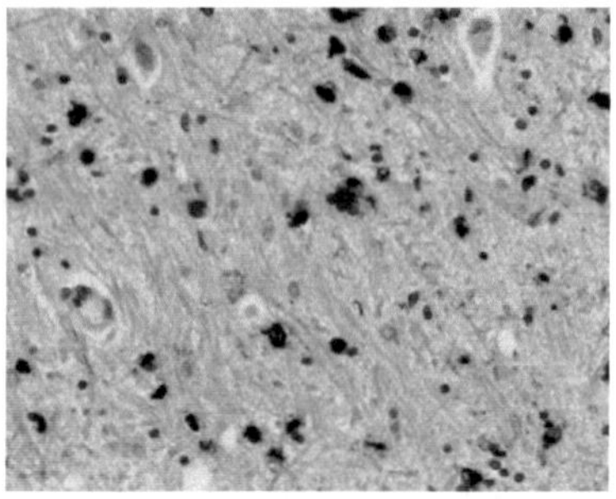

Donor information:

62-year-old female, MSA, late onset, rapid progression
Ion PGM™ Inherited Disease Panel sequencing did not reveal any known genetic mutations.

In the absence of α-synuclein deposits, what other PD-related phenotypes will be seen?

25

ThermoFisher SCIENTIFIC

First, isolated a heterozygous (SNCA +/-), clonal population in the MSA iPSC line

GeneArt® Genomic cleavage detection assay confirms cleavage at desired locus

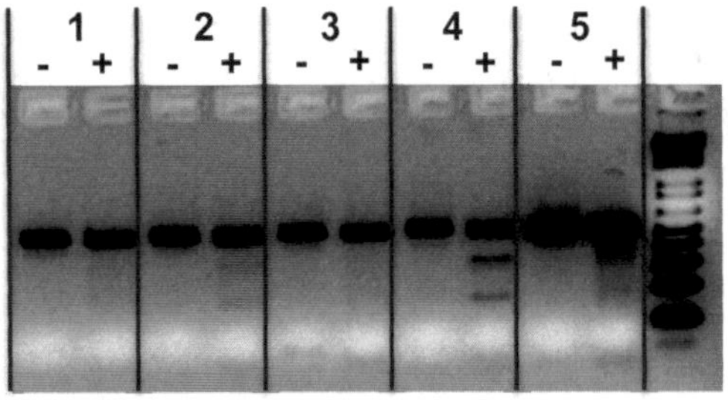

1 – 4 = single colonies
5 = multiple colonies

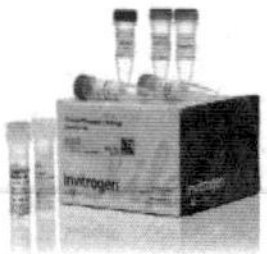

GeneArt® Genomic Cleavage Detection Kit

Next-gen (PGM-based) sequencing assay confirms clonality, heterozygosity, and deletion at desired loci

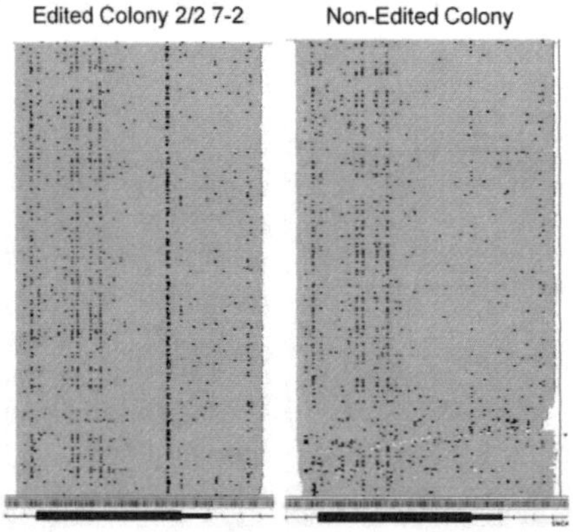

26

ThermoFisher SCIENTIFIC

Characterization of the edited MSA (SNCA +/-) iPSC line

Karyotypically normal

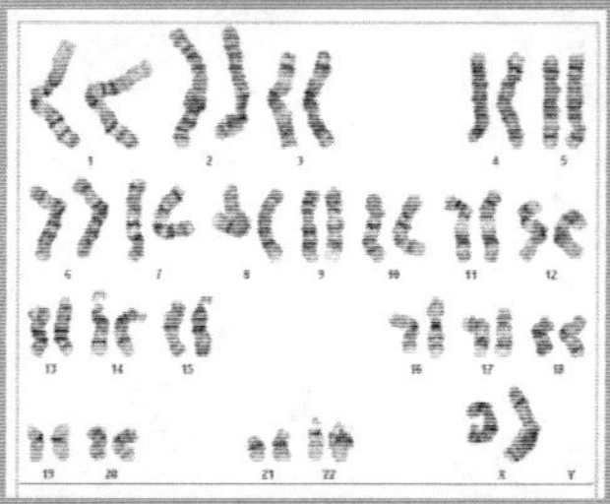

Express pluripotency markers

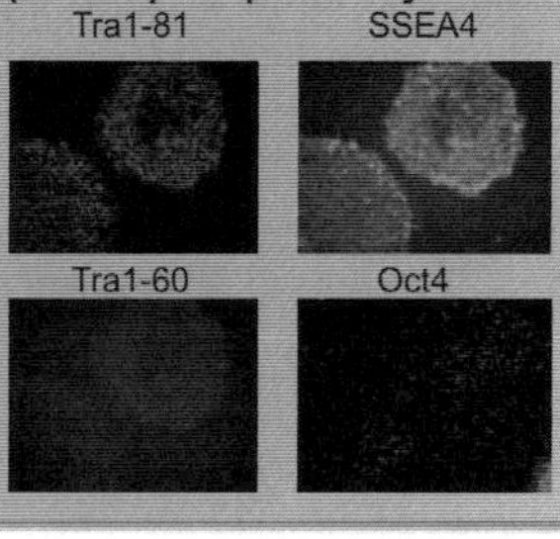

Decreased α-synuclein protein level in NSCs derived from the edited iPSCs

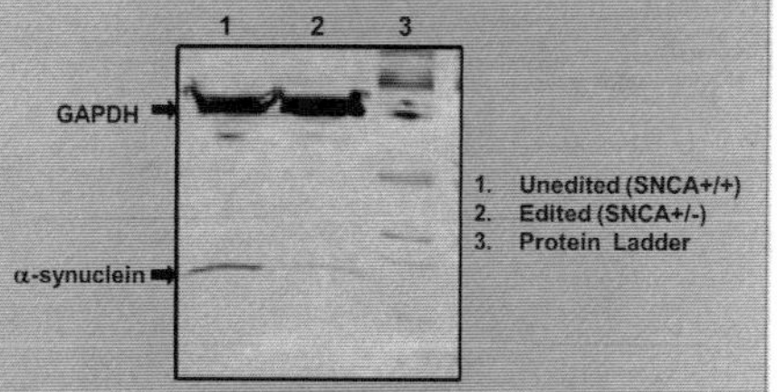

Western blot: anti-α-synuclein

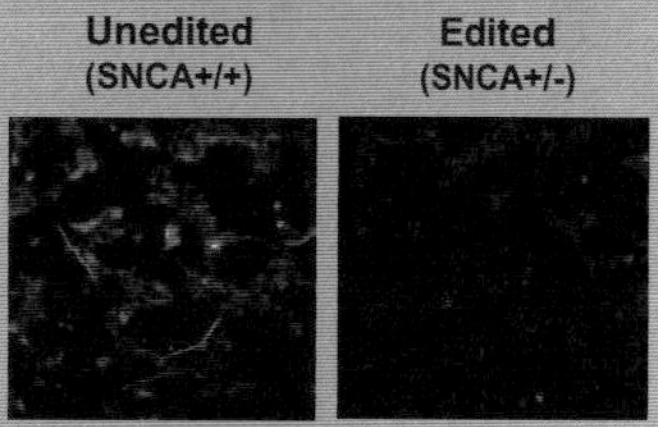

ICC: anti-α-synuclein

Thermo Fisher SCIENTIFIC

Re-editing to obtain Double Knockout via NHEJ repair

Generation of SNCA -/- from the MSA (SNCA +/-) iPSC line

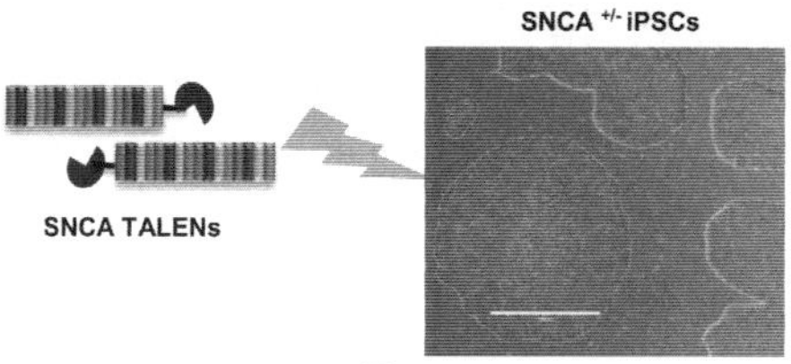

TaqMan® SNP Genotyping Assay

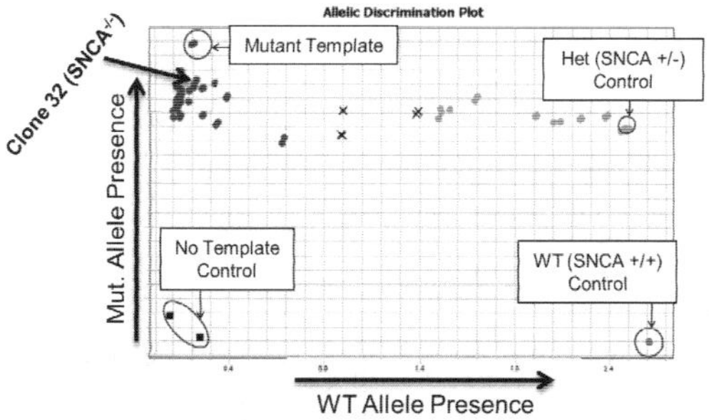

PGM™ Sequencing

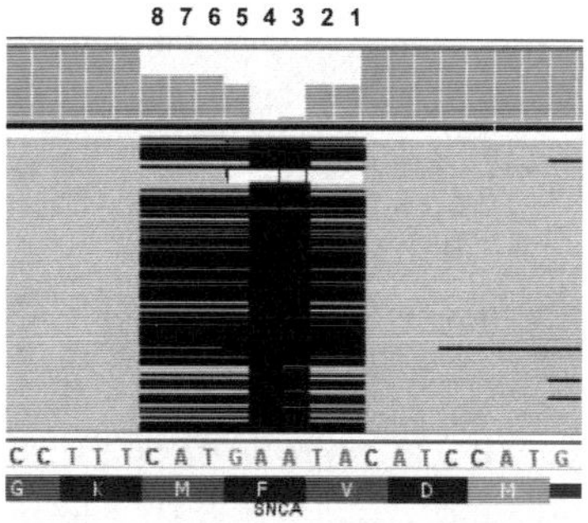

8bp deletion in the 2nd allele starting in 3rd codon to the end of 5th codon causing frame shift resulting in 16 aa then stop codon.

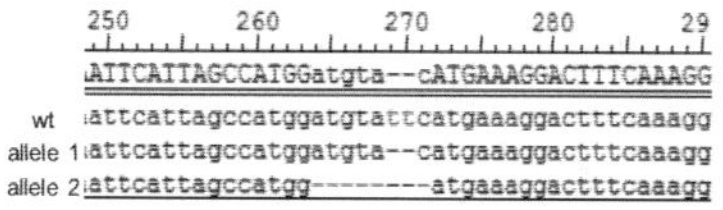

Thermo Fisher SCIENTIFIC

Verification of SNCA homozygous knockout

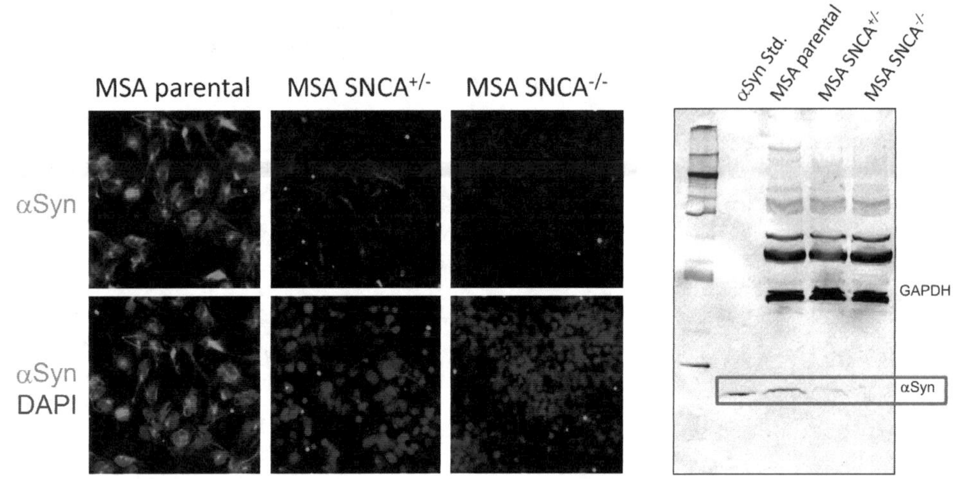

Absence of αSyn protein confirmed by ICC and Western blot

Editing via homologous recombination: LRRK2

Correcting mutation of LRRK2 G2019S to wild type

LRRK2 TAL pairs and donor fragment:

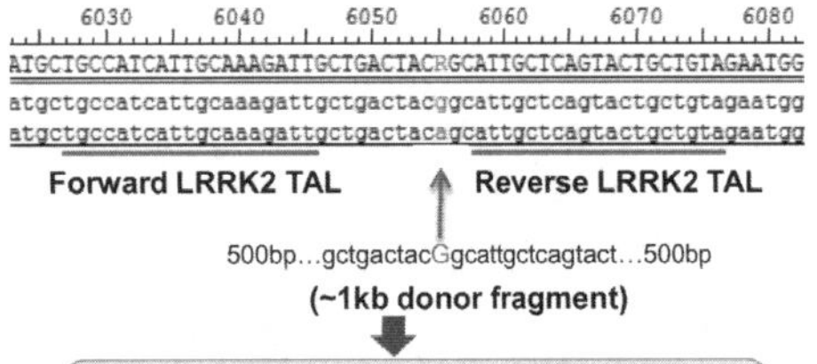

LRRK2 G2019S corrected to WT
Initial screen: 2 out of 140 (1.4%)

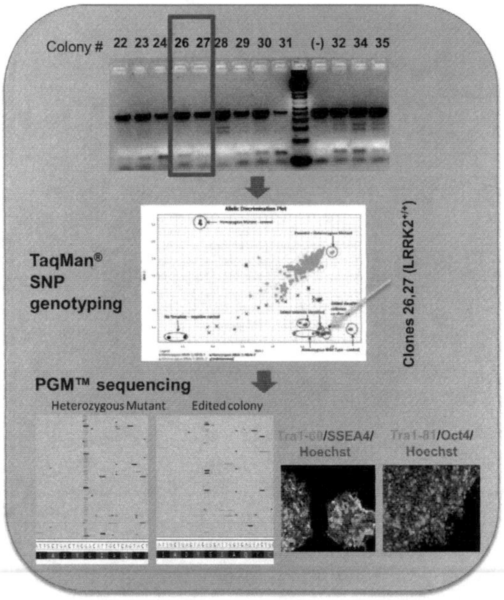

Editing via homologous recombination: GBA

Correcting mutation of GBA N370S to wild type

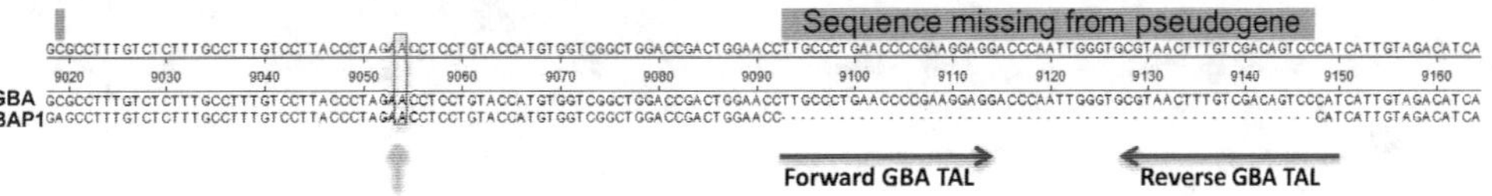

500 bp…gtccttaccctagaAcctcctgtaccatgtggtcggctggaccga…500 bp

(1 kb donor PCR fragment)

↓

Initial screen: 1 out of 48 (2%) GBA N370S corrected to WT (~77% corrected)

↓

Daughter colonies screen: 2 of 12 via sequencing; GBA N370S corrected to WT

↓

Clone expansion and characterization

Presence of nearby pseudogene with substantial homology necessitated "cutting" ~ 60 BP from site of edit

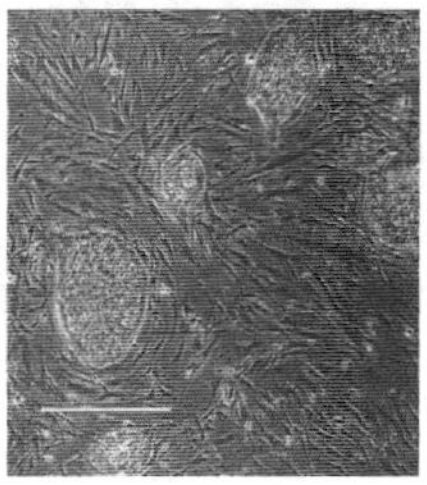

Pilot experiment using parental iPSC line and two TAL edited lines

No off-target effects detected for either edited line

Number of variants detected

Method	"Off-target" indels detected (avg)	After manual annotation
1 sample per chip	5	0
2 samples per chip	2	0
3 samples per chip	3	0

- Expected variants (2bp and 8bp deletions) detected in all run formats, detected in 14/14 runs.
- Manageable number of variants to manually annotate after Ion Reporter workflow and filter chain applied, <10 from any format.
- No off-target indels detected after Sanger confirmation.

Examples of manual annotation using IGV viewer

Expected variants | **Manually removed variant**

Genotype assigned: GAA/GAA — Genotype assigned: T/TCATGTA — Genotype assigned: G/G — Genotype assigned: GG/GG

Expected variants in edited cell line are clearly distinguished when compared to parental cells.

Sequencing errors in homopolymer regions require manual annotation and verification by Sanger sequencing.

Summary of iPSC lines generated

ID	Age at biopsy	Disease	Mutation	Line No.	iPSC Line Description
Ctrl-1	25	control		1	Ctrl-1
Ctrl-2	69	control		2	Ctrl-2
PD-1	19	YOPD	PARKIN	3	PD-1
PD-2	74	PD	LRRK2	4	PD-2
PD-3	61	PD	LRRK2/GBA	5	$LRRK2^{G2019S}/GBA^{N370S}$
	61	PD		6	$LRRK2^{WT}/GBA^{WT}$ Underway
	61	PD		7	$LRRK2^{WT}/GBA^{N370S}$
	61	PD		8	$LRRK2^{G2019S}/GBA^{WT}$
MSA	62	MSA		9	MSA: $SNCA^{+/+}$
	62	MSA		10	MSA: $SNCA^{+/-}$
	62	MSA		11	MSA: $SNCA^{-/-}$

In process: Monitor PD-relevant phenotypes using fluorescence-based HTS compatible assays

Cell Line	Cell Type	Compare under Untreated and Stress Conditions:
Parkinson's Disease Donor	iPSC	➢ **Colony Morphology**
	NSCs	➢ **Metabolic Activity and Neuronal Cell Health** *(PrestoBlue®, Neurite Outgrowth Staining Kit)*
	Neurons	
	Glial cells	➢ **Oxidative Stress** *(CellROX®)*
PD edited lines (gene correction, gene knockdown	iPSC	
	NSCs	➢ **Mitochondrial Stress** *(MitoSOX™)*
	Neurons	
	Glial cells	➢ **Apoptosis**
Age Matched Control Donors	iPSC	➢ **Protein aggregation** *(Western Blot, Immunocytochemistry, Autophagy Assays)*
	NSCs	
	Neurons	➢ **Gene Expression Analysis** *(QuantStudio™)*
	Glial cells	

Case Study Summary

1. Integration-free iPSCs from 5 Parkinson's disease and 3 control donors were created using the CytoTune®-iPS Sendai reprogramming kit, and fully characterized.
2. Neural Stem Cells from the above iPSCs were created using the Gibco® PSC Neural Induction Medium. Phenotypic assays have been developed to assess cellular processes in NSCs.
3. Isogenic cell lines were created using GeneArt® Precision TAL technology. These will be used in comparison against isogenic parental lines to study the impacts of SNCA, LRRK2, and GBA on PD-related phenotypes.

Thermo Fisher SCIENTIFIC

Acknowledgements

The Parkinson's Institute and Clinical Center

causecarecure

Birgitt Schuele
J. William Langston
Y. Anne Huang
Ramya Sundararajan
Malini Vangipuram
Jessica Westfall

Fibroblast derivation funded by CIRM TR1-01246

life technologies™

Thermo Fisher SCIENTIFIC

The world leader in serving science

Tori Barron
Kun Bi
Bonnie Hanson
Spencer Hermanson
Connie Lebakken
Marian Piekarczyk

David Piper
Chetana Revankar
Laurie Reichling
Steve Riddle
David Thompson
Kurt Vogel

Thermo Fisher SCIENTIFIC

Induction of Somatosensory Neural Lineage from Human and Mouse Fibroblasts

Blanchard, J.,
Eade, K., Scripps Research Institute
Baldwin, K., Scripps Research Institute
Originally presented on: 3/17/2014 0:40:00 - 1:00:00

Somatic sensory neurons of the dorsal root ganglion and trigeminal nerve mediate the sensation of chemical, thermal, painful, pruritogenic and mechanical stimuli based on cell type-specific expression of various receptor proteins and ion channels. Treatment of human pain, itch or inherited disorders affecting somatic sensory neurons would be aided by a means to produce these neurons *in vitro* from genetically diverse populations of humans. However, at present the only method to generate human sensory neurons *in vitro* is to differentiate induced pluripotent stem cells into a restricted sub-class of nociceptors that do not recapitulate the endogenous neuronal diversity of *in vivo* sensory neurons. Here we report that transient ectopic expression of two transcription factors rapidly and efficiently converts both human and mouse fibroblasts to mature neurons, without transiting through a proliferative intermediate. Using immunostaining and RT-PCR we show that this conversion is largely selective for the somatosensory lineage, with the majority of neurons(~90%) expressing one of the three major somatic sensory sub-lineages markers, TrkA, TrkB or TrkC. Remarkably, induced somatic sensory lineage neurons (iSLN) also exhibit pseudounipolar morphologies and subset-specific soma sizes that are characteristic of their *in vivo* counterparts, indicating that specific neuronal subtype morphology may be specified through direct reprogramming. Additionally, iSLNs display functional diversity that is appropriate for somatic sensory neurons. Through calcium imaging we show that subsets of iSLNs respond to distinct combinations of painful and pruritogenic compounds. These findings demonstrate that discrete neural lineages can be induced and patterned through transient ectopic expression of only two transcription factors, showing that neither cell division nor exogenous factors are required for lineage specification. These results also provide rapid and efficient methods for generating human sensory neurons, delivering new technology for therapy, investigating fundamentals of pain, somatosensory biology, and therapeutic screening.

See more Presentations from this session: Intracellular Programming & Reprogramming

See more Sessions in this Group: General Submissions

Drug-Loaded Nanoparticles Induce Gene Expression in Human Pluripotent Stem Cell Derivatives

Saha, K., Whitehead Institute for Biomedical Research
Harkness, T.,
Gong, S.,
Escalante, L.,
Gajbhiye, V.,
Chen, G.,
Laperle, A.,
Zheng, Q.,
Steyer, B.,
Originally presented on: 3/17/2014 1:00:00 - 1:20:00

Tissue engineering and advanced manufacturing of human stem cells requires a suite of tools to control gene expression spatiotemporally in culture. Inducible gene expression systems offer cell-extrinsic control, typically through addition of small molecules, but small molecule inducers typically contain few functional groups for further chemical modification. Doxycycline (DXC) is a potent small molecule inducer of tetracycline (Tet) transgene systems. To gain additional functional handles on drug release and thus gene expression, here we report the conjugation of DXC to a hyperbranched biodegradable and biocompatible polymer nanoparticles (e.g., Boltorn H40). The hyperbranched polymer nanoparticle core offers a large number of peripheral functional groups which can be used to conjugate a large number of DXC molecules as well as polyethylene glycol (PEG) arms, thereby making the drug-polymer conjugate water soluble. The resulting nanoparticle (e.g., PEG–H40–DXC) exhibited pH-sensitive drug release behavior and successfully controlled gene expression in stem-cell-derived fibroblasts with a Tet-On system. While free DXC inhibited fibroblast proliferation and matrix metalloproteinase (MMP) activity, the nanoparticles maintained higher fibroblast proliferation levels and MMP activity. The results demonstrate that the DXC-nanoparticle system provides an effective tool to controlling gene expression in human stem cell derivatives. Further work with these nanoparticle systems could produce patterned hPSC-derived cell cultures, offering researchers the ability to investigate these cells in environments that optimally recapitulate complexities found within developing and diseased human tissues.

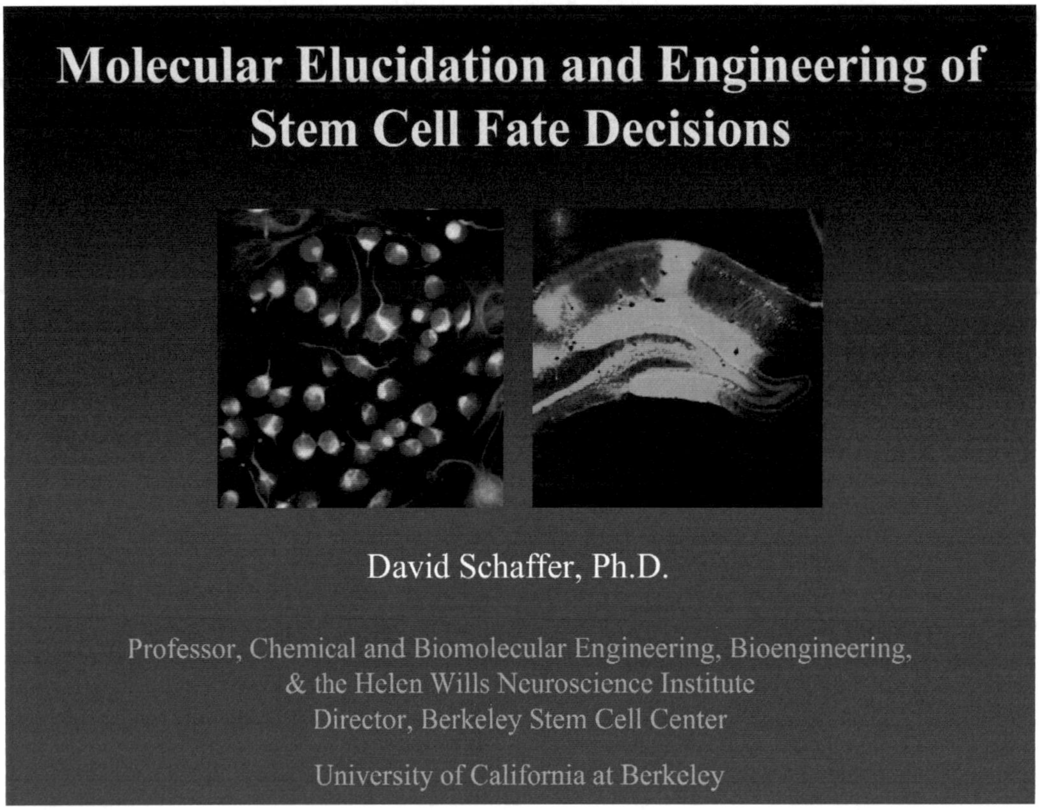
Molecular Elucidation and Engineering of Stem Cell Fate Decisions
David Schaffer, Ph.D.
Professor, Chemical and Biomolecular Engineering, Bioengineering, & the Helen Wills Neuroscience Institute
Director, Berkeley Stem Cell Center
University of California at Berkeley

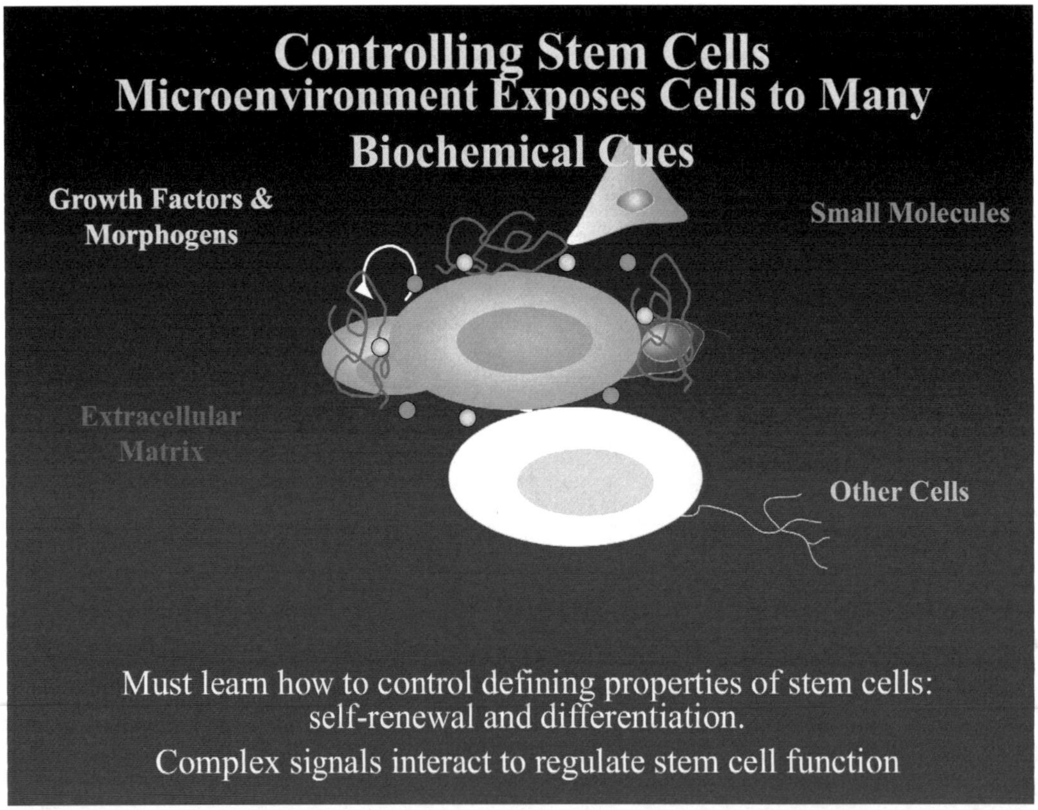
Controlling Stem Cells
Microenvironment Exposes Cells to Many Biochemical Cues
Growth Factors & Morphogens
Small Molecules
Extracellular Matrix
Other Cells
Must learn how to control defining properties of stem cells: self-renewal and differentiation.
Complex signals interact to regulate stem cell function

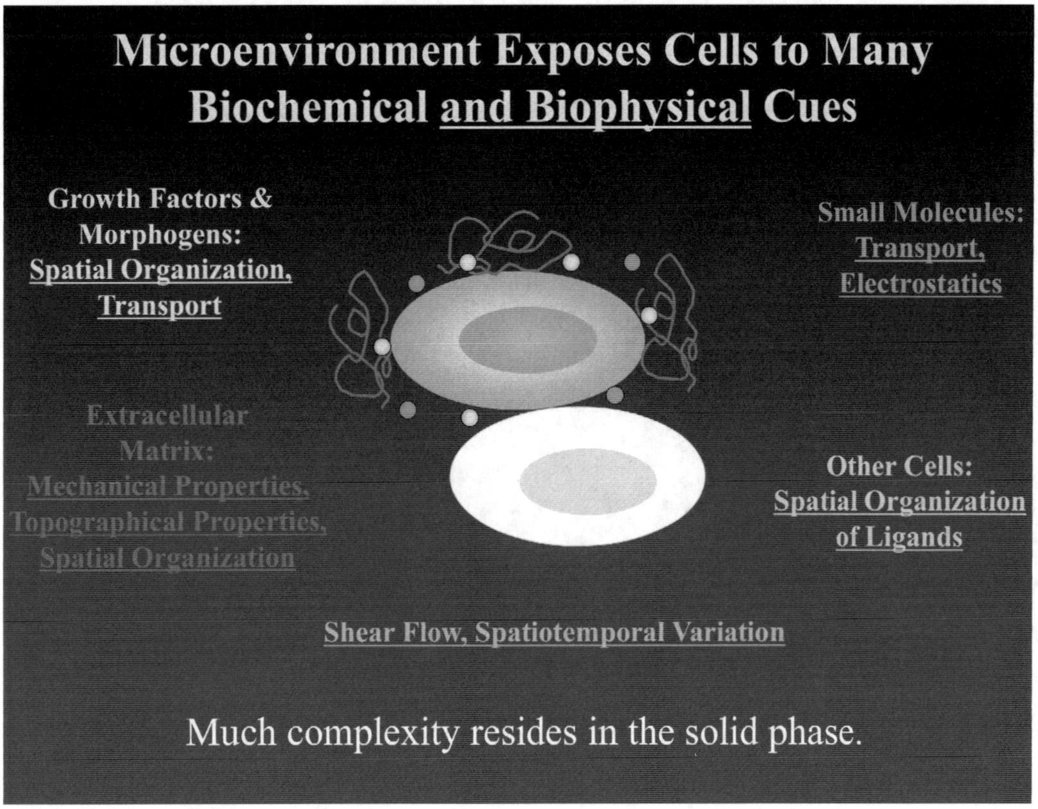
Microenvironment Exposes Cells to Many Biochemical and Biophysical Cues
Growth Factors & Morphogens: Spatial Organization, Transport
Small Molecules: Transport, Electrostatics
Extracellular Matrix: Mechanical Properties, Topographical Properties, Spatial Organization
Other Cells: Spatial Organization of Ligands
Shear Flow, Spatiotemporal Variation
Much complexity resides in the solid phase.

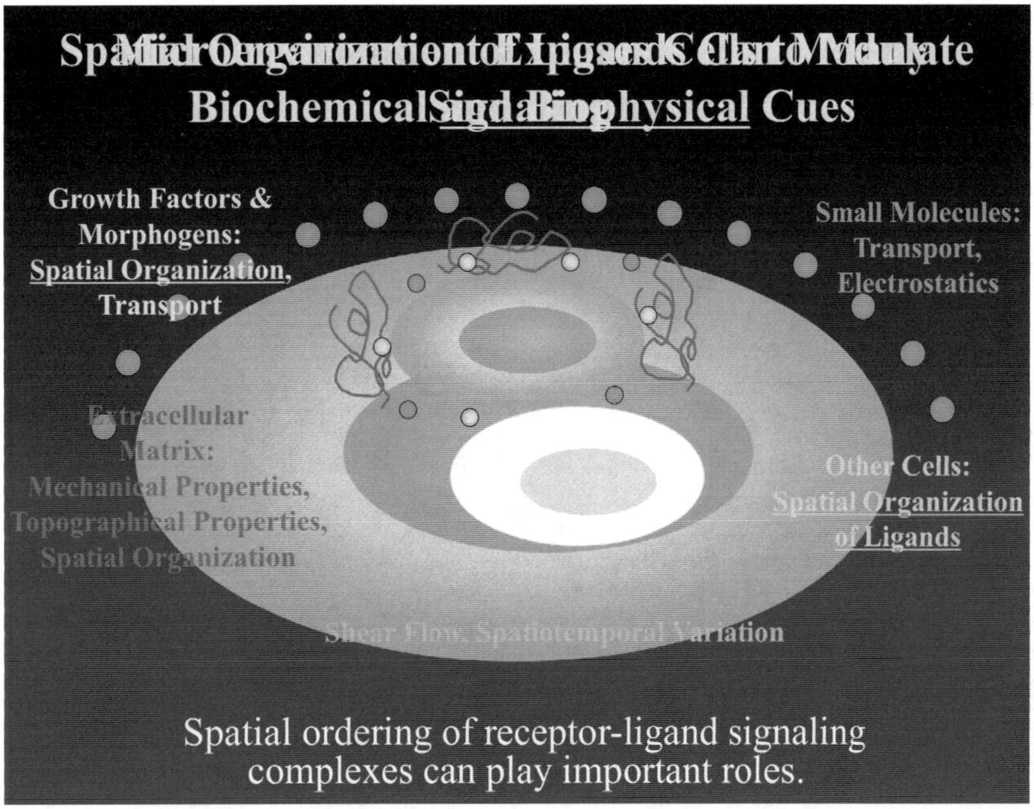
Growth Factors & Morphogens: Spatial Organization, Transport
Small Molecules: Transport, Electrostatics
Extracellular Matrix: Mechanical Properties, Topographical Properties, Spatial Organization
Other Cells: Spatial Organization of Ligands
Spatial ordering of receptor-ligand signaling complexes can play important roles.

Outline

- Discovery of signaling system that regulates stem cell differentiation

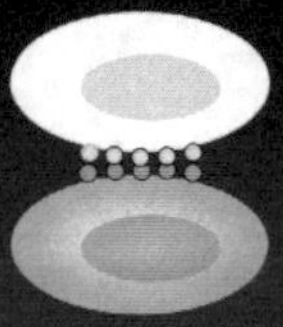

- Optogenetic system for spatiotemporal control of signaling

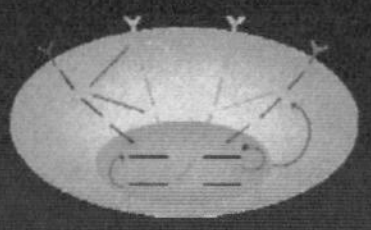

- Biomaterial system to control spatial organization in cellular signaling

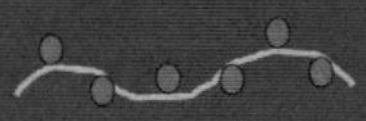

Stem Cells in the Adult Nervous System … ?

"Once development was ended, the fonts of growth and regeneration of the axons and dendrites dried up irrevocably. In adult centres the nerve paths are something fixed, ended, immutable. Everything may die, nothing may be regenerated. It is for the science of the future to change, if possible, this harsh decree."

S. Ramon y Cajal, *Degeneration and regeneration of the nervous system*, 1928

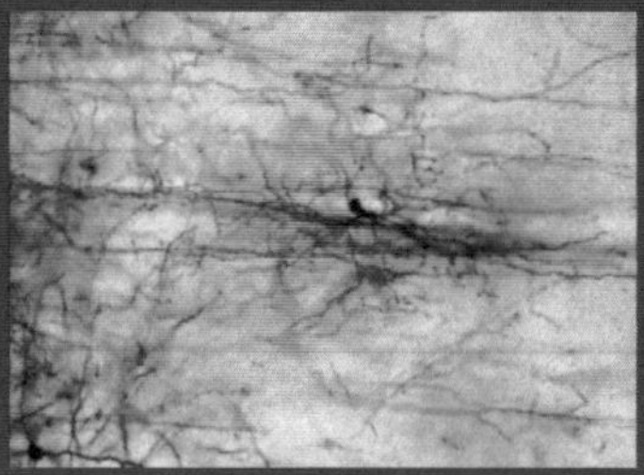

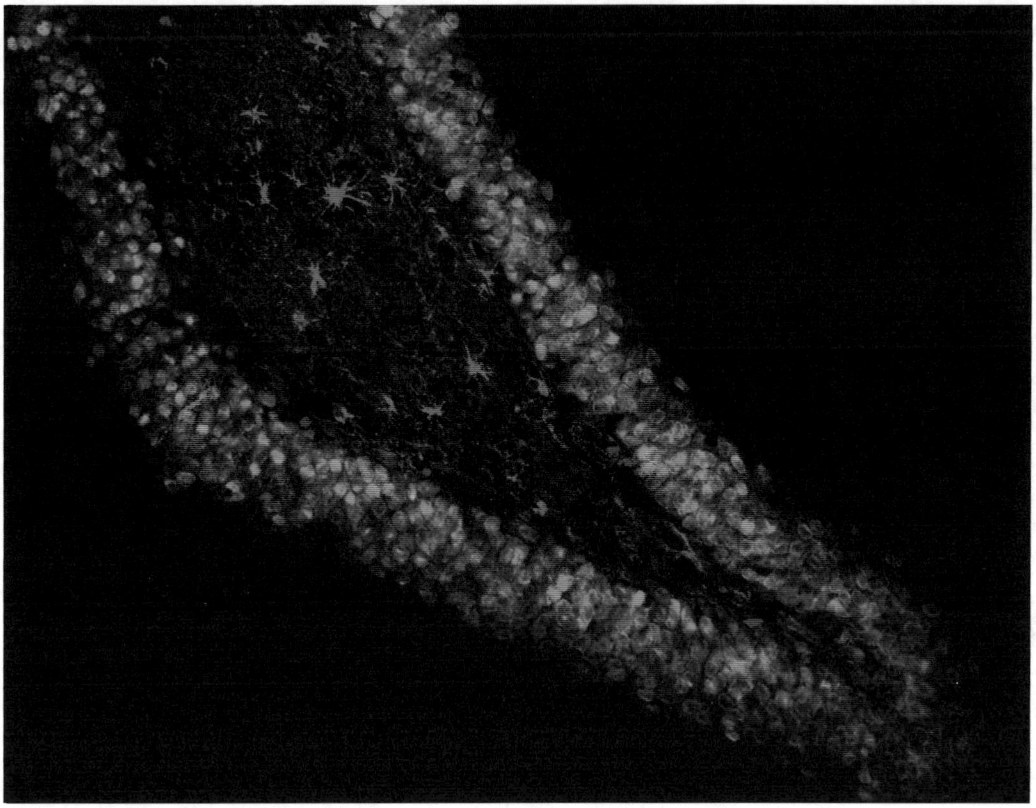

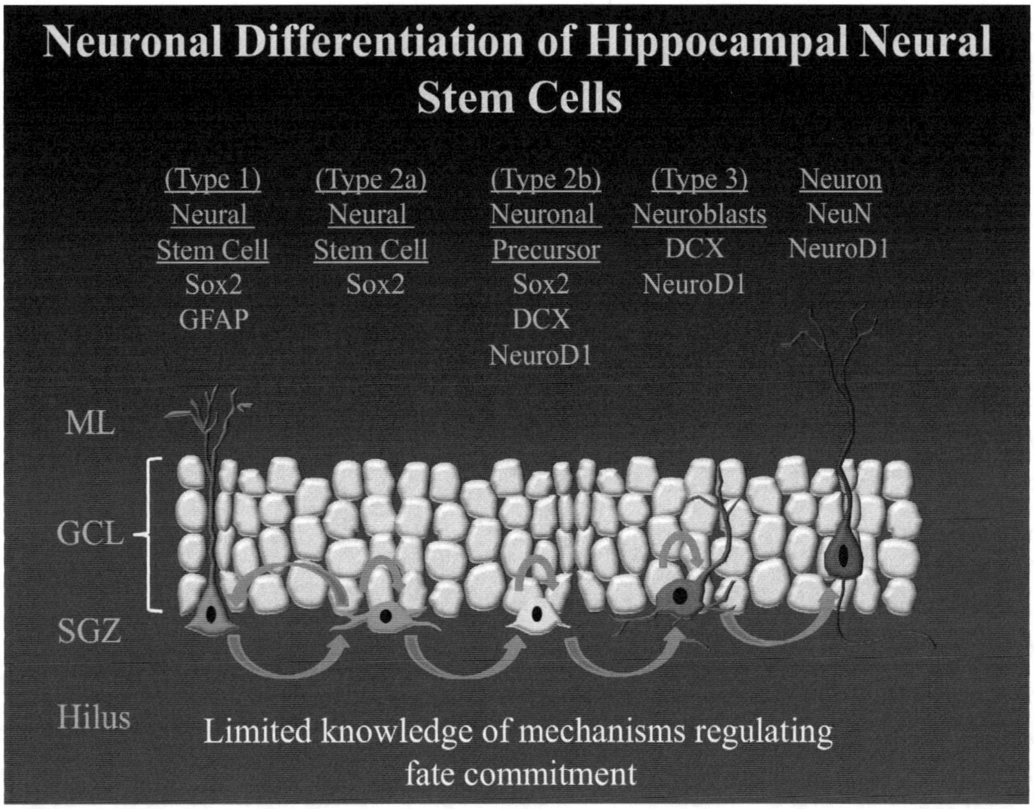

Neuronal Differentiation of Hippocampal Neural Stem Cells
(Type 1)
Neural
Stem Cell
Sox2
GFAP
(Type 2a)
Neural
Stem Cell
Sox2
(Type 2b)
Neuronal
Precursor
Sox2
DCX
NeuroD1
(Type 3)
Neuroblasts
DCX
NeuroD1
Neuron
NeuN
NeuroD1
ML
GCL
SGZ
Hilus
Limited knowledge of mechanisms regulating fate commitment

Eph/Ephrins

- Ephs are cell surface receptors, and Ephrins are cell-associated ligands
- Signaling system implicated in numerous developmental processes, and particularly axon pathfinding in the CNS
- Active only when clustered
- Signaling is typically bidirectional
- Signal via numerous pathways, most notably ones involved in cytoskeletal rearrangements

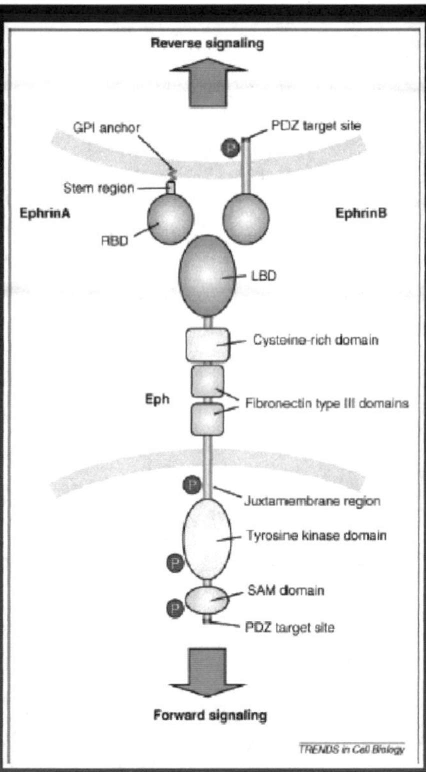

Egea *et al* (2005)

Ephrin-B2 and EphB4 Expression in the Brain

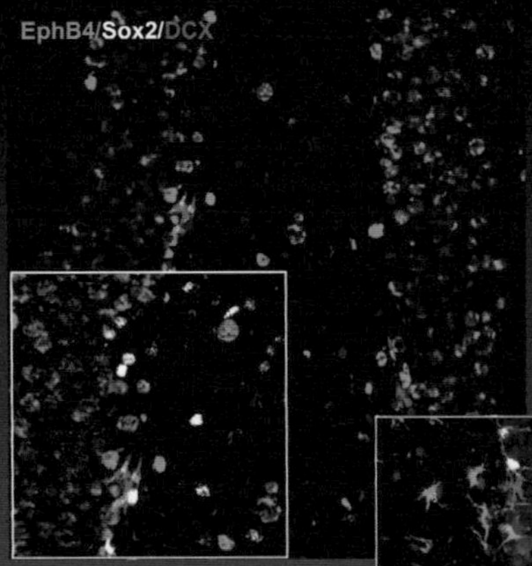

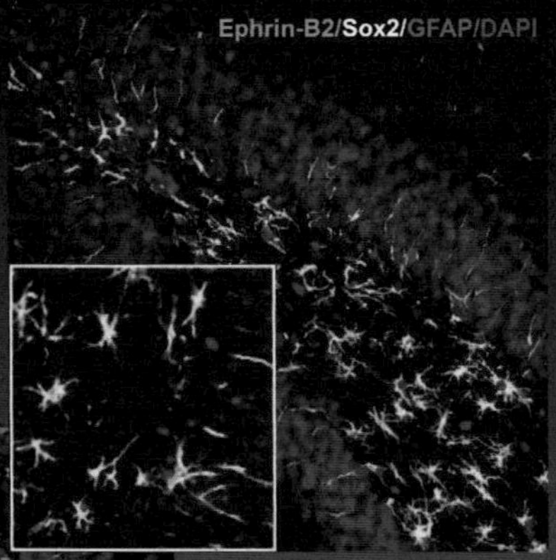

EphB4 receptor expressed by adult neural stem cells

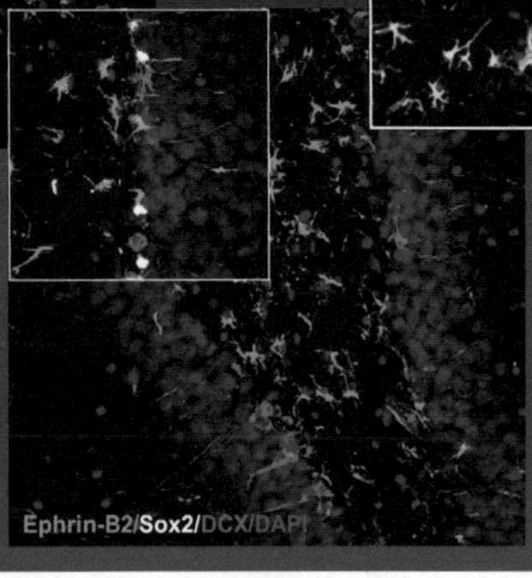

Ephrin-B2 ligand expressed by astrocytes that neighbor the NSCs

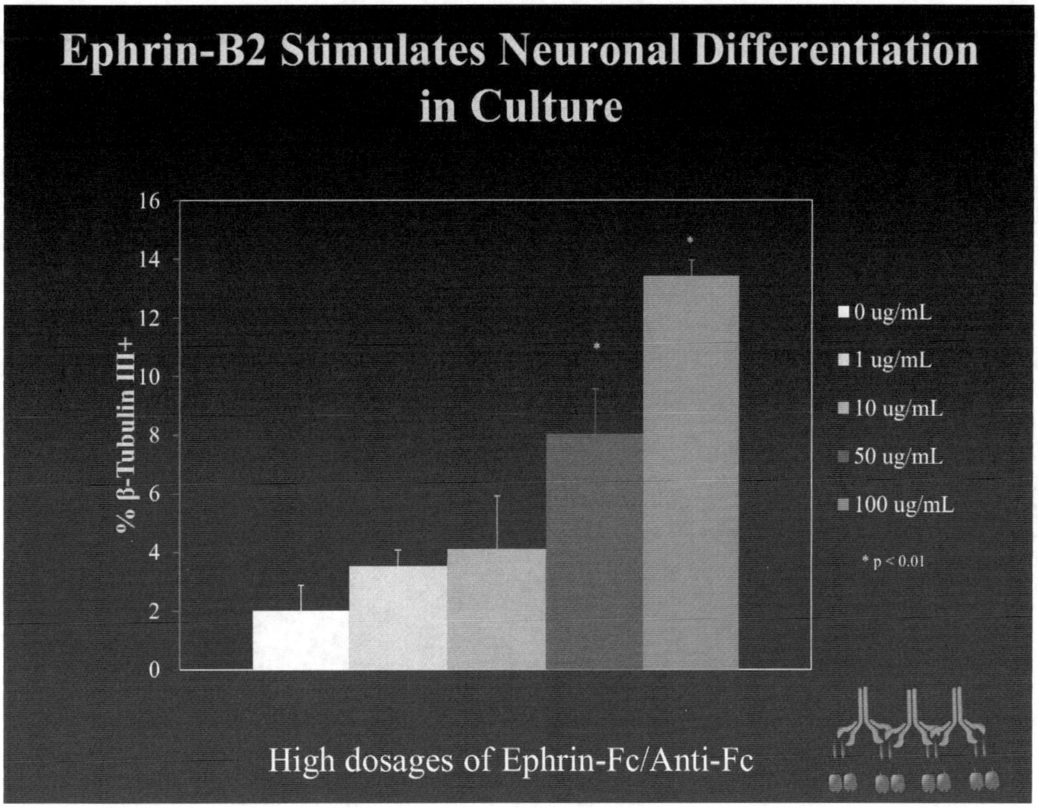
Ephrin-B2 Stimulates Neuronal Differentiation in Culture
% β-Tubulin III+
16
14
12
10
8
6
4
2
0
*
*
0 ug/mL
1 ug/mL
10 ug/mL
50 ug/mL
100 ug/mL
* p < 0.01
High dosages of Ephrin-Fc/Anti-Fc

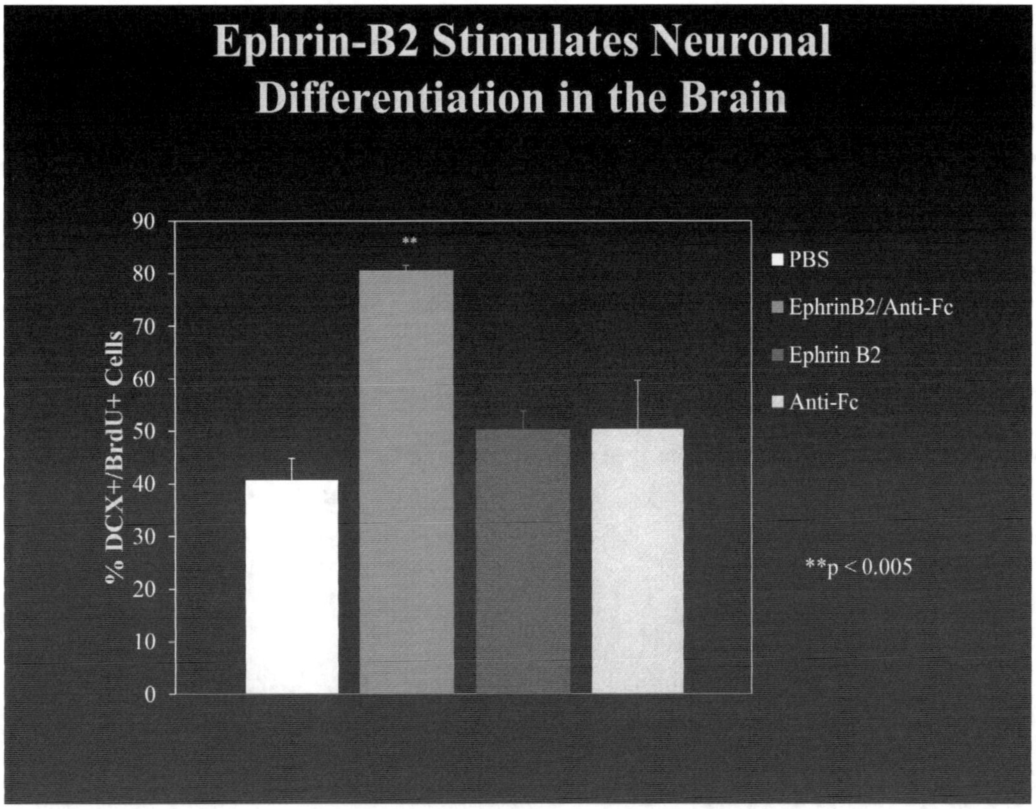
Ephrin-B2 Stimulates Neuronal Differentiation in the Brain
% DCX+/BrdU+ Cells
90
80
70
60
50
40
30
20
10
0
**
PBS
EphrinB2/Anti-Fc
Ephrin B2
Anti-Fc
**p < 0.005

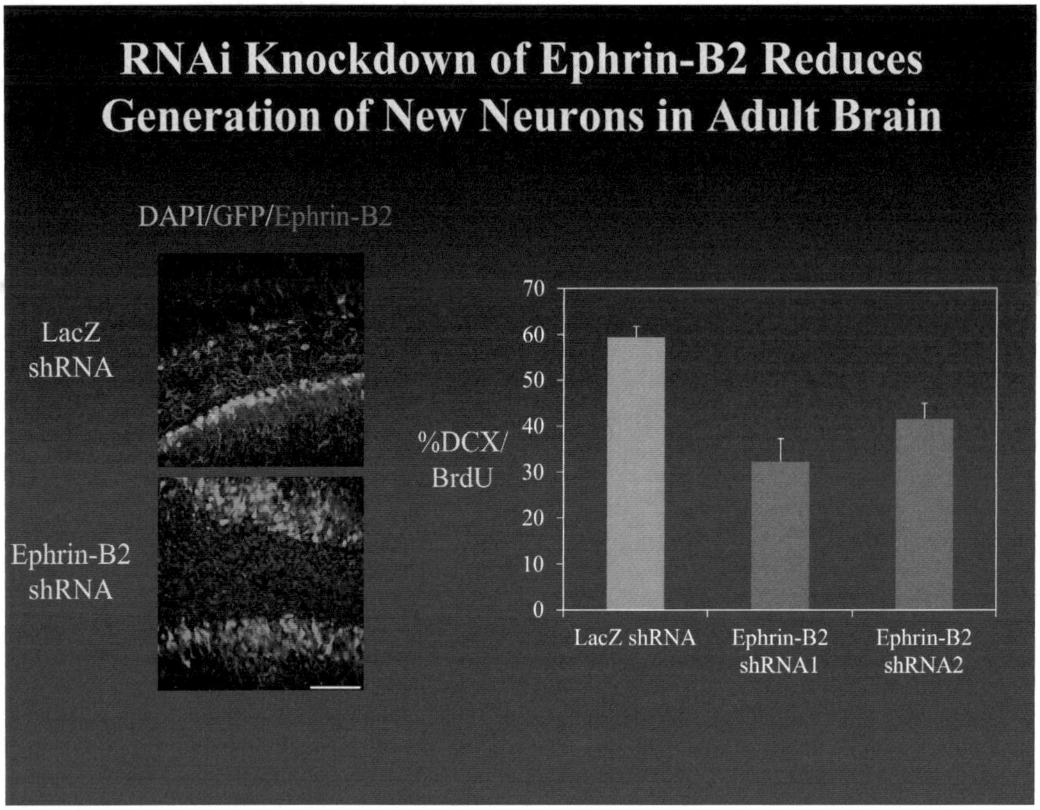
RNAi Knockdown of Ephrin-B2 Reduces Generation of New Neurons in Adult Brain
DAPI/GFP/Ephrin-B2
LacZ shRNA
Ephrin-B2 shRNA
%DCX/ BrdU
70
60
50
40
30
20
10
0
LacZ shRNA
Ephrin-B2 shRNA1
Ephrin-B2 shRNA2

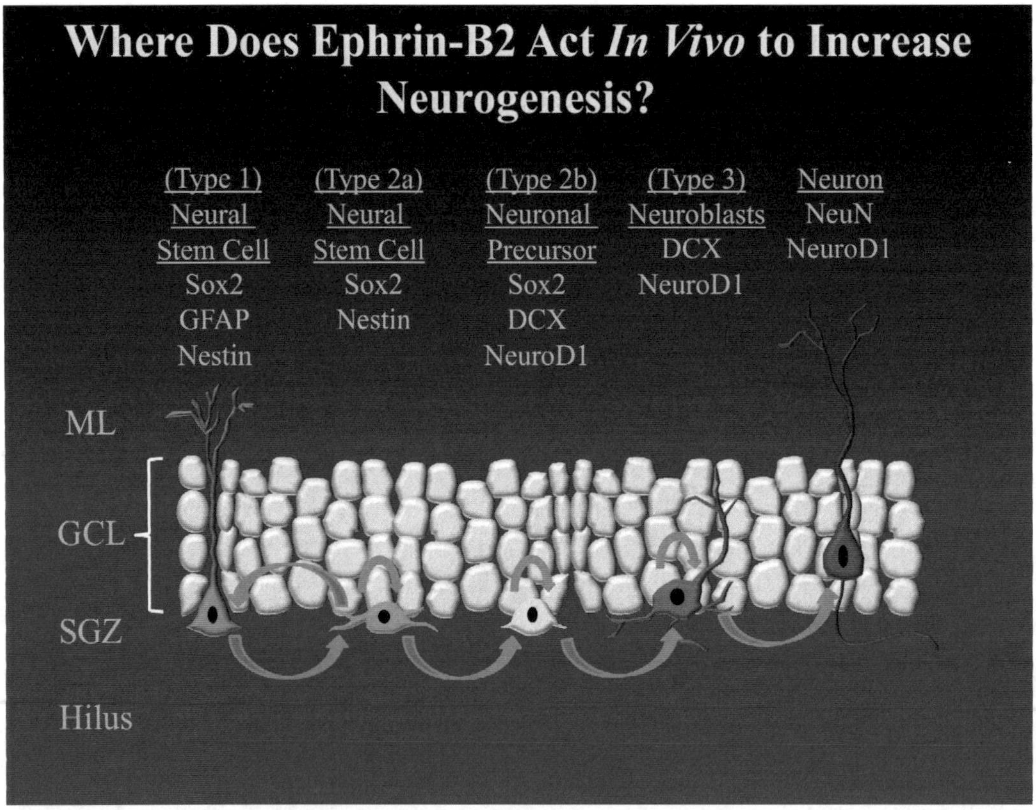
Where Does Ephrin-B2 Act *In Vivo* to Increase Neurogenesis?
(Type 1) Neural Stem Cell
Sox2
GFAP
Nestin
(Type 2a) Neural Stem Cell
Sox2
Nestin
(Type 2b) Neuronal Precursor
Sox2
DCX
NeuroD1
(Type 3) Neuroblasts
DCX
NeuroD1
Neuron
NeuN
NeuroD1
ML
GCL
SGZ
Hilus

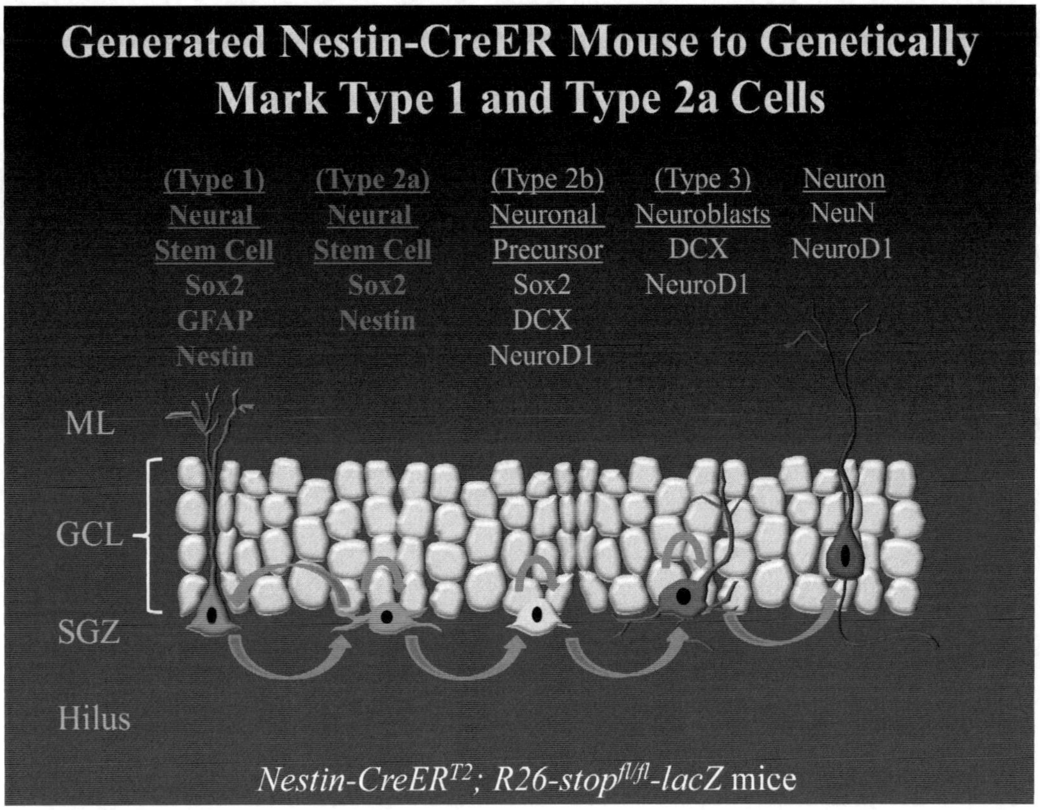
Generated Nestin-CreER Mouse to Genetically Mark Type 1 and Type 2a Cells
(Type 1) Neural Stem Cell Sox2 GFAP Nestin
(Type 2a) Neural Stem Cell Sox2 Nestin
(Type 2b) Neuronal Precursor Sox2 DCX NeuroD1
(Type 3) Neuroblasts DCX NeuroD1
Neuron NeuN NeuroD1
ML
GCL
SGZ
Hilus
Nestin-CreER^T2; R26-stop^fl/fl-lacZ mice

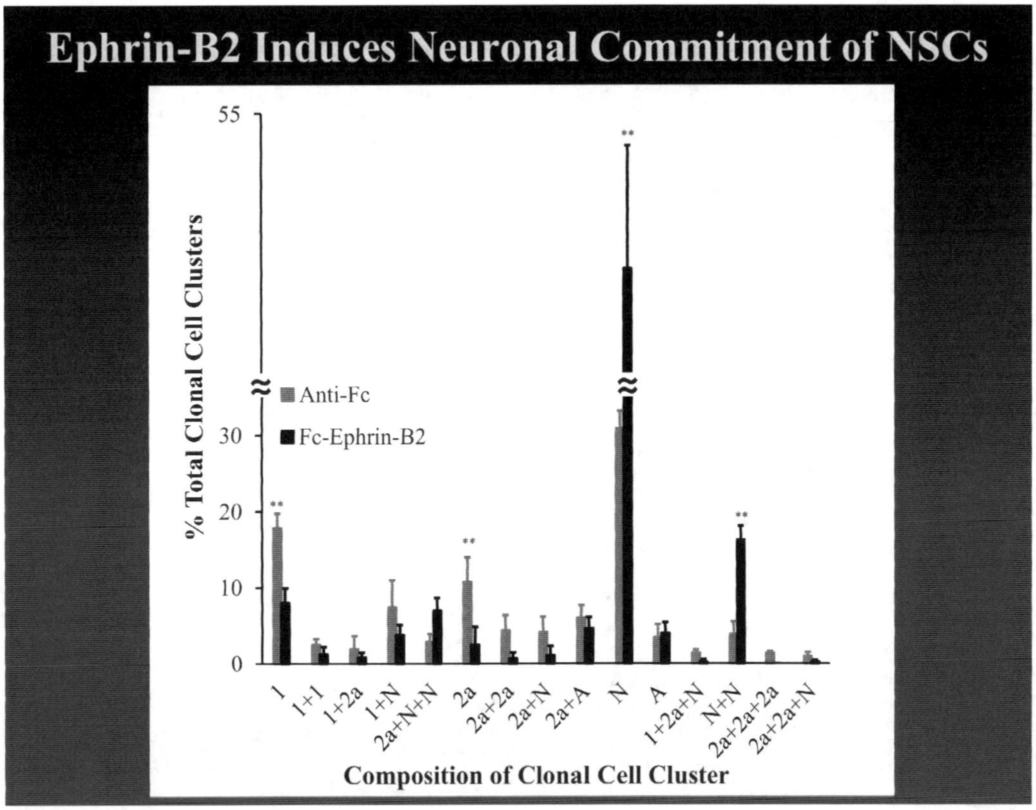
Ephrin-B2 Induces Neuronal Commitment of NSCs
% Total Clonal Cell Clusters
55
30
20
10
0
Anti-Fc
Fc-Ephrin-B2
**
1
1+1
1+2a
1+N
2a+N+N
2a
2a+2a
2a+N
2a+A
N
A
1+2a+N
N+N
2a+2a+2a
2a+2a+N
Composition of Clonal Cell Cluster

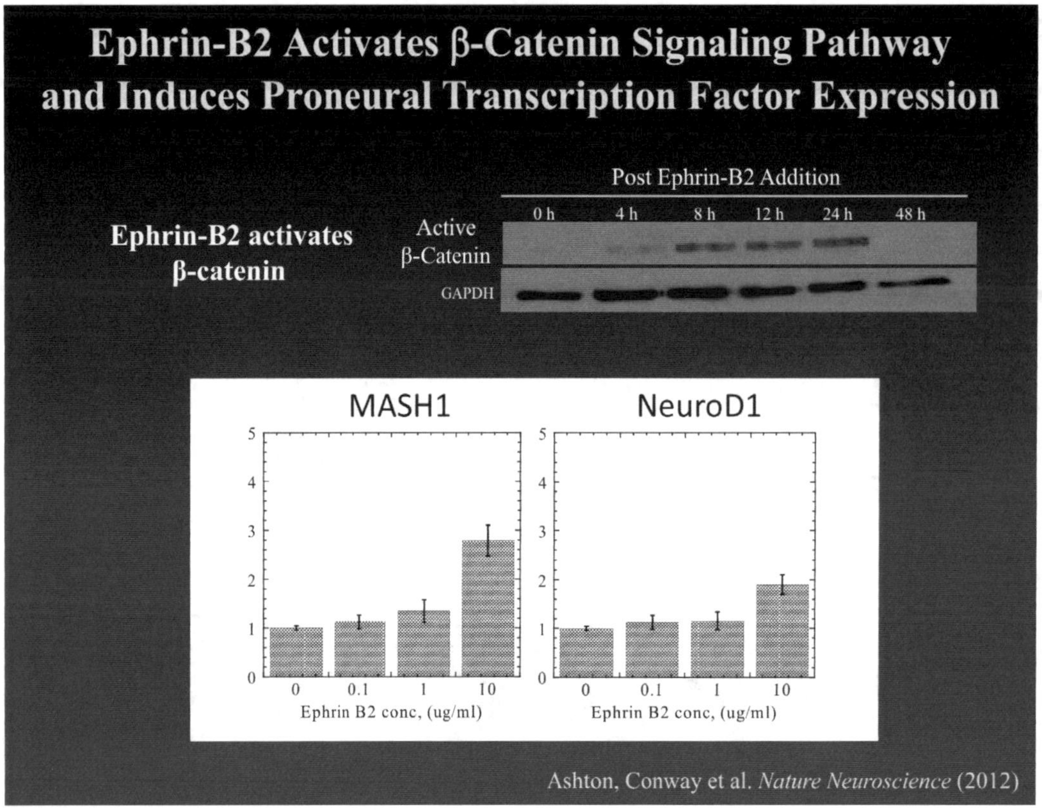
Ephrin-B2 Activates β-Catenin Signaling Pathway and Induces Proneural Transcription Factor Expression
Ephrin-B2 activates β-catenin
Post Ephrin-B2 Addition
0 h
4 h
8 h
12 h
24 h
48 h
Active β-Catenin
GAPDH
MASH1
NeuroD1
Ephrin B2 conc, (ug/ml)
Ephrin B2 conc, (ug/ml)
Ashton, Conway et al. Nature Neuroscience (2012)

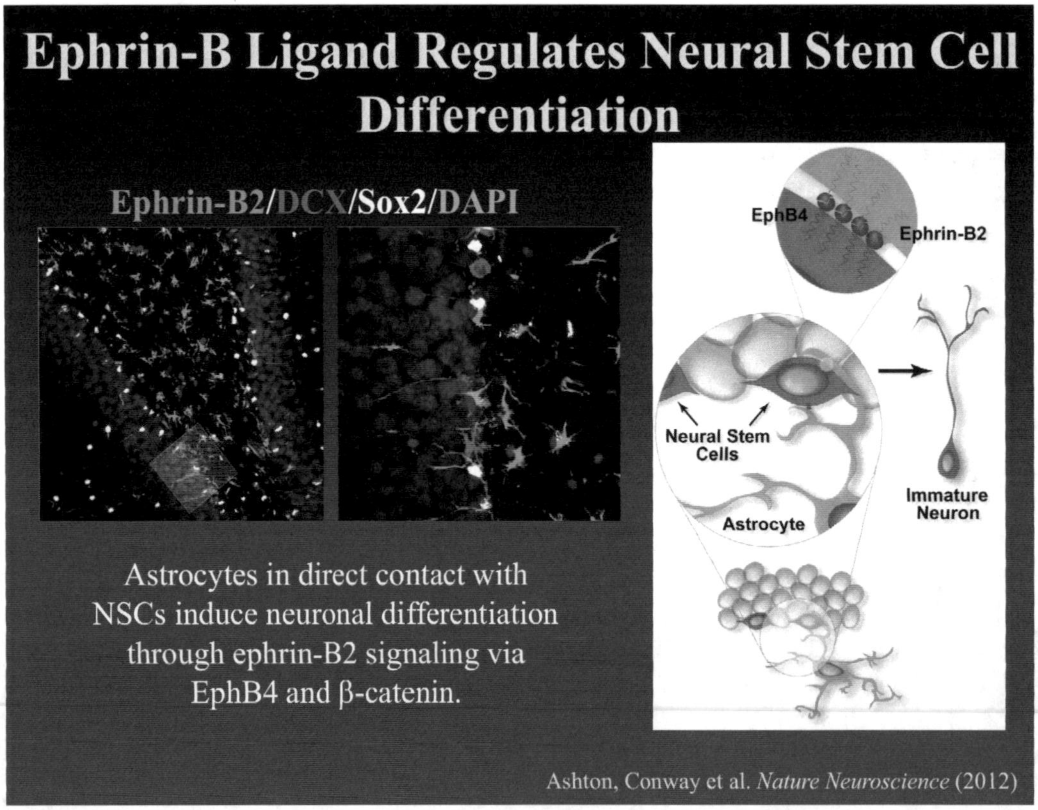
Ephrin-B Ligand Regulates Neural Stem Cell Differentiation
Ephrin-B2/DCX/Sox2/DAPI
EphB4
Ephrin-B2
Neural Stem Cells
Astrocyte
Immature Neuron
Astrocytes in direct contact with NSCs induce neuronal differentiation through ephrin-B2 signaling via EphB4 and β-catenin.
Ashton, Conway et al. Nature Neuroscience (2012)

Outline

- Discovery of signaling system that regulates stem cell differentiation
- **Optogenetic system for spatiotemporal control of signaling**
- Biomaterial system to control spatial organization in cellular signaling

Microenvironmental Inputs

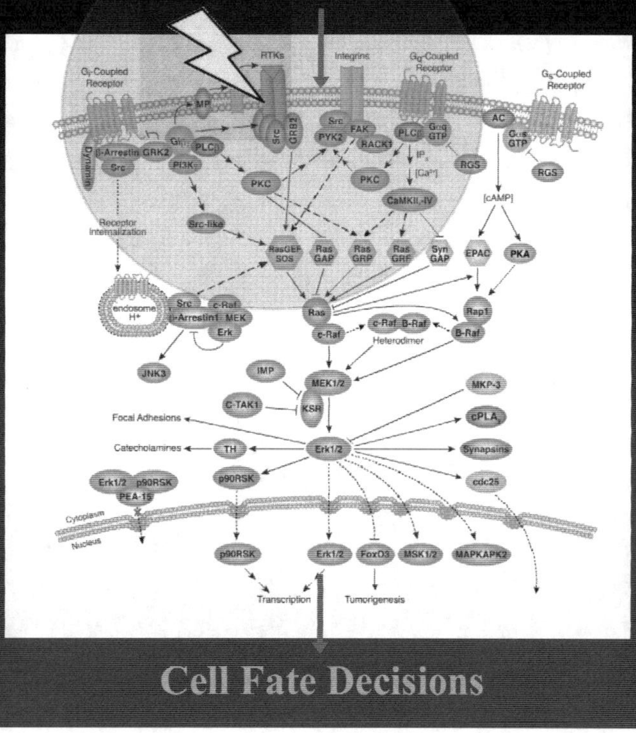

1) **Rapid & dynamic**
2) **Modular**

Cell Fate Decisions

Light as a Signaling Input

- Orthogonal signal for most cells
- Easy to "synthesize" and to sense
- Easy control of "concentration" (intensity)
- High degree of spatial and temporal control
- Electrophysiology has led applications, but has strong potential for protein-mediated signal transduction

β-Catenin Signaling Pathway

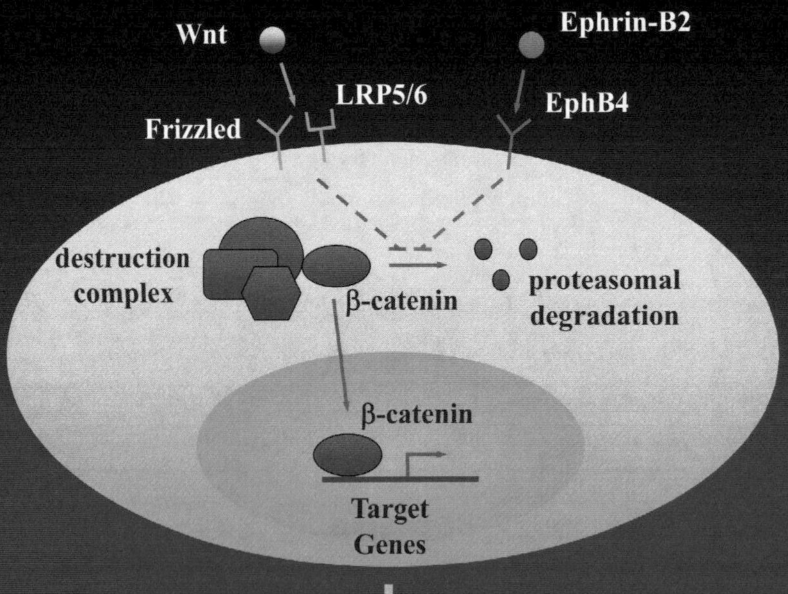

Signaling involved in stem cells, development, cancer …

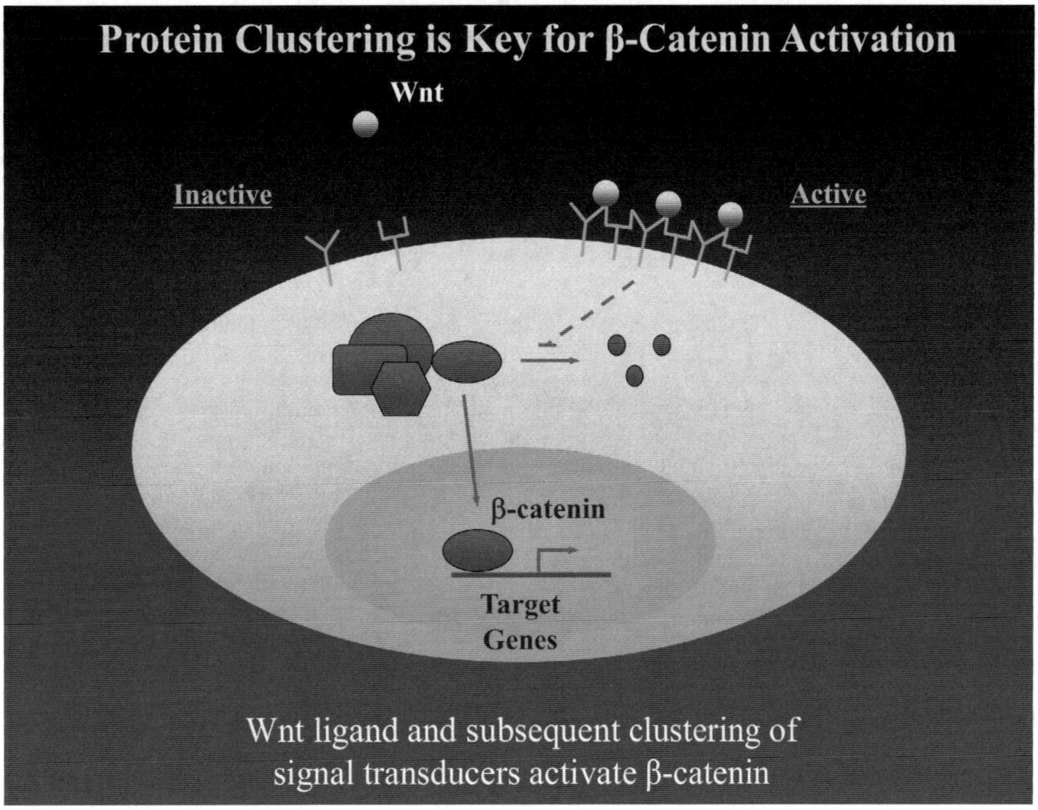
Protein Clustering is Key for β-Catenin Activation
Wnt
Inactive
Active
β-catenin
Target
Genes
Wnt ligand and subsequent clustering of
signal transducers activate β-catenin

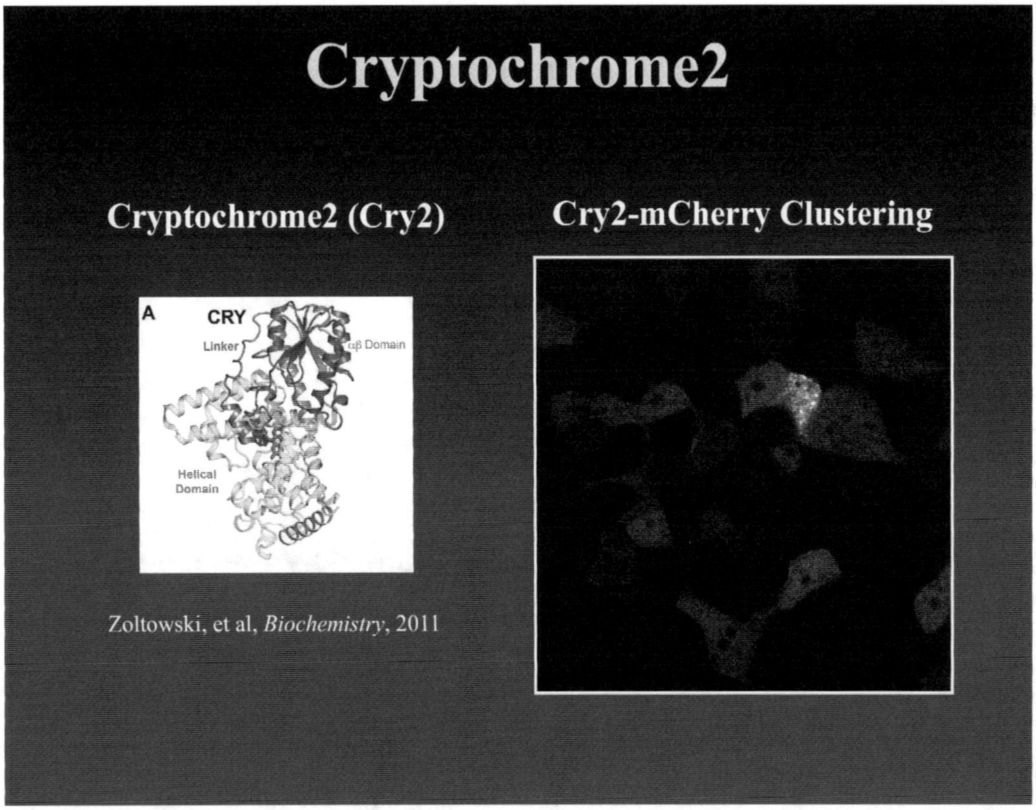
Cryptochrome2
Cryptochrome2 (Cry2)
Cry2-mCherry Clustering
A
CRY
Linker
αβ Domain
Helical
Domain
Zoltowski, et al, Biochemistry, 2011

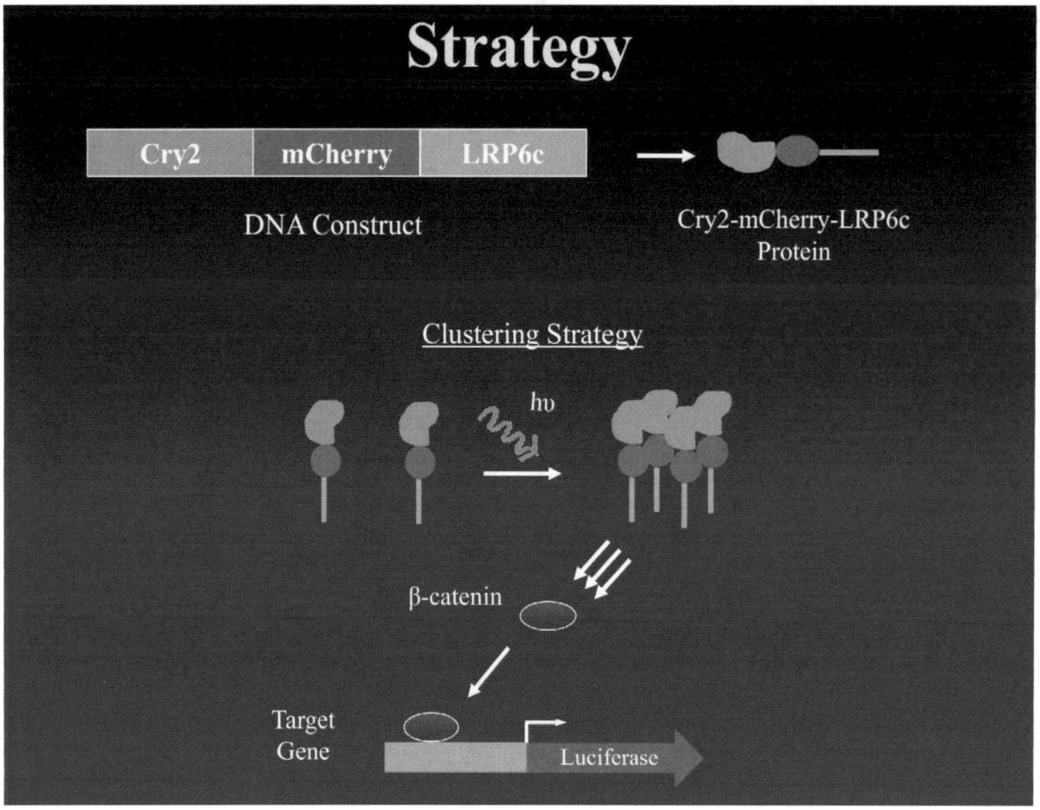
Strategy
Cry2
mCherry
LRP6c
DNA Construct
Cry2-mCherry-LRP6c
Protein
Clustering Strategy
hυ
β-catenin
Target
Gene
Luciferase

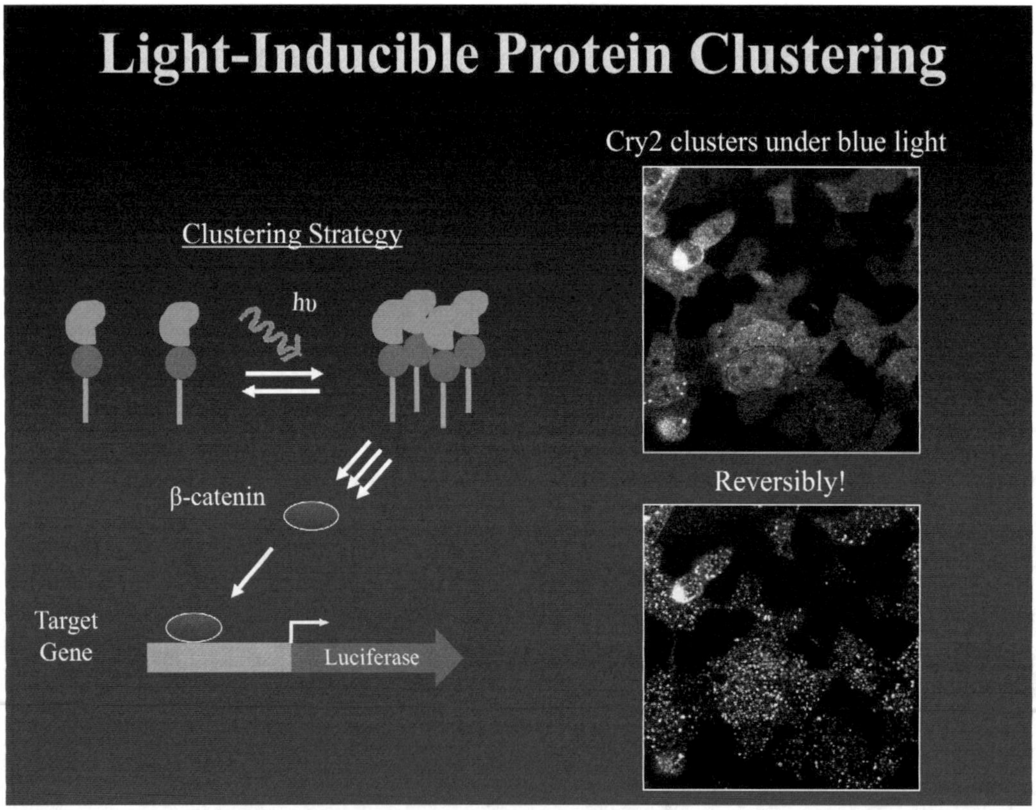
Light-Inducible Protein Clustering
Cry2 clusters under blue light
Clustering Strategy
hυ
β-catenin
Target
Gene
Luciferase
Reversibly!

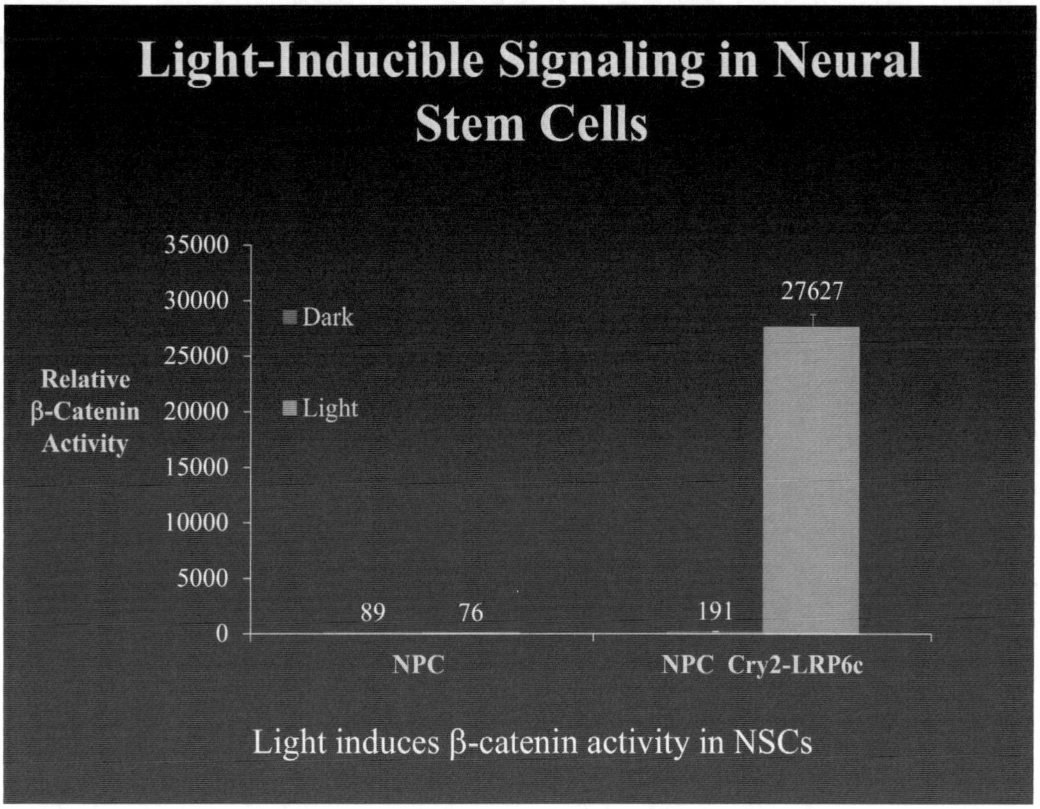

Light-Inducible Signaling in Neural Stem Cells
35000
30000
25000
20000
15000
10000
5000
0
Relative β-Catenin Activity
Dark
Light
27627
89
76
191
NPC
NPC Cry2-LRP6c
Light induces β-catenin activity in NSCs

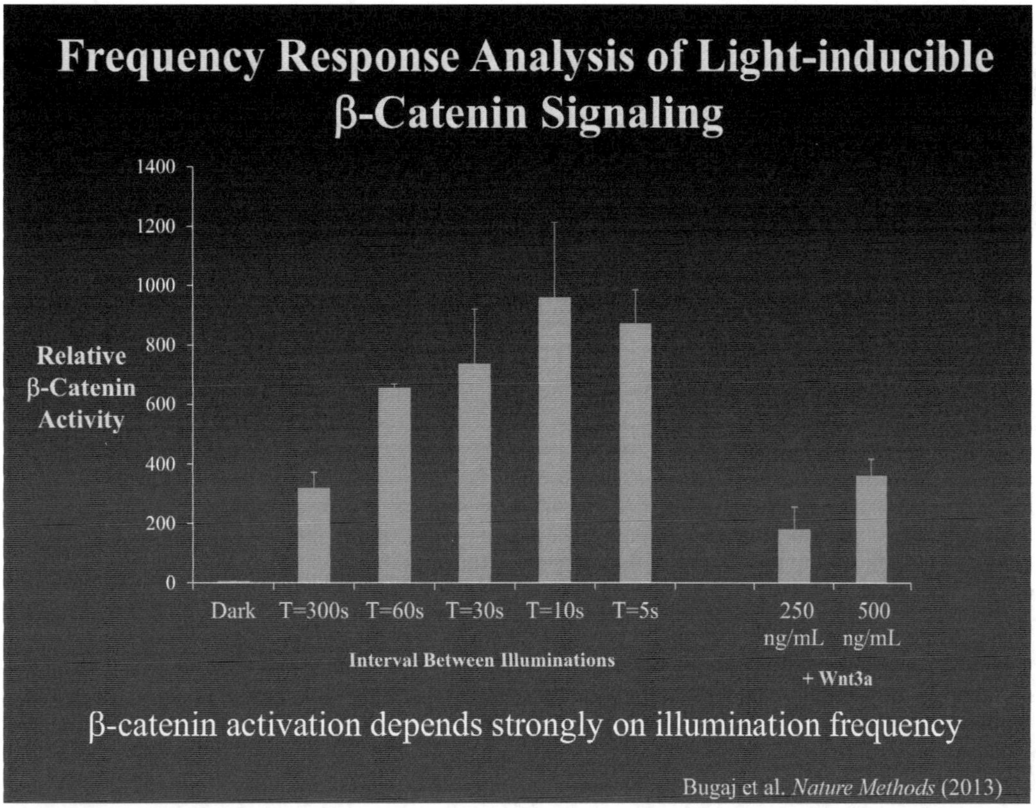

Frequency Response Analysis of Light-inducible β-Catenin Signaling
1400
1200
1000
800
600
400
200
0
Relative β-Catenin Activity
Dark
T=300s
T=60s
T=30s
T=10s
T=5s
250 ng/mL
500 ng/mL
Interval Between Illuminations
+ Wnt3a
β-catenin activation depends strongly on illumination frequency
Bugaj et al. Nature Methods (2013)

Outline

- Discovery of signaling system that regulates stem cell differentiation

- Optogenetic system for spatiotemporal control of signaling

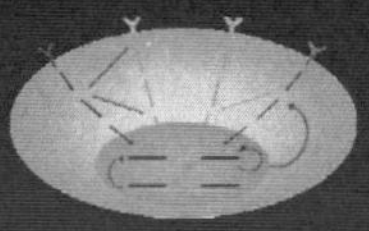

- **Biomaterial system to control spatial organization in cellular signaling**

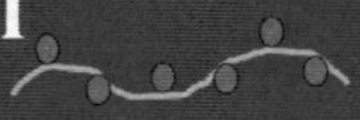

"Under physiological conditions, small changes in clustering could, perhaps, modulate Eph forward signaling. Therefore, it would be extremely useful to design ways to gradually change clustering behavior of Ephs and ephrins under physiological conditions."

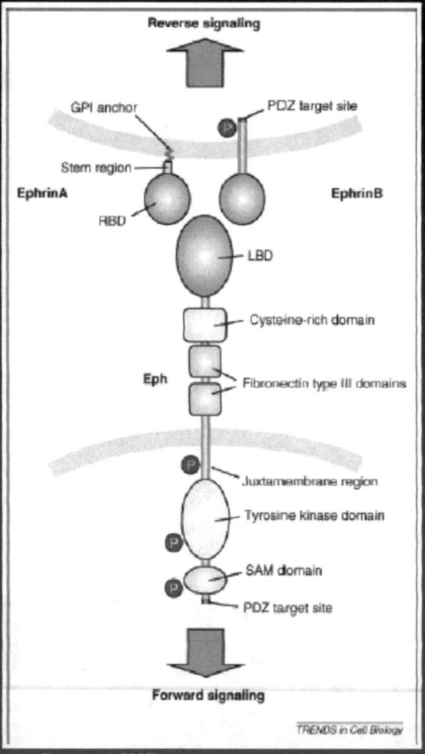

Egea *et al* (2007)

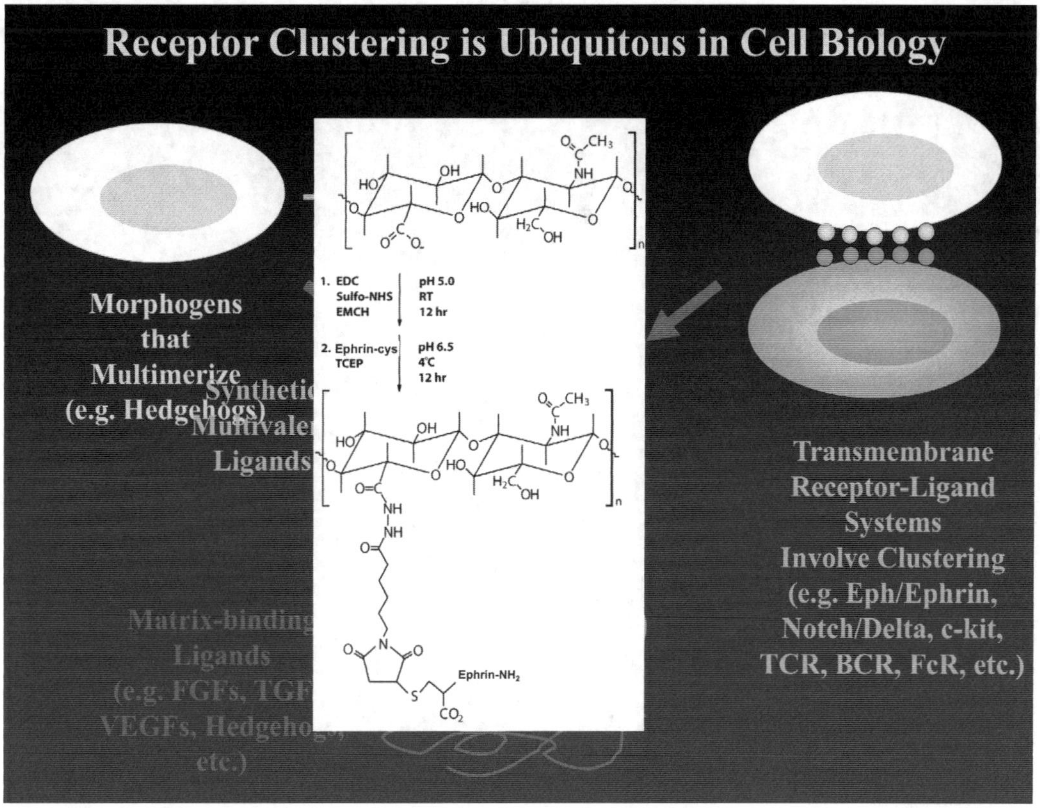
Receptor Clustering is Ubiquitous in Cell Biology
Morphogens that Multimerize (e.g. Hedgehogs)
Ligands
1. EDC Sulfo-NHS EMCH
pH 5.0 RT 12 hr
2. Ephrin-cys TCEP
pH 6.5 4°C 12 hr
Ephrin-NH2
Transmembrane Receptor-Ligand Systems Involve Clustering (e.g. Eph/Ephrin, Notch/Delta, c-kit, TCR, BCR, FcR, etc.)

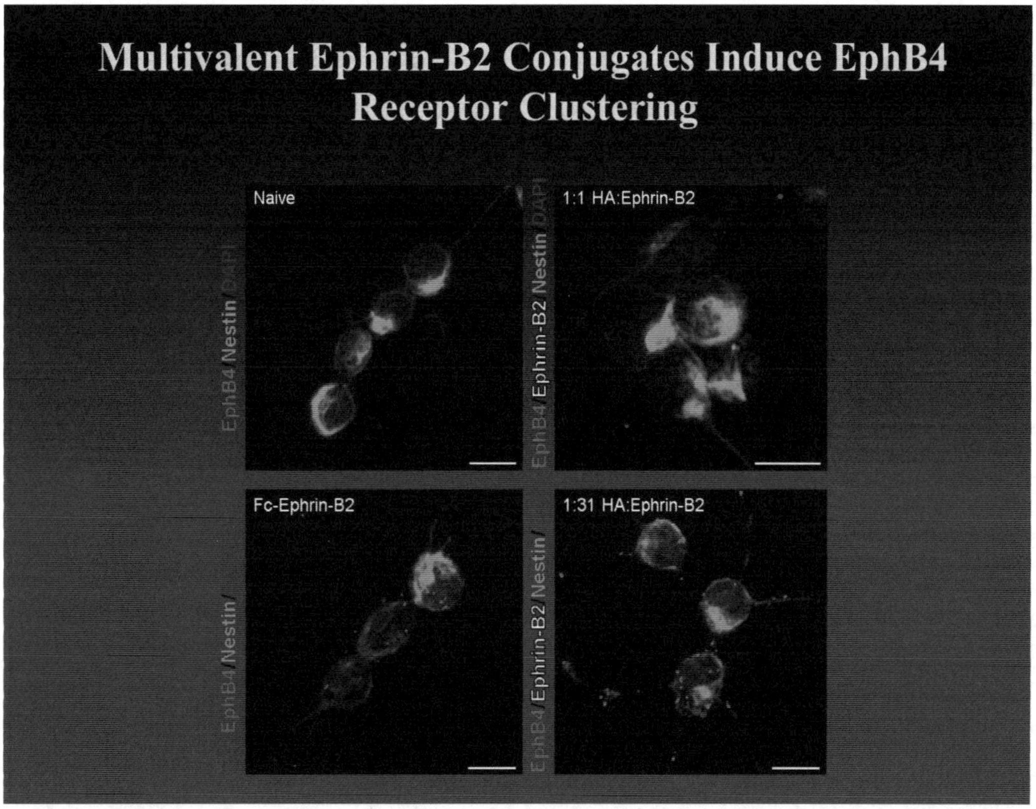
Multivalent Ephrin-B2 Conjugates Induce EphB4 Receptor Clustering
Naive
1:1 HA:Ephrin-B2
Fc-Ephrin-B2
1:31 HA:Ephrin-B2
EphB4/Nestin/DAPI
EphB4/Ephrin-B2 Nestin/DAPI
EphB4/Nestin/
EphB4/Ephrin-B2 Nestin/

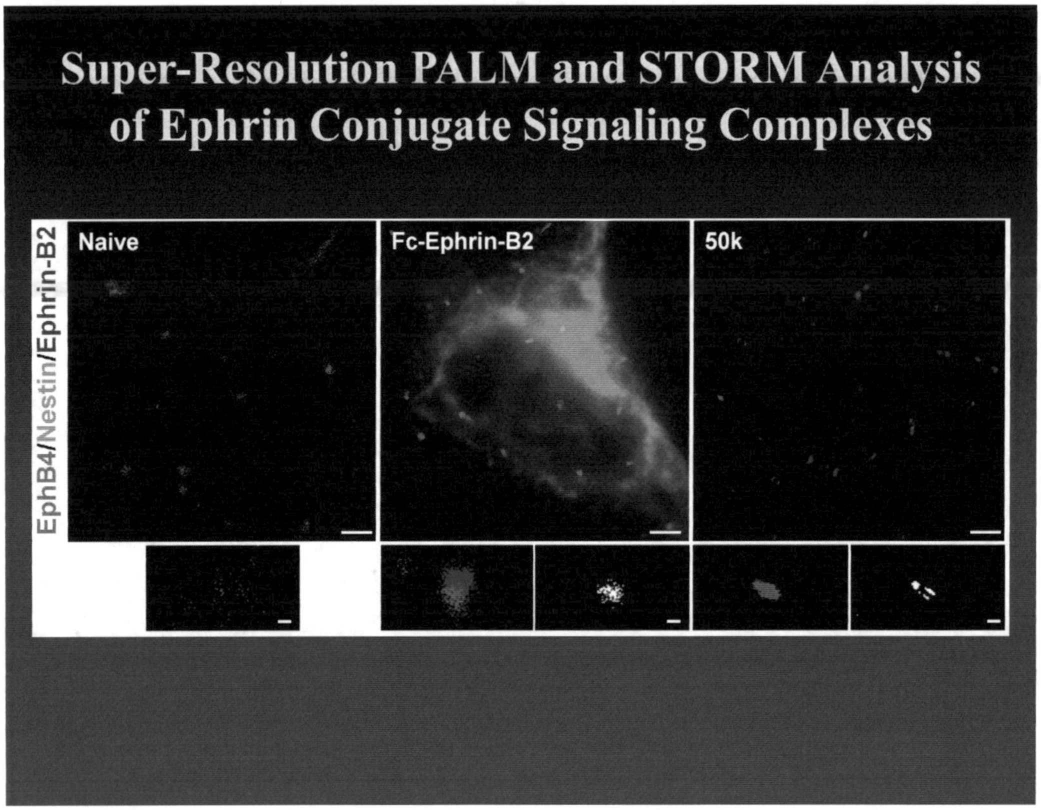
Super-Resolution PALM and STORM Analysis of Ephrin Conjugate Signaling Complexes
EphB4/Nestin/Ephrin-B2
Naive
Fc-Ephrin-B2
50k

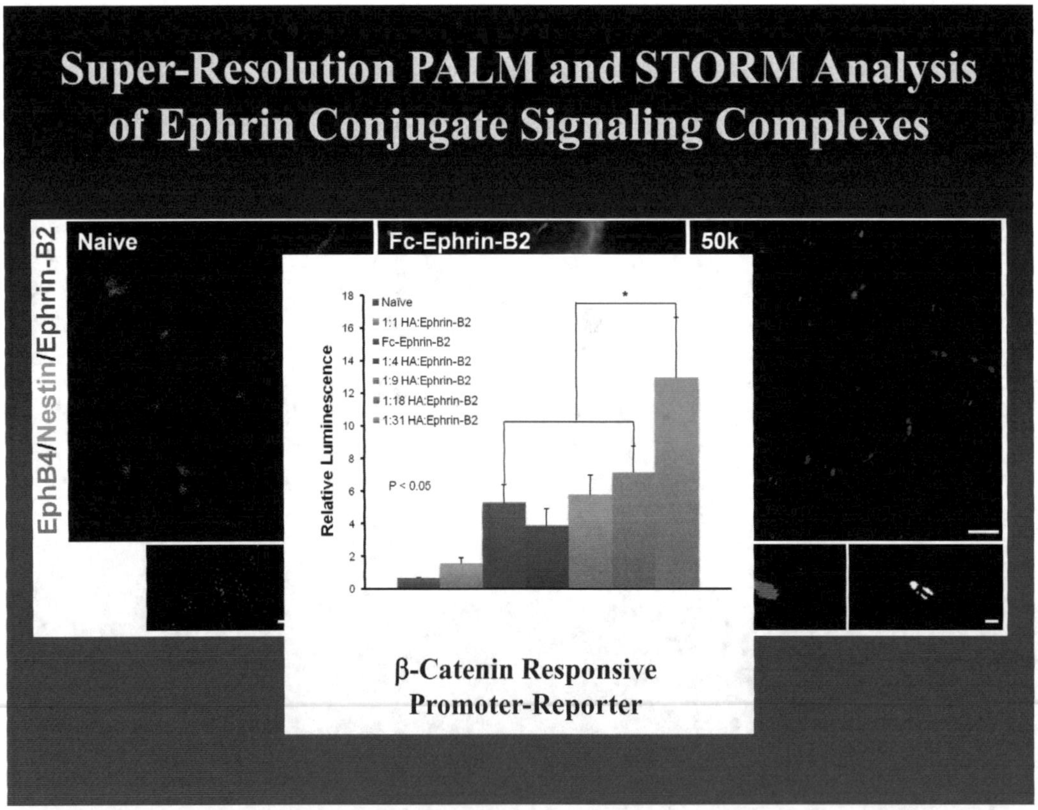
Super-Resolution PALM and STORM Analysis of Ephrin Conjugate Signaling Complexes
EphB4/Nestin/Ephrin-B2
Naive
Fc-Ephrin-B2
50k
Relative Luminescence
Naive
1:1 HA:Ephrin-B2
Fc-Ephrin-B2
1:4 HA:Ephrin-B2
1:9 HA:Ephrin-B2
1:18 HA:Ephrin-B2
1:31 HA:Ephrin-B2
P < 0.05
*
0
2
4
6
8
10
12
14
16
18
β-Catenin Responsive Promoter-Reporter

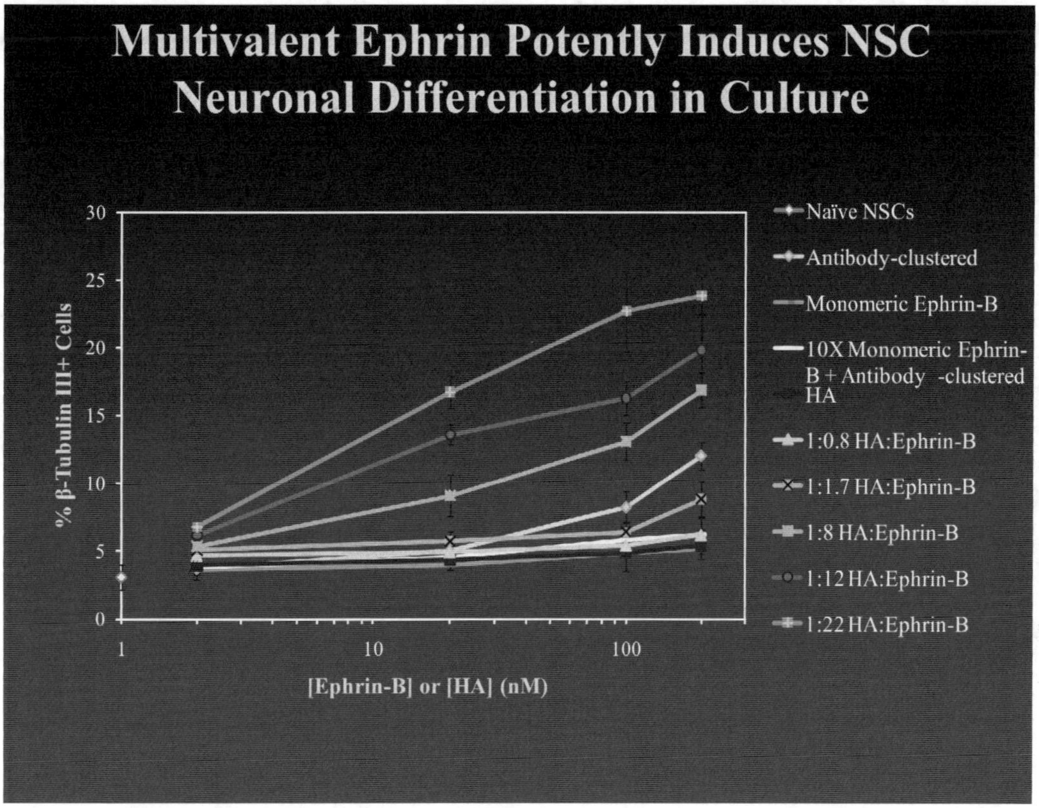
Multivalent Ephrin Potently Induces NSC Neuronal Differentiation in Culture
% β-Tubulin III+ Cells
[Ephrin-B] or [HA] (nM)
0
5
10
15
20
25
30
1
10
100
Naïve NSCs
Antibody-clustered
Monomeric Ephrin-B
10X Monomeric Ephrin-B + Antibody -clustered HA
1:0.8 HA:Ephrin-B
1:1.7 HA:Ephrin-B
1:8 HA:Ephrin-B
1:12 HA:Ephrin-B
1:22 HA:Ephrin-B

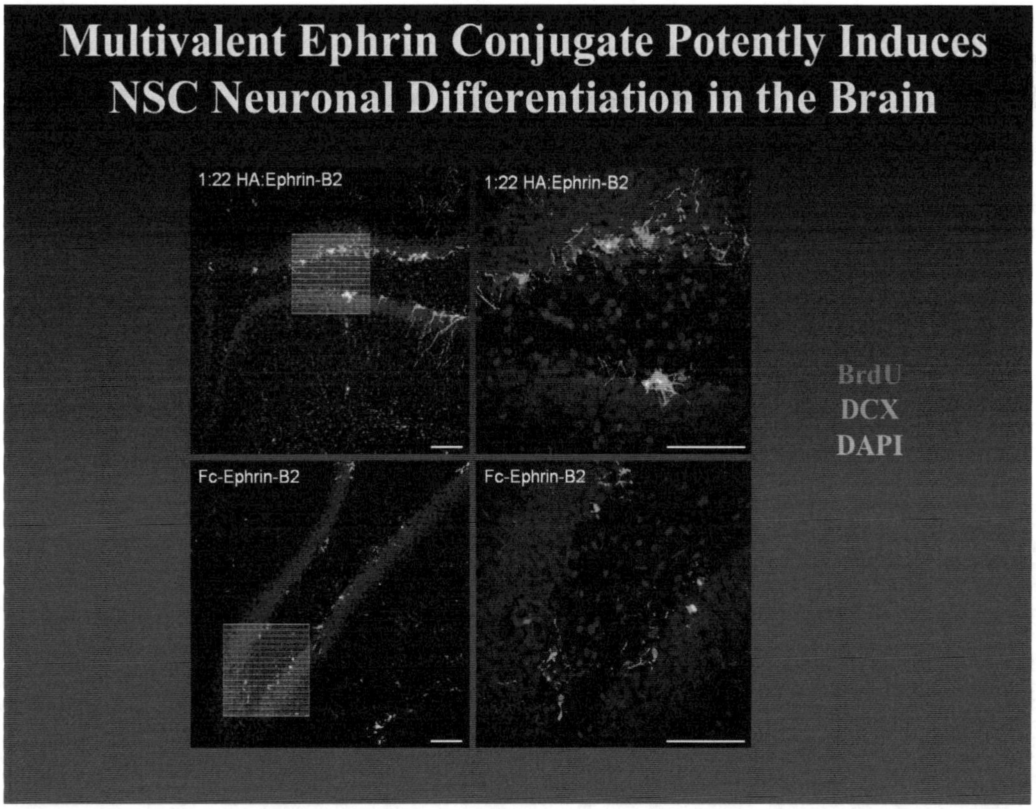
Multivalent Ephrin Conjugate Potently Induces NSC Neuronal Differentiation in the Brain
1:22 HA:Ephrin-B2
1:22 HA:Ephrin-B2
Fc-Ephrin-B2
Fc-Ephrin-B2
BrdU
DCX
DAPI

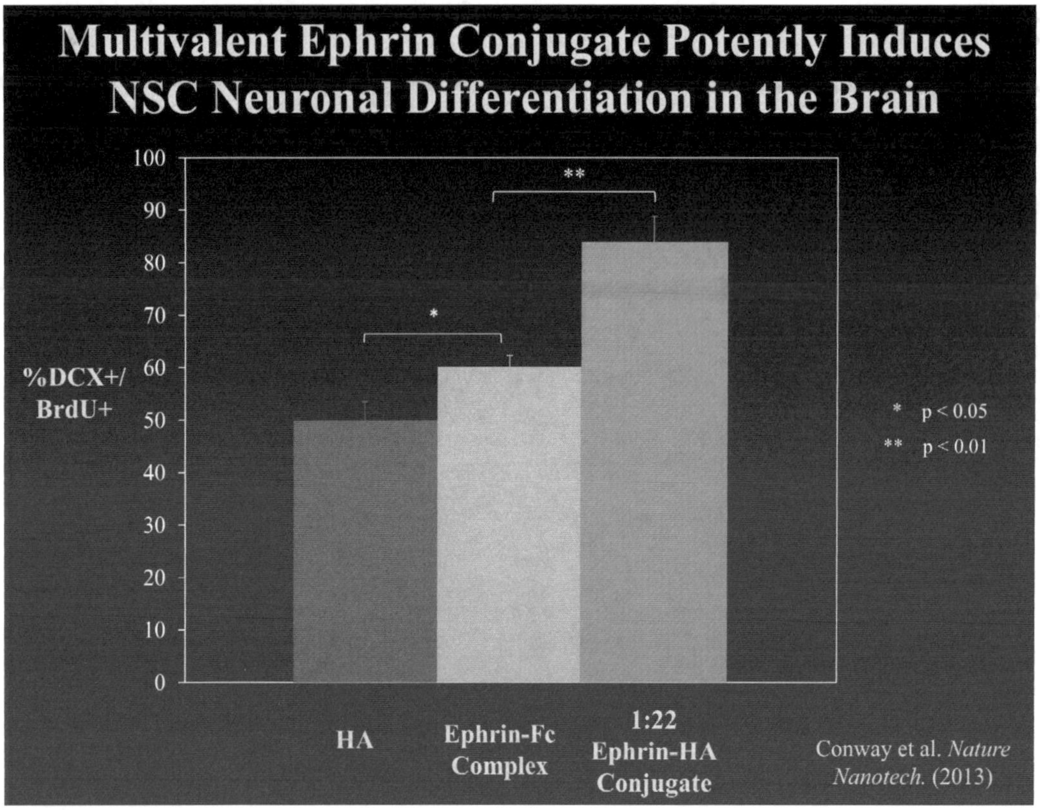
Multivalent Ephrin Conjugate Potently Induces NSC Neuronal Differentiation in the Brain
%DCX+/ BrdU+
100
90
80
70
60
50
40
30
20
10
0
**
*
* p < 0.05
** p < 0.01
HA
Ephrin-Fc Complex
1:22 Ephrin-HA Conjugate
Conway et al. Nature Nanotech. (2013)

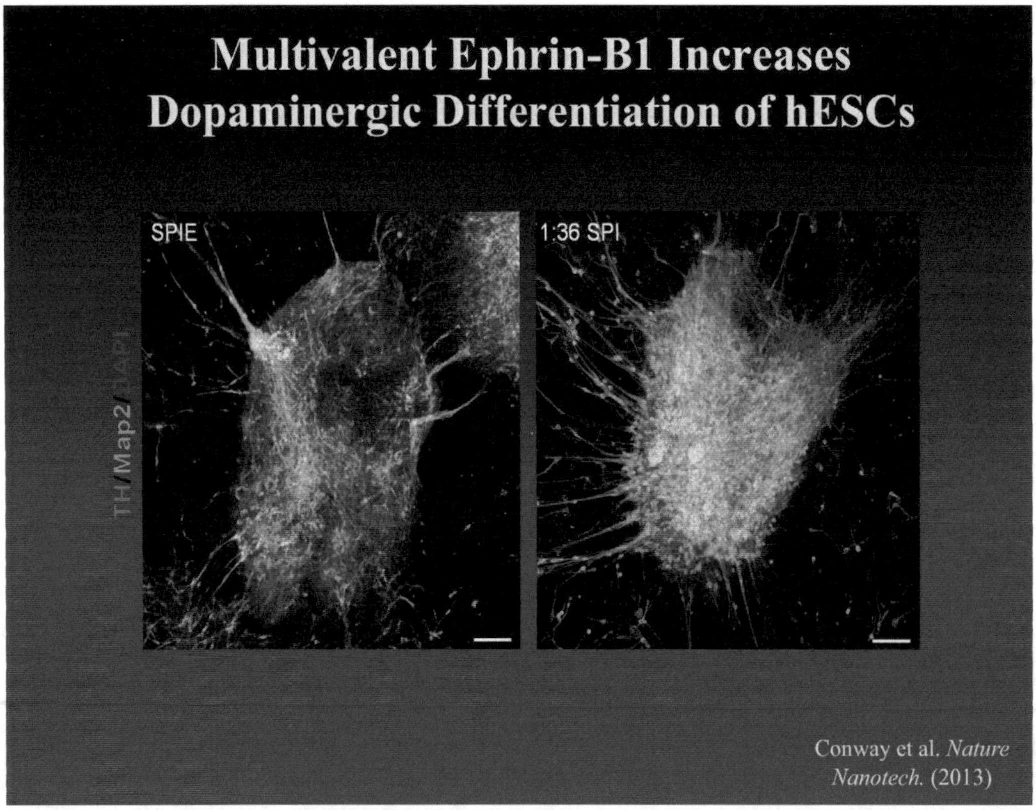
Multivalent Ephrin-B1 Increases Dopaminergic Differentiation of hESCs
SPIE
1:36 SPI
TH/Map2/
Conway et al. Nature Nanotech. (2013)

Multivalent Ephrin-B1 Increases Dopaminergic Differentiation of hESCs

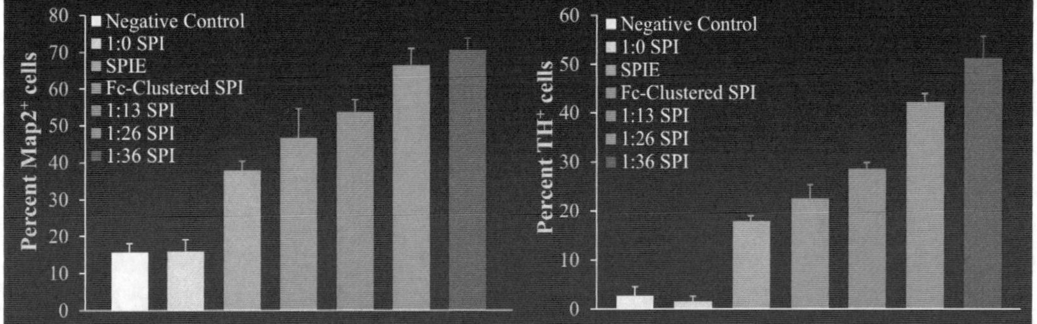

Conway et al. *Nature Nanotech.* (2013)

Summary

- Must better understand and engineer microenvironmental signaling mechanisms, both biochemical and biophysical, to control stem cell function.
- Ephrins regulate neuronal differentiation of adult neural stem cells and human pluripotent stem cells.
- Optogenetics is a valuable tool to elucidate spatiotemporal dynamics of signaling and cell fate decisions.
- Biomimetic materials to spatially organize cellular receptors can potently activate signaling and regulate cell fate in vitro and in vivo.

Schaffer Group

Melissa Bartel
Lukasz Bugaj
Leah Byrne, Ph.D.
Sisi Chen, Ph.D.
Anthony Conway
Tim Day
Siddharth Dey, Ph.D.
Barbara Ekerdt
Benjamin Epstein
Jonathan Foley
Ashley Fritz
Bum-Yeol Hwang, Ph.D.
Leo Lei, Ph.D.
Prajit Limsirichai
David Ojala
Anusuya Ramasubramaniam

Wanichaya Ramey
Sebastian Rammensee, Ph.D.
Jorge Santiago-Ortiz
Dawn Spelke
Marjan Varedi, Ph.D.
Tandis Vazin, Ph.D.
John Weinstein
Hanadie Yousef

Contributing Alums
Randy Ashton

Collaborators
Kate Excoffon (WSU)
Kevin Healy
Kathy High (CHOP)
John Flannery
Ravi Kane
Sanjay Kumar
Paul McCray (UI)
Bill Merigan (UR)

NIH, CIRM, NSF, DOE, Foundation Fighting Blindness, Project ALS, ALS Association, Alzheimer's Association, Cystic Fibrosis Foundation, American Parkinson's Disease Association, NSF CAREER Award, ONR Young Investigator Award, Whitaker Foundation,, Hellman Family, Merck & Co., Inc., UC Davis CRCC, UC Discovery, Bio-Rad, Geron, CDMRP TSC

Schaffer Group

Melissa Bartel
Lukasz Bugaj
Leah Byrne, Ph.D.
Sisi Chen, Ph.D.
Anthony Conway
Tim Day
Siddharth Dey, Ph.D.
Barbara Ekerdt
Benjamin Epstein
Jonathan Foley
Ashley Fritz
Bum-Yeol Hwang, Ph.D.
Leo Lei, Ph.D.
Prajit Limsirichai
David Ojala
Anusuya Ramasubramaniam

Wanichaya Ramey
Sebastian Rammensee, Ph.D.
Jorge Santiago-Ortiz
Dawn Spelke
Marjan Varedi, Ph.D.
Tandis Vazin, Ph.D.
John Weinstein
Hanadie Yousef

Contributing Alums
Randy Ashton

Collaborators
Kate Excoffon (WSU)
Kevin Healy
Kathy High (CHOP)
John Flannery
Ravi Kane
Sanjay Kumar
Paul McCray (UI)
Bill Merigan (UR)

NIH, CIRM, NSF, DOE, Foundation Fighting Blindness, Project ALS, ALS Association, Alzheimer's Association, Cystic Fibrosis Foundation, American Parkinson's Disease Association, NSF CAREER Award, ONR Young Investigator Award, Whitaker Foundation,, Hellman Family, Merck & Co., Inc., UC Davis CRCC, UC Discovery, Bio-Rad, Geron, CDMRP TSC

Schaffer Group

Melissa Bartel
Lukasz Bugaj
Leah Byrne, Ph.D.
Sisi Chen, Ph.D.
Anthony Conway
Tim Day
Siddharth Dey, Ph.D.
Barbara Ekerdt
Benjamin Epstein
Jonathan Foley
Ashley Fritz
Bum-Yeol Hwang, Ph.D.
Leo Lei, Ph.D.
Prajit Limsirichai
David Ojala
Anusuya Ramasubramaniam

Wanichaya Ramey
Sebastian Rammensee, Ph.D.
Jorge Santiago-Ortiz
Dawn Spelke
Marjan Varedi, Ph.D.
Tandis Vazin, Ph.D.
John Weinstein
Hanadie Yousef

Contributing Alums
Randy Ashton

Collaborators
Kate Excoffon (WSU)
Kevin Healy
Kathy High (CHOP)
John Flannery
Ravi Kane
Sanjay Kumar
Paul McCray (UI)
Bill Merigan (UR)

NIH, CIRM, NSF, DOE, Foundation Fighting Blindness, Project ALS, ALS Association, Alzheimer's Association, Cystic Fibrosis Foundation, American Parkinson's Disease Association, NSF CAREER Award, ONR Young Investigator Award, Whitaker Foundation,, Hellman Family, Merck & Co., Inc., UC Davis CRCC, UC Discovery, Bio-Rad, Geron, CDMRP TSC

Multiphoton-Based Fabrication of 3D ECM Scaffolds Via Modulated Raster Scanning
Ogle, B.,
Originally presented on: 3/17/2014 0:00:00 - 0:25:00

Multiphoton-based Fabrication of 3D ECM Scaffolds via Modulated Raster Scanning

Visar Ajeti,1 Ping-Jung Su,1 Jayne Squirrell,1 Brenda M. Ogle,1, 2 and Paul J. Campagnola1*

1Department of Biomedical Engineering, University of Wisconsin-Madison, 1550 Engineering

Dr., Madison WI, 53706, USA

2Department of Biomedical Engineering, University of Minnesota-Twin Cities, 312 Church Street

SE, Minneapolis, MN 55439, USA

Background. Extracellular matrices change dramatically with cardiac tissue development and maintenance. The oft-made conclusion is that these changes reflect passive accommodations to support differentiating cell types, including cardiomyocytes, smooth muscle cells and endothelial cells. We posit that, in addition or instead, changes of ECM actively direct cardiac differentiation. Approach. To test a portion of this hypothesis, that ECM distribution contributes to cardiomyocyte specification, we present here the development of a multiphoton-based fabrication technology to print 3D matrices based on tissue-specific ECM patterns and composed entirely of whole ECM proteins, with high spatial resolution. The method is analogous to multi-photon laser scanning microscopy (MPLSM) in that the excitation, and thus the photochemistry, is restricted to the focal volume. To this end, we implemented a new form of instrument control, termed modulated raster scanning, where rapid laser shuttering (40 MHz) is used to directly map the greyscale image data to the resulting protein concentration in the fabricated scaffold. Results. To assess the fidelity of the modulated raster scanning approach, individual optical sections of murine cardiac tissue immunostained for type IV collagen were compared to structures fabricated according to this template. Images of the immunostained tissue were overlain with images of fabricated structures and pixel by pixel co-localization tests using Fiji (ImageJ) yielded 95% or greater fidelity of both spatial localization and intensity. These results are substantially improved over more commonly used STL-based approaches that, in our hands, realize a fidelity of only 75%

or greater. We used this approach to fabricate

3D scaffolds from a mixture of BSA and FN where the design was derived directly from a 3D confocal immunofluorescence (FN) image stack of the left ventricle of mouse at postnatal day 2. The image stack, (taken near and around a blood vessel), was comprised of 22 optical sections, taken 1 micron apart (**Figure 1**). *Conclusion*. 3D, ECM-based structures fabricated using this approach preserved most of the microarchitecture of the original images stack with the resulting relative protein concentration corresponding to relative intensities of the original pattern. In
addition, we have seeded induced pluripotent stem cells and cardiac progenitor cells (Irx4+

cells) into the scaffolds and show that cells survive within and are capable of remodeling structures fabricated in this way. Future Work. Future studies will discern whether and to which extent ECM distribution, especially as it relates to the distribution of ECM with development, directs cardiac differentiation. Further, this approach could be utilized to study cell-ECM interactions of other tissue and organ systems and will likely influence the way we culture cells outside the body and our approaches to transplanting them back to the body.

Selection Mechanisms during Culture Adaptation in Stable Human Pluripotent Stem Cell Cultures

Rao, R. R., Virginia Commonwealth University

Originally presented on: 3/17/2014 0:25:00 - 0:50:00

Human pluripotent stem cells (hPSCs) have generated a lot of interest in the scientific community based on their potential applications in tissue engineering and regenerative medicine. However, numerous research groups have continued to report a propensity for genomic alterations during hPSC culture that poses concerns for basic research and clinical applications. To realize the clinical promise of stem cells, significant research in the development of propagation systems and bioprocessing strategies that generate stable stem cells is needed. Work from our laboratory and others have demonstrated that amplification of specific chromosomal regions is correlated with increased gene expression. To date, the phenotypic association of common genomic alterations remains unclear and is a cause for concern during hPSC propagation. Determining specific genes responsible for competitive advantage of abnormal hPSCs has been elusive; and progress on our molecular understanding of why hPSC single□cell dissociation increases aneuploid selection has been slow. In our study, we focus on a common genomic aberration and a list of candidate genes with increased gene expression to functionally validate a gene that may confer selective advantage when overexpressed. Our studies demonstrate that transduced hPSCs exhibited culture dominance in co-cultures of modified overexpression lines with non-overexpression lines. Furthermore, during low density seeding, we demonstrate increased clonality of our overexpression line against matched controls. Our results also indicate that this selective advantage could be reduced by varying the hPSC propagation and culture conditions. Our findings have important implications for hPSCs to reach their full potential in regenerative medicine, as it is critical to address the mechanisms that act as molecular switches during culture adaptation that lead to genomic instability in order to develop appropriate propagation strategies for generation of chromosomally stable hPSCs.

Cell Fate, Morphogenesis and Biophysical Dynamics within Pluripotent Microenvironments

Kinney, M. A., Georgia Institute of Technology and Emory University
Saeed, R., Georgia Institute of Technology and Emory University
McDevitt, T. C., Georgia Institute of Technology
Originally presented on: 3/17/2014 0:50:00 - 1:15:00

Cell fate, morphogenesis and biophysical dynamics within pluripotent microenvironments

M. A. Kinney[1], R. Saeed[1], and T. C. McDevitt[1,2]

[1]The Wallace H. Coulter Department of Biomedical Engineering

[2]The Parker H. Petit Institute for Bioengineering and Bioscience

Georgia Institute of Technology & Emory University, Atlanta, GA, USA

Stem cell fate and function are dynamically modulated by the interdependent relationships between biochemical and biophysical signals constituting the local 3D microenvironment. While approaches to recapitulate the stem cell niche have been explored for directing stem cell differentiation, a quantitative relationship between embryonic stem cell (ESC) morphogenesis and intrinsic biophysical cues within three-dimensional microtissues has not been established. The objective of this study, therefore, was to define the intrinsic cellular biophysical characteristics and phenotypic changes that arise simultaneously during cell fate specification and morphogenesis of three-dimensional pluripotent stem cell microenvironments.

Spheroids were formed from murine embryonic stem cells (D3) via forced aggregation (1000 cells per well in 400 μm diameter PDMS microwells) for 24 hours and were subsequently transferred to rotary orbital suspension (45 rpm), which together enable precise control of aggregation kinetics and spheroid size. ESCs were differentiated in serum free media (N2B27) in the presence or absence of BMP4, in order to direct differentiation trajectories toward mesoderm or ectoderm lineages, respectively. Biophysical characteristics were calculated using the linear viscoelastic model of creep compression, which was measured via a micron-scale material testing platform (Microsquisher, CellScale) that employs cantilever deformation to measure the force on individual aggregates. The multiparametric dynamics of morphogenesis was analyzed via histological, gene expression, and biomechanical changes in 3D microtissues, and correlative associations were assessed using multivariate data modeling.

Consistent with results from monolayer differentiation, pluripotent spheroids supplemented with BMP4 increased the expression of genes related to mesoderm lineages (*Hba-x*, *Gata2*), in contrast to the increased endoderm (*Foxa2*) and ectoderm (*Pax6*) expression in basal, serum-free biochemical environments. In addition, mesodermal aggregates exhibited loss of traditional epithelial characteristics, resulting in elongated, mesenchymal cells expressing alpha smooth muscle actin (α-SMA), suggesting that ESCs differentiated in 3D undergo an epithelial-to-mesenchymal transition (EMT) during mesoderm specification. Interestingly, the stiffness of stem cell spheroids (10^1-10^2 Pa) was several orders of magnitude less than that of native tissues (10^3-10^4 Pa), and was dynamically increased during differentiation, with increased stiffness (1.5 fold; $p=0.004$) concomitant with increasing mesoderm cell fate and mesenchymal morphogenesis. Microtissue stiffness was significantly ($p<0.03$ for all conditions) decreased upon inhibition of ROCK-mediated tension (Y27632; 10 μM) at all stages of differentiation, indicating that the mechanical properties were largely (~50%) attributed to structural characteristics of the actomyosin cytoskeleton. Moreover, while individual mechanical properties, such as stiffness, were not directly related to gene expression, the phenotypic profiles were accurately predicted by the combination of several biophysical properties, thus indicating that multiparameteric viscoelastic characteristics can describe pluripotent tissue composition.

Together, the dynamic remodeling during 3D pluripotent stem cell differentiation motivates the development of engineering approaches for controlling spatiotemporal administration of physical and chemical cues. Understanding biochemical and physical tissue morphogenesis, including the relationships between remodeling of cytoskeletal elements and the physical properties of the EB structure will inform approaches for the development of complex, functional tissue structures for regenerative therapies.

Intracellular Labeling of Pluripotent Stem Cell-Derived Neural Progenitor Aggregates with Micron-Sized Particles of Iron Oxide for Magnetic Resonance Imaging

Sart, S., Ecole Polytechnique
Calixto Bejarano, F., The National High Magnetic Field Laboratory
Baird, M., The National High Magnetic Field Laboratory
Yan, Y., Florida State University
Rosenberg, J., The National High Magnetic Field Laboratory
Ma, T., Florida State University
Grant, S., The National High Magnetic Field Laboratory
Li, Y., FAMU-FSU College of Engineering
Originally presented on: 3/17/2014 1:15:00 - 1:40:00

Magnetic resonance imaging (MRI) can be used to assess the fate of the transplanted stem cells, including the neural progenitor cells (NPC) derived from pluripotent stem cell (PSC), in the injured nervous system. NPCs are usually derived from PSCs through the formation of embryoid body (EB), and 3D multicellular structures have been maintained to enhance the function of PSC-derived neural cells. This study investigated the effect of labeling multicellular EB-NPC aggregates with micron-sized particles of iron oxides (MPIO) for MRI. While labeling and tracking single stem cells have been demonstrated for MPIO, the ability to label multicellular aggregates and consequences on cellular effects have yet to be characterized fully. In this study, intact and dissociated EB-NPC aggregates were labeled with various concentrations of MPIOs and analyzed by fluorescent imaging, flow cytometry and *in vitro* MRI for labeling efficiency and detectability. Moreover, the effects of MPIO on cell viability, cytotoxicity, proliferation and neural differentiation were evaluated. MPIO incorporation did not affect the proliferation and neural differentiation of EB-NPCs. However, intact EB-NPC aggregates showed higher cell proliferation and viability compared to the dissociated cells. Despite diffusion limitation at low MPIO concentration, higher concentrations of MPIO (*i.e.* 10 μg Fe/mL) were able to label EB-NPC aggregates at similar efficiency to the dissociated cells. *In vitro* MRI showed efficient MPIO detection over 2-3 population doublings. This study demonstrated the feasibility of using MPIO to label EB-NPC aggregates for future animal study.

Lysophosphatidic Acid Enhances Stromal Cell-Directed Angiogenesis

Binder, B., University of California, Davis
Sondergaard, C., School of Medicine, UC Davis
Nolta, J., School of Medicine, UC Davis
Leach, K., UC Davis

Statement of Purpose: Ischemic diseases such as peripheral vascular disease (PVD) affect more than 15% of the general population and are characterized by loss of blood flow to the extremities. While the delivery of cells or growth factors to promote angiogenesis has shown promise, the high cost of producing and purifying proteins such as vascular endothelial growth factor (VEGF) limits large-scale implementation. Lysophosphatidic acid (LPA) is a platelet derived growth factor that has been reported to promote the secretion of angiogenic factors by human adipose-derived stromal cells (hASC) and is implicated in promoting angiogenesis in metastatic tumors. Additionally, hASC are readily available, have outstanding proliferative capacity, and can function as pericytes *in vivo*, thus making the concomitant delivery of LPA with hASC an attractive target for investigation. We hypothesized that LPA would increase the secretion of angiogenic cytokines by hASC under ischemic conditions, and that this combination would synergistically promote revascularization and recovery in a mouse critical hindlimb ischemia model.

Methods: hASC were cultured in FBS-free α-MEM supplemented with 0.1% (w/v) fatty-acid free BSA and incubated in hypoxia (1% O_2) for 24 h to simulate the serum deprivation and hypoxia (SD/H) of *in vivo* ischemia. To explore the effect of LPA on the transcriptional regulation of angiogenic factors, hASC were cultured in SD/H with 25 μM LPA. When used, 10 μM $LPA_{1/3}$ receptor inhibitor (Ki16425) was added to medium containing LPA. qPCR was performed using TaqMan primers for VEGF and basic fibroblast growth factor (bFGF). To determine the functional effects of LPA treatment, conditioned media from hASC incubated with 25 μM LPA for 24 h was removed and used to assay migration of endothelial colony forming cells (ECFC). To study the ability of LPA and hASC to promote angiogenesis *in vivo*, 1e6 hASC were entrapped in an 8 mm x 1 mm fibrin gel containing 25 μM LPA and placed over the defect site of a ligated and segmentally resected femoral artery in NOD SCID gamma (NSG) mice.

Results: After 24 h in SD/H, hASC cultured with 25 μM LPA exhibited significantly higher levels of VEGF and bFGF expression. These increases were abrogated by the addition of 10 μM Ki16425. Conditioned media from LPA-treated hASC was also significantly more chemoattractive to endothelial cells in a transwell migration assay, but the addition of Ki16425 negated this effect. Similarly, media from hASC embedded in fibrin gels with 25 μM LPA in SD/H *in vitro* showed higher levels of VEGF production as measured by ELISA. Importantly, ischemic hindlimbs treated with both ASC and LPA had significantly increased numbers of blood vessels and functional recovery after 2 weeks compared to ASC or LPA alone.

Conclusions: These data demonstrate that LPA enhances the pro-angiogenic capacity of hASC under SD/H. This information will be valuable for optimizing cell-based treatments for PVD and ischemic disease *in vivo*.

Scalable Production of Human Mesenchymal STEM/Stromal CELLS in Microcarrier-Based Culture Systems

Carmelo, J. G., Department of Bioengineering and IBB - Institute for Biotechnology and Bioengineering, Instituto Superior Técnico, Lisboa, Portugal
Fernandes-Platzgummer, A., Department of Bioengineering and IBB - Institute for Biotechnology and Bioengineering, Instituto Superior Técnico, Lisboa, Portugal
Weber, J. L.,
Diogo, M. M., Department of Bioengineering and IBB - Institute for Biotechnology and Bioengineering, Instituto Superior Técnico, Universidade de Lisboa, Lisboa, Portugal
Lobato da Silva, C., Department of Bioengineering and IBB - Institute for Biotechnology and Bioengineering, Instituto Superior Técnico, Universidade de Lisboa, Lisboa, Portugal
Cabral, J. M. S., Department of Bioengineering and IBB - Institute for Biotechnology and Bioengineering, Instituto Superior Técnico, Universidade de Lisboa, Lisboa, Portugal

SCALABLE PRODUCTION OF HUMAN MESENCHYMAL STEM/STROMAL CELLS IN MICROCARRIER-BASED CULTURE SYSTEMS

Joana G. Carmelo, Department of Bioengineering and IBB - Institute for Biotechnology and Bioengineering, Instituto Superior Técnico, Universidade de Lisboa, Lisboa, Portugal

Ana Fernandes-Platzgummer, Department of Bioengineering and IBB - Institute for Biotechnology and Bioengineering, Instituto Superior Técnico, Universidade de Lisboa, Lisboa, Portugal

Jennifer L. Weber, Corning Life Sciences, Corning Incorporated, Corning, NY 14831, USA

Marie-Maud Bear, Corning SAS, CETC, Avon 77210, France

Martial Hervy, Corning SAS, CETC, Avon 77210, France

Maria Margarida Diogo, Department of Bioengineering and IBB - Institute for Biotechnology and Bioengineering, Instituto Superior Técnico, Universidade de Lisboa, Lisboa, Portugal

Cláudia Lobato da Silva, Department of Bioengineering and IBB - Institute for Biotechnology and Bioengineering, Instituto Superior Técnico, Universidade de Lisboa, Portugal

Joaquim M.S. Cabral, Department of Bioengineering and IBB - Institute for Biotechnology and Bioengineering, Instituto Superior Técnico, Universidade de Lisboa, Lisboa, Portugal

T: (+351) 218 419 063, joaquim.cabral@ist.utl.pt

The clinical application of human mesenchymal stem/stromal cells (MSC) for a wide range of diseases is at the front line of stem cell-based cellular therapies. Nevertheless, successful implementation of these cell-based therapies requires the ability to generate large numbers of cells with well-defined characteristics in a cost-effective way. To this end, significant research has been done on the development of microcarrier-based cultures in scalable stirred bioreactors alongside with the development and evaluation of well-defined serum-free and xeno-free media formulations which represent important milestones towards the production of MSC for cellular therapies.

In this work, a microcarrier-based suspension culture was explored for the scale-up of MSCexpansion in xeno-free medium using synthetic peptide acrylate surface beads. Cells were maintained on Corning® Synthemax® II polystyrene and CELLstart?-coated Solohill plastic microcarriers [1] for 14 days in xeno-free medium. Bone marrow (BM) derived- and adipose tissue (AT) derived-MSC were seeded at 50,000 cells/mL with 4.5 cm^2/mL of microcarriers in spinner flasks (80 mL volume). To maximize cell seeding, the adhesion step was performed in 50% final volume during the first 24 hours. The efficiency of initial cell adhesion to microcarriers was similar for both cell types; slightly better cell adhesion was observed on Synthemax II compared to CELLstart-coated microcarriers: 48% and 43% for AT MSC and 42% to 37% for BM MSC. The homogenous cell distribution on the first days of culture resulted in a higher expansion rate with exponential growth from day 4 to day 6 for AT MSC and day 4 to day 9 for BM MSC. The longer growth phase observed for BM MSC resulted in higher cell densities of 350,000 cells/mL compared to 250,000 cell/mL for AT MSC cultures. Expanded cells maintained their characteristic immunophenotype and multilineage differentiation potential. Ongoing work includes the translation of this culture system to a fully controlled stirred-tank bioreactor (1 L) and the study of the impact of different culture parameters namely feeding regime and dissolved oxygen concentration on cell productivity. This scalable xeno-free, microcarrier-based culture platform is anticipated to enable cost effective and robust production of MSC meeting the needs of the allogeneic ?off-the-shelf? MSC therapy sector.

[1] Dos Santos F, Andrade PZ, Abecasis MM, et al. Toward a clinical-grade expansion of mesenchymal stem cells from human sources: a microcarrier-based culture system under xeno-free conditions. Tissue engineering. Part C, Methods. 17(12), 1201?10 (2011).

The Effect of Polyethylene Glycol Linked Multi-Walled Carbon Nanotubes on Myogenic Differentiation of Human Mesenchymal Stem Cells for Skeletal Muscle Engineering

Zhao, C.,
Ho, H. K.,
Pastorin, G.,
Originally presented on: 3/17/2014 17:00:00 - 17:02:00

The effect of polyethylene glycol linked multi-walled carbon nanotubes on myogenic differentiation of human mesenchymal stem cells for skeletal muscle engineering

Chunyan Zhao, Giorgia Pastorin, Han Kiat Ho

Department of Pharmacy, National University of Singapore, Singapore

INTRODUCTION

The reconstruction of skeletal muscle defects has been a continuing challenge and the need for improved treatments has thus motivated the research on skeletal muscle tissue engineering. Human mesenchymal stem cells (hMSCs) have attracted much attention in tissue engineering due to their wide range of sources, unique capabilities of self-renewal in an undifferentiated state for prolonged time, and multi-lineage differentiation upon proper stimuli [1]. They are attractive, being readily isolated from bone marrow, umbilical cords, umbilical cords blood, lacrimal glands, glomeruli and other easily accessible sources such as adipose tissue and peripheral blood. Advances in stem cell biology have shown that hMSCs can differentiate into a variety of connective tissue cells including osteocytes, chondrocytes, adipocytes and myocytes, which envisage the versatile applications of hMSCs. Hence, to repair and regenerate skeletal muscle tissue, the employment of hMSCs is promising, triggering investigative works to improve its differentiation toward myogenic lineage.

Carbon nanotubes (CNTs), cylindrical tubes of rolled graphene sheets comprising of honeycomb structure of carbon atoms, have been at the forefront of nanotechnology due to their unique mechanic, thermal, electrical and chemical features. However, the pristine CNTs are extremely hydrophobic and rapidly precipitate in aqueous solutions, mitigating their physiological relevance. To improve the property of CNTs, polyethylene glycol (PEG) is often used to modify CNTs¡¯ surfaces and increase their hydrophilicity and biocompatibility. For the first time ever reported, our group found that the thin films of PEG linked multi-walled CNTs (PEG-CNTs) were not cytotoxic and accelerated the osteogenic differentiation of hMSCs, to a similar extent of hMSCs cultured with a commonly used growth factor, the bone morphogenetic protein-2 (BMP-2) [2]. Similarly, we found that graphene sheets also accelerated osteogenic differentiation of hMSCs, which was comparable to the one achieved with BMP-2 [3]. It was postulated that the intrinsic nanostructure of PEG-CNTs and the honeycomb structure of carbon atoms in the outer layer of PEG-CNTs helped in accelerating the osteogenic differentiation of hMSCs. Therefore, this study is driven by exploiting the breadth of this effect with investigation of PEG-CNT films¡¯ role in the myogenic differentiation of hMSCs, with the hope of improvements in skeletal muscle engineering.

METHODS

The PEG-CNT films were prepared by dropping PEG-CNTs suspension onto pre-heated cover slips. When dried, the films were characterized for surface roughness (helium ion microscopy, HIM; and atomic force microscopy, AFM), thickness (scanning electron microscopy, SEM) and stiffness (AFM). Subsequently, the hMSCs were cultured on PEG-CNT films and were induced for myogenic differentiation by incubating with normal medium for one week and myogenic medium (normal medium with dexamethasone and hydrocortisone) for another two weeks. The myogenically-induced and non-induced hMSCs were examined by cell morphology (fluorescent rhodamine-phalloidin staining) and cell viability (CellTiter-Glo assay). The quantitative real-time polymerase chain reaction (RT-PCR) was then employed to determine the extent of myogenesis by monitoring changes of hMSCs specific genes (CD73, CD90 and CD105), myogenic genes (Myf-5, MyoD, desmin and myosin heavy chain, MHC) and skeletal muscle specific genes (fast skeletal troponin C, TnC; and ryanodine receptor 1, Ryr) in the differentiated hMSCs. To verify the skeletal myogenic differentiation of hMSCs, the non-induced hMSCs on PEG-CNT films were also characterized with osteogenic genes (collagen-1, Col-1; osteocalcin, OCN; alkaline phosphatase, ALP; osteopontin, OPN).

RESULTS

The HIM image of PEG-CNT films showed that the PEG-CNT film exhibited smooth surface and orderly fashion of single PEG-CNT (Figure 1 A). The AFM analysis substantiated the HIM result and the PEG-CNT films had small values in the roughness parameters (Table 1). The estimated thickness of PEG-CNT films was around 5¦Ìm (Figure 1 B) while the average elastic modulus for the PEG-CNT films was 944¡À474MPa, indicative of high stiffness.

Over the three weeks of incubation, rhodamine-phalloidin staining of hMSCs showed that PEG-CNT films did not alter the hMSCs morphology (Figure 2). The non-induced hMSCs retained a spindle shape, which is similar with that observed in human skeletal muscle cells (SKMCs). However, the induced hMSCs adopted rounder shape, which is different from the shape of SKMCs.

The cell viability assay showed a slight decrease in the viability after the myogenic induction of hMSCs on cover slips (Figure 3). Significant inhibition of cell viability in hMSCs on PEG-CNT films was observed compared to the control (non-induced hMSCs on cover slips), regardless of presence or absence of myogenic induction. As cells differentiate, their rate of proliferation usually decreases, thus cell proliferation potential decreased with increasing differentiation [4, 5]. Coupling to the lack of overt cell death based on fluorescent staining described earlier, the decrease of cell viability on PEG-CNT films might be due to myogenic differentiation of hMSCs and hence suppression of proliferation.

The non-induced hMSCs on cover slips and PEG-CNT films presented a set of highly expressed CD genes (CD73, CD90, CD105) (Figure 4 A). With myogenic induction, it was shown that the hMSCs seeded on both cover slips and PEG-CNT films had a reduction of these CD genes. The suppressed hMSCs feature gene expression indicated that the induced hMSCs has diminished hMSC characteristics, thus may be more prone to myogenic differentiation. Comparing to the control (non-induced hMSCs on cover slips), the expression of myogenic genes and skeletal muscle specific genes was weakly increased in the myogenically-induced hMSCs on cover slips and highly increased on PEG-CNT films (Figure 4 B). The up-regulation of myogenic markers verified the myogenic phenotype of investigated cell culture and indicated that the differentiation protocol was capable of guiding myogenic differentiation of hMSCs. The TnC and Ryr are primarily expressed in skeletal muscle among the three types of muscle (skeletal muscle, cardiac muscle and smooth muscle) [6]. Therefore, the higher expression of TnC and Ryr in myogenically-induced hMSCs indicated that the hMSCs derived myobalsts may be committed towards skeletal muscle cells. Interestingly, these myogenic genes and skeletal muscle specific genes were also significantly increased in the non-induced hMSCs on PEG-CNT films compared to the control. This occurrence suggests that PEG-CNT films alone, in the absence of myogenic inducers, such as dexamethasone and hydrocortisone, could trigger the myogenesis of hMSCs. The advantage of this effect is the possible removal of potentially noxious dexamethasone and hydrocortisone from the induction protocol. Today, there is no well recognized way of controlling the optimal concentrations of the inducers for efficient differentiation with reduced or no side effects *in vivo*. Therefore, the observed myogenic commitment of hMSCs by PEG-CNT films in the absence of myogenic inducers is a major advancement of hMSCs application in skeletal muscle engineering. Furthermore, the non-induced hMSCs on PEG-CNT films did not demonstrate higher expression of Col-1, OCN and ALP compared to the control, indicating a lack of concurrent osteogenesis during the skeletal myogenesis (Figure 4 C).

CONCLUSION

In this study, we created PEG-CNT films with slight surface roughness with the PEG-CNTs aligned in orderly fashion on the surface. The PEG-CNT film thickness was approximately 5¦Ìm with exhibition of significant stiffness (944¡À474MPa). The films supported the culturing of hMSCs and preserved the hMSCs morphology as compared to cover slip controls. Upon myogenic induction, hMSCs adopted new morphology with rounder shape while the non-induced hMSCs retained the spindle shape. Based on RT-PCR results, myogenic induction of hMSCs was successful on cover slips and PEG-CNT films with myogenic medium. Moreover, the PEG-CNT films alone could specifically trigger the myogenic differentiation of hMSCs (without concurrent osteogenesis), although somewhat at the expense of the viability of hMSCs, an expected secondary effect during differentiation process. Therefore, the PEG-CNT films show promise as a potential scaffold in guiding myogenic differentiation of hMSCs for skeletal muscle tissue engineering. This was the first study to investigate the CNTs¡¯ influence in the myogenesis of hMSCs. Further studies will be conducted to substantiate this finding with myogenic protein expression and physiological function tests.

REFERENCE

[1] C. Zhao, et. al, Biotechnology advances, 31(2013) 654-668.

[2] T.R. Nayak, et. al, ACS nano, 4 (2010) 7717-7725.

[3] T.R. Nayak, et. al, ACS nano, 5 (2011) 4670-4678.

[4] M. Olbrich, et. al, PloS one, 7 (2012) e47176.

[5] C. GM, The Cell: A Molecular Approach. 2nd edition. , 2000.

[6] R. Gahlmann, et. al, Journal of molecular biology, 201 (1988) 379-391.

Extracellular Matrices Derived from Pluripotent Stem Cell Aggregates for Tissue Development

Sart, S., Florida State University
Yan, Y., Florida State University
Ma, T., Florida State University
Li, Y., FAMU-FSU College of Engineering
Originally presented on: 3/17/2014 17:02:00 - 17:04:00

Pluripotent stem cells (PSCs) secrete large amounts of endogenous extracellular matrices (ECMs), which play an important role in regulating PSC self-renewal, lineage commitment, and tissue morphogenesis. ECMs derived from PSCs have broader signaling capacity compared to somatic ECMs and contain the unique signaling networks that regulate the self-renewal and lineage specification during tissue development. In this study, ECMs from undifferentiated PSC aggregates or differentiated embryoid bodies at different developmental stages and lineage specifications were decellularized and their capacities to direct PSC proliferation and differentiation were characterized. The results demonstrate that the PSC-derived ECMs were able to influence PSC proliferation and differentiation by direct interactions with the cells and by influencing the signaling functions of the regulatory macromolecules. Proteomic analysis (by LC-MS) indicated the shared and distinct proteins in ECMs derived from undifferentiated and different types of differentiated aggregates. To enhance the stability of PSC-derived ECMs, ECMs were crosslinked by a natural cross-linker which increased the stiffness of ECMs. The effect of biomechanical properties of ECMs on neural differentiation was also evaluated. Taken together, the PSC-derived ECMs have the potential to direct lineage- and development-specific cellular responses for in vitro applications or in vivo cell delivery.

Autologous Non-Activated Platelet Rich Plasma (PRP) a Proliferation Promoting Substitute for Xenogenic Culture Media

Atashi, F., Geneva university

Originally presented on: 3/17/2014 17:04:00 - 17:06:00

Introduction.

Adipose derived Mesenchymal stem cells (AT-MSC) are promising cells for cell therapy and tissue engineering. However, their expansion needs in vitro culture. Actually non-autologous culture media (e.g fetal bovine serum (FBS), plasma from blood bank) that brings potential immunological reactions and are used. In this study we investigate the effects of non-activated autologous platelet-rich plasma (nPRP) and autologous thrombin activated platelet rich plasma (tPRP) on AT-MSCs proliferation, as a FBS substitute.

Methods.

AT-MSCs of 10 donors were isolated and cultured for 10 days under different conditions: (a) 10% FBS as control, (b) 1%, (c) 5%, (d) 10%, (e) 20%, (f) 40% and (g) 60% of tPRP or nPRP. Platelet viability was assessed by Calcein. The proliferation capacity of nPRP and FBS cultured cells was quantified by EDU assay. MSCs phenotype markers were analyzed by FACS calibur. Differentiation was proved by corresponding staining.

Results.

57% of platelets were viable until 10 days of culture. PRP increased AT-MSC proliferation dose dependently: 20% nPRP provided a higher cell proliferation rate than tPRP and FBS while MSCs phenotype and differentiation potency were not modified.

Conclusion.

20% nPRP could serve as a potent substitute for non-autologous media in human cell therapies and tissue engineering researches based on GMP protocols.

Serum-Free Co-Culture of Human Mesenchymal Stem Cells and Articular Chondrocytes in Suspension Bioreactors for Cartilage Repair

Khurshid, M., University of Calgary
Mulet-Sierra, A., University of Alberta
Adesida, A., University of Alberta
Sen, A., Pharmaceutical Production Research Facility (PPRF), Schulich School of Engineering, University of Calgary
Originally presented on: 3/17/2014 17:06:00 - 17:08:00

Background:
Defects in articular cartilage can lead to a chronic and painful condition called osteoarthritis1,2. A proposed cell therapy is the implantation of human articular chondrocytes (hACs) into the defect site. A major drawback of this technique is that hACs do not proliferate extensively in culture without losing their native phenotype2,3. Recently, it has been discovered that when hACs and human mesenchymal stem cells (hMSCs) are co-expanded in static culture, hACs proliferate extensively and hMSCs exhibit enhanced chondrogenesis compared to monoculture of each cell type2. Thus, the co-culture of hACs and MSCs represents a feasible method of generating cells capable of cartilage repair.

Static culture is an inefficient means of generating cells that lacks reproducibility and the ability to control key aspects of the culture environment. However, suspension bioreactors can reproducibly generate large quantities of cells under controlled conditions. Thus, the aim of this project was to determine if it is possible to co-expand populations of hACs and hMSCs in bioreactors as aggregates under serum-free conditions.

Methods:
First, two serum-free media (PPRF-msc64 and TheraPEAK) and a serum-containing medium were screened to determine if they could support the proliferation of co-expanded cells in static culture. Next, bone marrow-derived hMSCs and hACs were inoculated in a 1:3 ratio2 at 20,000 cells/mL in 125 mL stirred suspension bioreactors. Samples of aggregates were enzymatically dissociated and counted to determine the cell density. Cell and matrix characteristics, such as gene expression, the amount of matrix produced and the distribution of each cell type in the aggregates were also analyzed. We have examined several variables to improve cell proliferation in the bioreactors including: agitation rate, feeding schedule and oxygen tension.

Results:
The cells reached the highest cell densities in a serum-free medium, PPRF-msc6 (540,000±30,000 cells/mL on day 6), as compared to TheraPEAK (9,000±2,000 cells/mL), and serum-containing medium (36,000±2,000 cells/mL). Thus, from this point forward, all experiments were conducted with PPRF-msc6.

We have demonstrated, for the first time, that hACs and hMSCs can be co-expanded in suspension bioreactors as aggregates of cells as shown in Figure 1, producing more matrix (3.3 μg glycosaminoglycans (GAG)/μg DNA) than in static vessels (1.5 μg

GAG/μg DNA). Agitation rate had a significant impact on cell proliferation. Specifically, the average cell density was greater in the 80 rpm condition, peaking at 89,000±1,000 cells/mL at day 12, as compared to the 60 rpm (48,000±1,700 cells/mL on day 10) and the 100 rpm conditions (68,000±0 cells/mL on day 12). Additionally, bioreactor operation as cyclic fed-batch, with 50% medium changes on day 8 and 12, resulted in higher cell densities (95,000±7,000 cells/mL on day 16) than batch operation (40,000±1,600 cells/mL). Also, hypoxic conditions (3% O2 tension5) resulted in more than double the cell growth compared to normoxic conditions.

Significance:
We have developed protocols to rapidly and reproducibly co-culture human stem cells and chondrocytes to clinically relevant quantities. The large-scale production of hACs and hMSCs in bioreactors with serum free medium represents a novel strategy for the generation of cells for cartilage tissue-engineering.

References:
1. Gao, J. et al. 2007. J Eng Med 221:441?450. 2. Acharya, C. et al. 2012. J Cell Physiol. 227:88?97. 3. Dhinsa, B. & Adesida, AB. 2012. Curr Stem Cell Res Ther. 7:143?148. 4. Jung, S. et al. 2011. J Tissue Eng Regen Med 6:391-403. 5. Adesida, A. et al. 2012. Stem Cell Res & Ther, 3:9.

Efficient Differentiation of Human Pluripotent Stem Cells to Angioblasts and Endothelial Cells Via Small Molecule Activation of Wnt Signaling

Bao, X., University of Wisconsin Madison

Originally presented on: 3/17/2014 17:08:00 - 17:10:00

Heart disease is the leading cause of morbidity and mortality in industrialized countries. One possible treatment for heart disease involves producing heart cells in vitro and tranplanting these cells to restore the function of damaged tissue. We and others have previously described mthods to generate large numbers of heart muscle cells from human pluripotent stem cells (hPSCs), which have the potential to facilitate heart regeneration. However, the heart is not merely composed of muscle cells, but also contains vascular endothelial, smooth muscle cells, and cardiac fibroblasts. Thus, engineering heart tissue patches which contain both cardiomyocytes and support cells representing a more desirable approach for cell-based therapies to treat heart disease. Here we describe a method for the simple and efficient conversion of hPSCs to CD34+CD31+ angioblasts with bipotent differentiation potential, capable of generating smooth muscle and endothelial cells. We systematically optimized Wnt, FGF2, and VEGF signaling activity during angioblast differentiation and then illustrate that appropriate temporal activation of regulators of Wnt signaling alone, in the absence of exogenous FGF2 and VEGF signaling, is sufficient to drive multiple hPSC lines to differentiate to greater than 50% CD34+CD31+ angioblasts. These angioblasts can be further enriched to 99% purity with a single step of simple magnetic separation based on CD31 or CD34 expression. Single-cell clonal differentiation assays revealed that CD34+CD31+ angioblasts generated by Wnt pathway activation were bipotent and could differentiate to functional endothelial and smooth muscle cells.

Analysis of Prostate Fibroblast Derived Extracellular Matrix Proteins Unveils the Microenvironment Niche of Prostate Epithelial Stem Cells

Ojalill, M.,
Rappu, P.,
Heino, J., Turku University
Originally presented on: 3/17/2014 17:12:00 - 17:14:00

The composition and the organization of extracellular matrix (ECM) participate in the formation of the niche for epithelial stem cells. During tumor development the stroma surrounding the cancer cells is activated and remodeled, which may have a significant influence on cancer stem cells, tumor progression and cell invasion. Fibroblasts and cancer associated fibroblasts (CAFs) are the main producers of ECM. In prostate epithelial/cancer stem cells are identified as $CD44^{+}$, $CD133^{+}$, $\alpha 2\beta 1$ integrin ^{high} and androgen receptor negative cells (Collins, *et al.,* 2005, Cancer Res.). Integrin $\alpha 2\beta 1$ is a collagen receptor proposing that cell–collagen interaction is important for stem cell biology. Still, the function of specific ECM molecules and receptors in regulating prostate stem cells remains poorly understood.

Here, we have used primary cultures of prostate-derived fibroblastic cells, allowed them to generate ECM in the presence of ascorbic acid and analyzed the composition of the ECM using mass spectrometry. Fibroblasts originated from either normal (n=18) or cancerous tissue area (n=10). In both cases the most abundant proteins were tenascin, fibronectin and collagens VI, XII and XIV. Significantly increased levels of collagen III ($p< 0,05$; Student's t-test) and Periostin ($p< 0,05$; Student's t-test) were observed in CAF related ECM when compared to normal fibroblasts.

Further studies on the function of altered matrix proteins to integrin signaling in prostate stem cells are currently in progress.

Pluripotent Stem Cells Derived Mesenchymal Stromal Cells Improve Therapies for Advanced Cancers

Liu, F.,
Zhao, Q.,
Originally presented on: 3/17/2014 17:14:00 - 17:16:00

Bone marrow Mesenchymal stromal cells (BM-MSCs) have a unique tumor-homing capacity and are a promising vehicle of gene therapy for advanced cancers. However, BM-MSCs are difficult to genetically engineer due to their limited expandability and may promote cancer progression. Large amount of MSC-like cells can be derived from embryonic stem cells (ESCs) or iPS cells, which will greatly facilitate engineering and expansion of MSCs carrying anti-cancer genes. Starting from a human iPS cell line CY2 free of reprogramming transgenes, we derived adherent cells expressing typical MSC surface markers (PSC-MSCs). These PSC-MSCs have the same karyotypes as the parental line and did not form teratoma in our initial studies, whereas the parent line did so robustly. Similar to BM-MSCs, these PSC-MSCs differentiated into bone and adipocytes in vitro, and showed homing property to cancer cells in vitro with comparable expression of genes related with tumor tropism. Notably, these PSC-MSCs express much low level of receptors for TGF-beta and Interleukin-1, two major cytokines inducing the pro-tumor activities of BM-MSCs. Consistently, after stimulation with TGF-beta, Interleukin-1 or tumor-conditioned medium, the expression of multiple pro-tumor factors was much lower in PSC-MSCs than that in BM-MSCs. Moreover, when co-cultured with various cancer cells, PSC-MSCs did not promote the epithelial-mesenchymal transition of cancer cells or expand the cancer stem cell population as BM-MSCs did. On the other hand, when stimulated with TNF-alpha and poly dA/dT, PSC-MSCs expressed high level of Trail and were capable to induce apoptosis of cancer cells in vitro. These data suggest that PSC-MSCs are a promising safer alternate of BM-MSCs with better expandability and genetic engineering potential.

Microposts As a Means to Measure the Contractile Properties of hiPSC Derived Cardiomyocytes

Rodriguez, M. L., University of Washington
Murry, C. E., University of Washington
Sniadecki, N. J., University of Washington

Originally presented on: 3/17/2014 17:18:00 - 17:20:00

INTRODUCTION

Cardiomyocytes derived from human stem cells (hSC-CMs) are an invaluable research tool in the fight against heart disease. Recent research demonstrates that these cells have the ability to help elucidate the mechanisms that govern heart development and lead to heart disease, they can be used to determine the effectiveness of candidate drugs in treating heart disease, and they have even been shown to help reverse some of the damage caused by heart disease [1, 2]. All of these applications can be improved by assessing the contractility of hSC-CMs at the single-cell level, as one of the most important functional characteristics of a cardiomyocyte is its ability to produce contractile force. Additionally, out of the currently available sources of human stem cells, human induced pluripotent stem cells (hiPSCs) are very promising: the number of cell lines that can be induced to the pluripotent state is extremely vast, they serve as a potential source for patient-specific cardiomyocytes, and their use is non-controversial.

In this work, we discuss the development of an in vitro micropost platform that is able to assess the contractile characteristics of individual hiPSC-CMs. Here, this platform was used to determine the contractile properties of iPS-CMs of different ages, to gauge their maturation over time. In conjunction, immunofluorescent staining was used to assess their structural maturation. We found that, over the course of 80 days, hiPSC-CMs demonstrated significant increases in twitch force, velocity, and power, as well as cell spread area, perimeter, and Z-band width.

METHODS

Cell Culture

Fibroblasts from the human IMR-90 cell line (James A. Thomson, U.Wisconsin-Madison) were de-differentiated to the pluripotent state via treatment with transcription factors OCT4, SOX2, NANOG, LIN28 [3]. Monolayers of these stem cells were then differentiated into cardiomyocytes by sequential treatment of activin A and bone morphogenetic protein 4 (BMP4) [4], supplemented with CHIR 99021 and Xav 939. These cells were cultured on Matrigel-coated plates in RPMI medium: RPMI 1640 with L-glutamine (Gibco), supplemented with B-27 (Gibco), and 1% penicillin/streptomyosin (Cellgro). At 20, 70, and 100 days post-differentiation, the hiPSC-CMs were dissociated from their culture surfaces, and seeded onto arrays of microposts in RPMI medium supplemented with %5 FBS. Following seeding, the cells were again cultured in RPMI medium, which was henceforth exchanged every other day.

Substrate Preparation

Arrays of silicone microposts were fabricated by casting polydimethylsiloxane (PDMS) off of silicon wafers with patterned SU-8 structures, onto #1 25mm round coverslips, as previously described [5, 6]. The tips of these microposts were coated via microcontact printing with either 50 μg/ml of mouse laminin (Life Technologies), human fibronectin (BD Biosciences), or human collagen IV (Millipore), while the remaining surfaces of the micropost array were fluorescently stained with bovine serum albumin conjugated with Alexa Fluor 594 and blocked with 0.2% Pluronic F-127 in phosphate buffer solution [7].

Cell Analysis

One week following cell seeding, spontaneously-beating, hiPSC-CMs were recorded using high-speed video microscopy (Fig. 1A). A phase contrast video was taken in the optical plane at the tips of the posts to track the movement of individual posts during twitch contractions (Fig. 1B), and a fluorescent image was taken at a plane where the base of the posts was in focus to establish the reference position for each post (Fig. 1C). To determine the deflection of each post over time, a custom-written MATLAB code was used to find and track the movement of the posts by their centroids. The position of the posts in each image frame was then compared to the position of the centroids of the base of the posts (Fig. 1D). The deflection of each post over time can then be determined with the same MATLAB code, and twitch force can be calculated by multiplying this deflection by the bending stiffness of the micropost (Fig. 1E). Using this same code, the contraction velocity and power of each of the posts beneath every measured cell were determined by calculating the change in the micropost?s deflection over time (Fig. 1F), and from the product of the force and velocity measured at that post over time (Fig. 1G), respectively. Finally, to obtain the total force or power produced by an individual cell, the absolute magnitude of these entities were summed across all the posts beneath an individual cell. Alternatively, for velocity, each of the individual post traces was analyzed to determine the maximum velocity that each cell was able to produce.

After live experiments, the cells were fixed and stained for nuclei and sarcomeric α-actinin in order to image their myofibril structure while still on the posts. Once fluorescent images were taken, NIS Elements software was used to manually measure sarcomere lengths and Z-band widths, as well as circularity and cell area.

RESULTS

Contractile Properties

Our analysis revealed that there were no significant changes in passive force (Fig. 2A), force per area (Fig. 2C), or spontaneous beating frequency (Fig. 2D) for hiPSC-CMs that were 20, 70, and 100 days post-differentiation. Alternatively, the twitch force (Fig. 2B), velocity (Fig. 2E), and power (Fig. 2F) were all found to significantly increase over time, in the absence of any external stimuli. However, these quantities are still relatively small when compared to those measured for adult cardiomyocytes.

Cytoskeletal Properties

Upon analyzing α-actinin structure of hiPSC-CMs, we found significant increases in the spread area (Fig. 2G), perimeter (Fig. 2H), and Z-band width (Fig. 2K), as well as a significant decrease in circularity (Fig. 2I), over time. However, there were no significant differences in sarcomere length between any of the cell age groups (Fig. 2J), and there was an increase in circularity between day 40 and day 70 cells. Additionally, these values are still relatively immature in nature when compared to those found for adult cardiomyocytes.

CONCLUSIONS

In this work, we reported on the use of a micropost-based technology for the assessment of hiPSC-CM contractility. We demonstrated that this platform is able to effectively measure the twitch force, velocity, and power at all points of adhesion underneath a cell, and that is compatible with immunofluorescent staining. Upon measuring the contractile and cytoskeletal properties of individual iPS-CMs that were 20, 70, and 100 days post-differentiation, we found significant changes in numerous different endpoints, consistent with cardiomyocyte maturation. These results indicate that hiPSC-CMs observably develop in culture over 80 days, but that these cells are still underdeveloped when compared to adult cardiomyocytes. Thus, if hiPSC-CMs are to be used as a model for adult human cardiomyocytes shortly after differentiation, they require additional stimuli to attain the same level of contractile and cytoskeletal maturation as an adult cardiomyocyte. Therefore, future studies employing this platform will focus on determining which stimuli are essential for enhancing hiPSC-CM maturation in culture.

ACKNOWLEDGEMENTS

This work was funded by the NSF Graduate Research Fellowship Program. Additionally, part of this work was conducted at the University of Washington Microfabrication/Nanotechnology User Facility, a member of the NSF National Nanotechnology Infrastructure Network.

REFERENCES

1. Mercola, M., Circ Res, 2013. 112(3): p. 534-48.
2. Bellin, M., Nat Rev Mol Cell Biol, 2012. 13(11): p. 713-26.
3. Yu, J., Science, 2007. 318(5858): p. 1917-20.
4. Laflamme, M.A., Nat Biotechnol, 2007. 25(9): p. 1015-24.
5. Tan, J.L., Proc Natl Acad Sci U S A, 2003. 100(4): p. 1484-9.
6. Rodriguez, M.L., J Biomech Eng, 2014 (In Review).
7. Sniadecki, N.J. Methods in Cell Biology: Cell Mechanics, ed. 2007. 83: p. 313-328.

Figure 1 The twitch characteristics of a single hiPSC-CM is determined by taking a high-speed video of the tips of the posts and a reference image of the base of the posts (A). This video is taken in the phase contrast setting (B), while the reference image is taken in the red fluorescent channel (C). A custom MATLAB code is then used to determine the location of each post?s centroid in the reference plane (red dot), as well as it is used to track the location of the post?s centroid in the video plane (blue dot) (D). The difference in location between these two centroids over time gives the deflection of a post over multiple twitch events, which can be multiplied by the post stiffness to yield the twitch force (E). This deflection data can then be used to determine the twitch velocity (F), and power (G) of that post. Scale bars represent 6 μm.

Figure 2 Upon investigating the structural and contractile maturation of 20, 70, and 100 day old hiPSC-CMs, we found no statistical changes in passive force (A), force per area (C), frequency (D), or sarcomere length (J). However, significant differences in twitch force (B), upstroke (black) and relaxation (grey) velocity (E), upstroke (black) and relaxation (grey) power (F), spread area (G), cell perimeter (H), circularity (I), sarcomere length (J), and Z-band width (K) were found. A representative hiPSC-CM is shown for reference (L). Scale bar indicated 6 μm.

Assessing Whether the Adult Neocortex Can Incorporate New Projection Neurons

McKeehan, N.,
Belalcazar, H.,
Hebert, J.,

Originally presented on: 3/17/2014 17:22:00 - 17:24:00

The neocortex is the seat of our highest cognitive functions. Neocortical projection neurons can be lost due to neurodegenerative diseases and once lost they are not normally replaced, leading to permanent functional deficits. There are a number of therapies in development to alleviate the symptoms and retard the onset of neurodegenerative diseases, such as Alzheimer's, in which neocortical projection neurons are lost. However, additional treatment, such as the replacement of cortical projection neurons, will eventually be necessary to return patients to pre-disease states. Developing strategies to replace lost projection neurons is a daunting task because of the complexity and size of the neocortex. Previous attempts at replacing projection neurons in the neocortex using several types of transplanted neural stem or precursor cells have not provided viable strategies. One reason for the limited integration of new neurons is that thus far transplanted cells have remained primarily in a clump at the transplant site. To achieve the goal of functionally integrating new projection neurons throughout broad areas of the neocortex, a novel strategy is required that takes into account cell dispersion. We are developing an approach for introducing new, widely dispersed, projection neurons in the adult neocortex, providing a paradigm for testing whether they can functionally integrate. Our preliminary data suggest that we can achieve the wide dispersion of precursors in the adult neocortex and that new neurons can extend long projections along their normal tracks and to their normal targets. We are using engineering precursors, which have a natural ability to disperse in the adult neocortex, with doxycycline-inducible transcription factors that could, once the precursors have dispersed, allow their reprogramming to a cortical projection neuron fate. We are also assess the ability of transplanted neuronal cells to disperse and integrate in a healthy environment compared with one in which projection neurons are degenerating.

3D Culturing of Stem Cells and Progenitors for Reproducible Differentiation and Tissue Engineering

Tomov, M. L., SUNY-CNSE
Olmsted, Z. T., SUNY-CNSE
Paluh, J. L., SUNY-CNSE

Originally presented on: 3/17/2014 17:24:00 - 17:26:00

@font-face { "Calibri"; }@font-face { "Cambria"; }p.MsoNormal, li.MsoNormal, div.MsoNormal { margin: 0in 0in 10pt; line-height: 115%; font-size: 11pt; ""; }p.MsoNoSpacing, li.MsoNoSpacing, div.MsoNoSpacing { margin: 0in 0in 0.0001pt; text-align: justify; font-size: 12pt; ""; }div.Section1 { page: Section1; }

Generation and control of three dimensional (3D) niche environments in a highly reproducible platform is necessary to optimize tissue engineering with stem cells for translational biomedical applications. The differentiation of human pluripotent stem cells through embryoid body (EB) formation is used to evaluate pluripotency potential in vitro and provides a common precursor for lineage differentiation and organoid formation. Although in the stem cell field the current norm is to allow EBs to spontaneously form by non-templated methods, their formation by different groups is far from uniform and results in EB populations that are often of mixed or inappropriate size and shape dependent on the degree of EB fusion events. Applying tools of nanotechnology we generated lithography-templated arrays coated with polydimethylsiloxane (LTA-PDMS grids) as a seeding platform for human embryonic stem cells (hESCs) as 2D cell clusters or as 3D early aggregates. A lithography-based templating method offers high throughput generation of EBs of controllable shape and size and a means to track 3D aggregate cell formation through time-lapse analysis in this optically clear template. The size of human EBs is critical to differentiation timing and efficiency, which has also been observed with mouse ESCs. By further combining EB pre-templating with an additional post-patterning step we are able to ensure uniform settling of EBs, an additional parameter that can affect statistical analysis, thus providing a powerful set of tools to evaluate and optimize differentiation factors and gradients using more homogeneously formed EBs. The internal core structure of EBs varies dependent on their size and is revealed by confocal microscopy and serial microtome sectioning. We are applying these tools and lithography-based approaches to optimize in vitro platforms for tumor formation and reprogramming studies in cancer research, in tissue engineering with 3D-initiated lineage progenitors, and in multicellular co-cultures. Our approach offers an exciting new strategy for studying cell aggregates by enabling better monitoring and control of single and multi-cellular 3D environments.

Mechanotransduction Via p190RhoGAP Regulates a Switch Between Cardiomyogenic and Endothelial Lineages in Adult Cardiac Progenitors

Kz, K.,
Hubbi, M.,
Afzal, J.,
Kim, D. H., Johns Hopkins University
Ahn, E.,
Abraham, R.,
Levchenko, A.,
Originally presented on: 3/17/2014 17:26:00 - 17:28:00

Mechanical cues can have pleiotropic influence on stem cell shape, proliferation, differentiation, and morphogenesis, and are increasingly realized to play an instructive role in regeneration and maintenance of tissue structure and functions. To explore the putative effects of mechanical cues in regeneration of the cardiac tissue, we investigated therapeutically important cardiosphere-derived cells (CDCs), a heterogeneous patient- or animal-specific cell population containing c-Kit$^+$ multipotent stem cells. We showed that mechanical cues can instruct c-Kit$^+$ cell differentiation along two lineages with corresponding morphogenic changes, while also serving to amplify the initial c-Kit$^+$ sub-population. In particular, mechanical cues mimicking the structure of myocardial extracellular matrix specify cardiomyogenic fate, while cues mimicking myocardium rigidity specify endothelial fates. Further, we found that these cues dynamically regulate the same molecular species, p190RhoGAP, which then acts through both RhoA-dependent and independent mechanisms. Thus, differential regulation of p190RhoGAP molecule by either mechanical inputs or genetic manipulation can determine lineage type specification. Since human CDCs are already in phase II clinical trials, the potential therapeutic use of mechanical or genetic manipulation of the cell fate could enhance effectiveness of these progenitor cells in cardiac repair, and shed new light on differentiation mechanisms in cardiac and other tissues.

Photoreversible Protein Patterning to Modulate 4D Stem Cell Fate

DeForest, C., University of Washington

Originally presented on: 3/17/2014 17:28:00 - 17:39:00

Synthetic hydrogels have emerged as a unique class of biomaterials that enable stem cells to be cultured in three-dimensions within near-physiological, simplified microenvironments. Recent strategies have been developed that permit bioepitopes (*e.g.*, peptides, full-length proteins) to be introduced at any point in time and space to affect cell function spatiotemporally within user-defined subvolumes of the bulk material. While these techniques have been successfully utilized to direct a variety of basic cellular functions, advanced platforms that permit biological cues to be both *introduced* and subsequently *removed*would be beneficial in recapitulating the dynamic abundance of signaling biomolecules in the native, temporally-variable niche. Moreover, as proteins provide high substrate specificity in comparison with simple chemical moieties and peptides, a system that enables dynamic presentation of full-length proteins would be of great interest for use in modulating complex cellular behavior. In this work, we demonstrate that the combination of two bioorthogonal light-based chemistries provides for the reversible immobilization of protein cues spatially within a hydrogel. Results further highlight the versatility of such dynamic biochemical signal presentation in better understanding basic cell physiology and in directing 4D progenitor cell fate.

'Deterministic' Biomaterial Platforms for Directing Stem Cell Fate Choices

Cameron, A., The University of Queensland
Frith, J. E., The University of Queensland
Rowlands, A., The University of Queensland
George, P.,
Gomez, G.,
Yap, A.,
Cooper-White, J., The University of Queensland
Originally presented on: 3/17/2014 17:32:00 - 17:34:00

Mesenchymal stem cells (MSCs) and pluripotent stem cells (hESC, iPSC) cells, due to their potential to turn into most if not all cell types in our body, represent ideal cell sources for regenerative medicine applications. This presentation will focus on recent outcomes from our lab aimed at elucidating the biophysical properties of a substrate that may be tailored to manipulate MSC (and hESC) fate decisions. Firstly, the complex interplay of substrate 'rigidity' and ligand type and availability on MSC fate decisions will be described. We show that substrate stiffness can be completely overridden by ligand-integrin interactions, and that substrate stiffness is not necessarily deterministic of stem cell fate. Secondly, the influence of the 'creepiness' of a substrate will be detailed. We have demonstrated that MSCs are explicitly sensitive to variations in the viscous properties of a viscoelastic substrate, not just the elastic properties. The 'creepier' the substrate, the more 'primed' the cells are for a whole host of fate decisions. More recently, using a Rac1 biosensor, we have shown that this is due to these substrates upregulating Rac1 activity, in contrast to previous studies showing RhoA dominated outcomes. Lastly, the impact of the nano-spatial arrangement of cell adhesion ligands on self assembled functional block copolymer surfaces (changes in lateral spacing of adhesive peptides from ~30 to 60 nm) on MSC fate choice will be described. We demonstrate that the adhesion of MSCs to the block copolymer surfaces is through specific attachment to the presented adhesion motif and with increased lateral spacing of this peptide, we observe significant changes in cell morphology, motility and the ability of MSCs to undergo osteogenesis and adipogenesis - lateral spacing of ligands can be tuned to preference lineage fate choices. We also show that the type of ligand presented to MSCs, at a set lateral spacing, can further specify lineage choice. Each of these 'deterministic' biomaterial platforms has provided new insight into permissive cues for differentiation of MSCs, the complexity of intracellular pathways that are regulated by cell-surface interactions and the ultra-sensitivity of stem cells to changes or modulations in their extracellular environment.

Scalable Expansion of Human Induced Pluripotent Stem Cells in Xeno-Free Microcarriers

Badenes, S. M.,
Fernandes, T. G., Instituto Superior Técnico, University of Lisbon
Rodrigues, C. A. V., Instituto Superior Técnico, University of Lisbon
Pusch, A., LIFE&BRAIN GmbH
Haupt, S., LIFE&BRAIN GmbH
Bear, M. M., CETC
Hervy, M., CETC
Diogo, M. M., Department of Bioengineering and IBB - Institute for Biotechnology and Bioengineering, Instituto Superior Técnico, Universidade de Lisboa, Lisboa, Portugal
Brüstle, O., LIFE&BRAIN GmbH
Cabral, J. M. S., Department of Bioengineering and IBB - Institute for Biotechnology and Bioengineering, Instituto Superior Técnico, Universidade de Lisboa, Lisboa, Portugal
Originally presented on: 3/17/2014 17:34:00 - 17:36:00

A microcarrier-based suspension culture system was developed for scaling-up the expansion of human induced pluripotent stem cells (hiPSCs) in serum-free medium and using xeno-free synthetic beads based on a peptide-acrylate surface designed for long-term support of hiPSCs self-renewal [1]. Envisaging a homogeneous cell distribution, single-cell inoculation of hiPSCs on microcarriers was optimized using a chemical inhibitor of the Rho-associated kinase pathway. Under static conditions a 5-fold expansion of the initially inoculated cells was achieved after 5 days of culture. This microcarrier-based system was then successfully scaled-up to a dynamic culture system using 15 mL-spinner flasks. After optimization of the agitation speed, feeding regimen and impeller geometry a maximum cell fold increase of 14 was obtained after 7 days of culture in relation to the number of the initially attached cells. Envisaging the improvement of the scalability of the culture, multipassage expansion on the microcarriers was attained both using single cells and confluent microcarriers as the inoculum by taking advantage of the bead-to-bead cell transfer. In the latter case, after 4 passages, a cumulative 214-fold expansion was attained over 15 days of continuous culture. In static as well as in dynamic cultures, cells maintained their typical colony morphology and pluripotency-associated marker-expression. Moreover, their potential for spontaneous differentiation into cells of the three embryonic germ layers was demonstrated through formation of embryoid bodies containing cells expressing typical markers of endoderm, ectoderm and mesoderm. Importantly, the directed differentiation of the expanded hiPSCs into neural precursors expressing Pax6 and Nestin markers was successfully achieved both onto the xeno-free microcarriers and also after cell re-plating in tissue culture plates, by means of the dual-SMAD inhibition method. Overall, we expect these technologies to facilitate the standardized and automated scaling-up of hiPSCs expansion and integrated controlled neural differentiation for further downstream applications including potential cell replacement therapies.

[1] Melkoumian Z, Weber JL, Weber DM, Fadeev AG, Zhou Y, Dolley-Sonneville P, Yang J, Qiu L, Priest CA, Shogbon C, Martin AW, Nelson J, West P, Beltzer JP, Pal S, Brandenberger R, "Synthetic peptide-acrylate surfaces for long-term self-renewal and

cardiomyocyte differentiation of human embryonic stem cells", Nat Biotechnol, (2010), 28(6): 606-10

Aqueous Two-Phase System Strategies: In Route for the Potential Isolation of Stem Cells

González-González, M., Tecnológico de Monterrey. Campus Monterrey
Rito-Palomares, M., Tecnologico de Monterrey
Originally presented on: 3/17/2014 17:36:00 - 17:38:00

Aqueous two-phase system strategies: In route for the potential isolation of stem cells

Mirna González-González and Marco Rito-Palomares

Centro de Biotecnología-FEMSA, Tecnológico de Monterrey. Campus Monterrey,

Ave. Eugenio Garza Sada 2501 Sur, Monterrey, NL 64849, México.

Tel: (52) 81 8328-4132, Fax: (52) 81 8328-4136, *e-mail:* mrito@itesm.mx

A fast, scalable and cost-effective cell isolation technique is required to promote the application of stem cells therapies. Aqueous two-phase system (ATPS) is a liquid-liquid extraction technique that offers high biocompatibility, scalability, short processing times, and a mild environment for the recovery of stem cells. In this context, we have implemented multiple bioengineering strategies based on ATPS for this purpose. In a first approach, the use of novel and traditional polymer-polymer ATPS composed of ficoll, polyethylene glycol (PEG), dextran (DEX) or UCON have been explored utilizing human umbilical cord blood as the experimental matrix to study the partition behavior and viability of $CD133^+$ stem cells. Additionally, the comparison between these conventional systems and the ones complemented with antibodies, known as immunoaffinity ATPS, has also been carried out. From these experiments, it can be concluded that a different fractionation profile of the $CD133^+$ cells and the contaminants can be observed within the three studied ATPS, highlighting that the $CD133^+$ stem cells partition to the top phase of the ficoll 400,000-DEX 70,000 systems, meanwhile the contaminants showed a greater preference for the dextran rich bottom phase. On the other hand, an absolute preference for the bottom phase is obtained for the PEG 8,000-DEX 500,000 and Ucon-DEX 75,000 systems, for both the cell of interest and the contaminants. Furthermore, no significant differences were observed when compared the immunoaffinity and the conventional ATPS (without antibody addition). It is important to mention that in all experiments, cell viability was at least 98% after ATPS recovery, confirming the suitability of ATPS for stem cell recovery. In a further attempt to reach the objective, novel strategies including the chemical modification of the antibodies introduced into the immunoaffinity ATPS are being developed to enhance the partitioning of the antibody of interest into the desired phase. An example of this strategy is the site-specific PEGylation of the CD133 antibody via streptavidin-biotin conjugation. For this, two modified PEGs were developed:

COOH-PEG-Biotin and NH_2-PEG-Biotin.

The present investigation aims to contribute as an inspiration point for the development of novel, robust and scalable strategies for the purification of stem cells, which may find application in clinical settings in the near future.

Acknowledgements

The authors wish to acknowledge the financial support of Tecnológico de Monterrey, Bioprocess research chair (Grant CAT161), Zambrano-Hellion Foundation and CONACyT for the fellowships of M. González-González No. 223963 and Grant 179775.

Development of a Combinatorial Search Algorithm for Optimizing Stem Cell Culture Conditions

Kim, M.,
Audet, J.,
Originally presented on: 3/17/2014 17:38:00 - 17:40:00

The ability to accurately manipulate stem and progenitor (?precursor?) cell cultures has become critical with the increasing interest of novel cell-based therapies. Efficiently optimizing cell culture systems is challenging due to the complexity generated by numerous factors in various dose combinations, non-linear cell response, interacting factors, and donor-to-donor variability. The conventional experimental approach to identify an optimal condition is by factorial design which is costly and unrealistic as the number of experiments (N) increases exponentially (N=k^n), dependent on the number of factors (n) and dose levels (k). A strategy to efficiently identify the optimal conditions for precursor cell expansion would facilitate the production of precursor cell populations and the discovery of potentially more effective complex conditions. This study proposes the development of a combinatorial search algorithm to optimize culture conditions for precursor cell expansion based on direct measurement of cellular response. Preliminary algorithms based on the combinatorial approach integrating aspects from conventional search methodologies have been developed and assessed using simulated hematopoietic stem cell culture data inspired by experimental findings, consisting of 5 factors over 5 dose levels including an inhibitory factor and a pair of interacting factors. A deterministic algorithm required testing of a similar number (~33) of combinations as in a central composite design experiment but interexperimental controls resulted in additional experimental cost, reducing efficiency. A purely stochastic algorithm struggled to recognize and respond to effects of interacting factors and was unable to converge to an optimal condition. In both cases, increasing variability simulated in the data posed a challenge to an efficient optimization process. To address these limitations, a differential evolution-based algorithm is being developed to enable the algorithm to accommodate the high degree of complexity of an expansion culture system resulting from factor interactions and the non-linearity in factor effects. The algorithm will aim to address the interexperimental variability associated with cell culture data as part of the decision-making process of the algorithm. Such algorithm will provide a robust optimization process that efficiently identifies the optimal culture condition of precursor expansion culture systems, facilitating the development and translation of precursor cell-based applications in research and industry.

Integrated Multi-Stage Tissue on a Chip Generation from Human Pluripotent Stem Cells

Michielin, F., University of Padova
Giobbe, G. G., Venetian Institute of Molecular Medicine - VIMM
Martewicz, S., Venetian Institute of Molecular Medicine - VIMM
Giulitti, S., University of Padova
Luni, C., University of Padova
Floreani, A., University of Padova
Elvassore, N., University of Padova
Originally presented on: 3/17/2014 17:40:00 - 17:42:00

The development of human organs-on-chips, in which the microscale engineering technologies are combined with cultured human cells to recapitulate whole living organ microenvironment, offers a unique opportunity to study human physiology and pathophysiology in an organ-specific context[1,2,3]. Recently, successful examples of organs-on-chips development have been provided[4,5,6,7], mainly using primary animal cells and, in few cases, primary human cells. The possibility of developing direct organogenesis-on-chip from human pluripotent stem cells (hPSCs) could overcome the limited availability of human primary cells, such as hepatocytes and cardiomyocytes. Human embryonic stem cells (hESCs) and induced pluripotent stem cells (hiPSCs) open a wide perspective for multi-organ generation on a chip. Integrated functional hPSCs differentiation on a chip has not yet been developed.

In this work, we demonstrate how to control stem cell expansion, selective germ layer commitment and derive functional tissue-specific cells on a chip from both hESCs and hiPSCs through a multi-stage microfluidic-based technology.

Experimentally, we showed that a discontinuous periodic medium delivery with stage-dependent frequency, f, (number of cycles of medium change per day) is an effective strategy for modulating stem cell niche specification *in vitro,* ensuring optimal delivery of exogenous factors and removal of endogenous cell-secreted factors. HPSCs homogeneously express pluripotency markers, such as Oct4, Sox2 and Tra1-60, along the microfluidic channel, as revealed by immunostaining. QRT-PCR analysis of Oct4 and Nanog after 5 days of culture, showed an optimal frequency of f=2d^{-1} and a significant three-fold higher expression compared to lower and higher frequencies was observed.

HPSCs were induced to spontaneously differentiate within the microfluidic platform and a frequency-dependent germ layer enrichment was observed. Compared to standard culture in Petri dish, we obtained a significant increase in ectoderm markers OTX2 and TUBB3 expression by adopting f=1d^{-1} and a dramatic reduction with f=8d^{-1}, confirming that ectoderm commitment is mainly driven by endogenous factors accumulation. Conversely, by adopting f=8d^{-1}, we observed a significant increase of meso-endoderm T and GATA4 markers and early endoderm FOXA2, EOMES and AFP markers expression. More selective germ layer commitment was successfully obtained following specific differentiation protocols and an optimal f for each germ layer commitment was established.

Cardiac cells on a chip were derived in 15 days of frequency-dependent multi-stage differentiation protocol, consisting on mesoderm induction, early cardiac commitment and functional maturation. 60% Troponin-T positive cells showed defined sarcomeric organization, spontaneous contractile activity, excitation contraction coupling up to 2 Hz and proper calcium dynamics. Cells were also responsive for caffeine and verapamil stimulation.

Similarly, hepatic-like cells on a chip were obtained, starting from endoderm commitment, up to hepatic maturation. Cells express cytokeratin 18, CYP-3A, which is the most relevant marker for drug toxicity assesment, and albumin. Cells are also responsive for indocyanine green digestion, glycogen storage (75% of total cells) and show 40% increase in albumin secretion, compared to static conditions.

Functionally differentiated cells derived in the microfluidic channels can be directly used for dynamic multi-parametric and large-scale drug screening or for developing micro-engineered human organ models, overcoming issues related to the limited availability of human primary cell sources.

1. Huh, D., Torisawa, Y., Hamilton, G. A., Kim, H. J. & Ingber, D. E. *Lab. Chip* **12,** 2156–2164 (2012).

2. Huh, D. *et al. Sci. Transl. Med.* **4,** 159ra147–159ra147 (2012).

3. Ghaemmaghami, A. M., Hancock, M. J., Harrington, H., Kaji, H. & Khademhosseini, A. *Drug Discov. Today* **17,** 173–181 (2012).

4. Huh, D. *et al. Science* **328,** 1662–1668 (2010).

5. Lee, S.-A. *et al. Lab. Chip* (2013). doi:10.1039/C3LC50197C

6. Jang, K.-J. *et al. Integr. Biol. Quant. Biosci. Nano Macro* (2013). doi:10.1039/c3ib40049b

7. Grosberg, A., Alford, P. W., McCain, M. L. & Parker, K. K. *Lab. Chip* **11,** 4165–4173 (2011).

8. Sasai, Y. *Nature* **493,** 318–326 (2013).

9. Eiraku, M. *et al. Nature* **472,** 51–56 (2011).

10. Eiraku, M. *et al. Cell Stem Cell* **3,** 519–532 (2008).

Expansion and Neural Commitment of Human Pluripotent Stem Cells As 3D Suspension Aggregates

Miranda, C. C., Instituto Superior Técnico, University of Lisbon
Pascoal, J. F., Instituto Superior Técnico, University of Lisbon
Fernandes, T. G., Instituto Superior Técnico, University of Lisbon
Diogo, M. M., Department of Bioengineering and IBB - Institute for Biotechnology and Bioengineering, Instituto Superior Técnico, Universidade de Lisboa, Lisboa, Portugal
Cabral, J. M. S., Department of Bioengineering and IBB - Institute for Biotechnology and Bioengineering, Instituto Superior Técnico, Universidade de Lisboa, Lisboa, Portugal
Originally presented on: 3/17/2014 17:44:00 - 17:46:00

Human pluripotent stem cells (hPSC) hold great promise to enable cell therapies, tissue regeneration and a better understanding of embryonic development and cell differentiation. Though great advances have been made in the study of these cells, their translation to the clinic is being hampered by technical issues, particularly related with their large scale expansion and the efficiency of the *in vitro* differentiation protocols. Therefore, more efficient and cost-effective methodologies need to be developed.

In order to better mimic the *in vivo* microenvironment, 3D suspension conditions are starting to emerge as a promising alternative to perform the efficient *in vitro* expansion and controlled differentiation of hPSC. This culture system is based on the formation of cell aggregates initiated by single-cell suspensions, manually dissociated cell clumps or cultured in microwells that allow the formation of aggregates with an homogeneous and controlled size distribution. These culture systems were evaluated and compared in this work to perform the expansion and neural commitment of hPSC lines.

When cultured in the presence of expansion medium, hPSC aggregates derived from single-cell suspensions exhibited a narrow range of size distribution after seven days of culture whereas manual dissociation resulted in a wider range of diameters. To facilitate the study of the effect of aggregate size in the expansion of hPSCs, aggregates ranging from 200 to 5,000 cells per aggregate were formed by changing the initial cell concentration, and forcing the cells to aggregate in AggreWell™ plates. After one week in culture, the fold increase in total cell number changed in an inverse proportion to the aggregate size and, therefore, 200 initial cells per aggregate reached the highest value (16.0±2.7). Manually dissociated aggregates achieved intermediate fold increase values (3.8±1.7), possibly due to aggregate heterogeneity. After three weeks in culture, cells exposed to the three different conditions were analyzed for the expression of pluripotency markers by flow cytometry and at least 90% of the cells were found to retain the expression of OCT4 and TRA-1-60.

Several methodologies have been applied to induce the neural commitment of hPSC, including the Neural Sphere [Steiner *et al.,* 2010] and the Dual Smad Inhibition [Chambers *et al.*, 2009] methods. Nevertheless, their performance has never been compared side by side. Since the neural sphere induction method requires cells to be cultured for four weeks, the growth of the aggregates is limited to the point where their diameter will condition the diffusion of nutrients, oxygen and metabolites to and from the inner volume of the aggregate. As a result, aggregates with smaller initial sizes, namely 200 and 1,000 cells per aggregate, were shown to lead to higher cell yields (7.6±0.15 and 13.1±1.4, respectively), whereas the final expression of OCT4 for all sizes of aggregates remained below 3% after neural specification. On the other hand, the dual Smad Inhibition method, initially designed to induce neural commitment in monolayer cultures, appears to require a higher initial cell density, attaining higher cell yields when using 5,000 or 10,000 initial cells per aggregate (5.0±0.4 and 6.6±0.5, respectively). Also, when using N2B27 chemically-defined neural induction media supplemented with LDN193189 and SB431542 small molecules, instead of the traditional Knockout serum replacement (KOSR)-supplemented medium, it was possible to decrease the percentage of OCT4-expressing cells from >95% at day 0 to ~30%, after twelve days of neural commitment. Moreover, after replating the cells it was possible to give rise to neural rosette colonies that were tested positive for the expression of the neural markers NESTIN, PAX6, FoxG1 and SOX2, confirming that this process is also suitable to induce hPSC neural commitment in a 3D suspension culture.

Conventional and Naive Human Pluripotent Stem Cell Genome Engineering Via Cas9

Lian, X., University of Wisconsin-Madison
Originally presented on: 3/17/2014 17:46:00 - 17:48:00

Conventional and naive human pluripotent stem cell genome engineering via Cas9

Because human pluripotent stem cells (hPSCs) can be propagated indefinitely while still retaining the capacity to differentiate into all somatic cell types, they are a potentially inexhaustible supply of clinical relevant somatic cells. Reporter or site-specific mutation cell lines generated through targeted gene modifications are expected to play an important role in developing optimal cell-specific differentiation protocols or disease models. Over the past 15 years of hPSC research, there were limited successful genes targeting examples in hPSCs partially due to the extremely low efficiency of traditional homologous recombination. Here we reported that the clustered regularly interspaced short palindromic repeats (CRISPR)/CRISPR-associated (Cas) system can be used to improve gene targeting efficiency in hPSCs compared to the traditional homologous recombination approach. In addition, conventional hPSCs can be converted to the mouse embryonic stem cell-like naive state and the gene targeting efficiency can be further increased with naive hPSCs.

Chemically Defined and Small Molecule-Based Generation of Cardiomyocytes

Burridge, P., Stanford University
Matsa, E., Stanford University
Shukla, P.,
Lin, Z.,
Churko, J.,
Ebert, A.,
Lan, F., Stanford Univerisity
Diecke, S.,
Mordwinkin, N.,
Gold, J.,
Wu, J., Stanford Univeristy
Originally presented on: 3/17/2014 17:48:00 - 17:50:00

Existing methodologies for human induced pluripotent stem cell (hiPSC) cardiac differentiation are efficient but require the use of complex undefined media constituents that hinder further elucidation of the molecular mechanisms of cardiomyogenesis. Using hiPSCs derived under chemically defined conditions on synthetic matrices, we systematically developed a highly optimized cardiac differentiation strategy employing a chemically defined medium, consisting of just three components: RPMI 1640, L-ascorbic acid 2-phosphate, and rice-derived recombinant human albumin. Along with small molecule-based differentiation induction, this produced contractile sheets of up to 95% TNNT2+ cardiomyocytes at a yield of up to 100 cardiomyocytes for every input pluripotent cell. This methodology was effective in 11 hiPSC lines tested repeatedly from p20 to p83, representing >600 differentiations. Cardiomyocytes produced this way were demonstrated to be of a predominantly atrial phenotype that progressively matured to a ventricular phenotype. The complete absence of any undefined or animal derived product in the entire hiPSC culture and differentiation platform makes it ideal for production of large numbers (>1E9) of cardiomyocytes to cGMP standards. This is the first fully chemically defined platform for cardiac specification of hiPSCs and permits elucidation of cardiomyocyte macromolecular/metabolic requirements, providing a minimally complex system for the study of maturation and subtype specification.

Current Status and Future Directions of MED-TVC Technology

Wazeer, I.,

Originally presented on: 3/17/2014 17:52:00 - 17:54:00

Desalination has now become one of the major water treatment process in several countries around the world where shortage of water is a serious problem. Energy consumption is a vital economic factor in selecting the type of desalination processes because current desalination processes require large amount of energy which is costly.

Multi-effect desalination system with thermal vapor compression (MED-TVC) is particularly more attractive than other thermal desalination systems due to its low energy consumption. MED-TVC is characterized by high performance ratio (PR), easier operation, low maintenance requirements and simple geometry. These attractive features make MED-TVC highly competitive to other well established desalination techniques that include the reverse osmosis (RO) and *multi-stage flash* desalination (MSF).

The primary goal of this paper is to present a preview of some aspects related with the theory of the technology, parametric study of the MED-TVC systems and its development. It will analyze the current and future aspects of the MED-TVC technology in view of latest installed plants.

Microfabricated Cell Culture Device for Stem Cell Bioprocessing with Quantative Online Monitoring

Szita, N., University College London

Originally presented on: 3/17/2014 17:54:00 - 17:56:00

Microfabricated Cell Culture Device for Stem Cell Bioprocessing with Quantative Online Monitoring

Microfabricated bioreactors have been successfully applied to the fields of microbial fermentation and mammalian suspension cell cultures. Key advantages that make microfabricated bioreactors a cost-effective proposition for early bioprocess development include: significant reduction in reagent use, real-time monitoring and control of process variables, ease of sterilisation via disposable polymer technology, reduced labour due to automation, and the capability to rapidly test different processing conditions. Clearly, a microfabricated, adherent culture device that could translate these advantages to stem cell culture would be of great value to the fields of regenerative medicine and cell therapy where more insight into cell culture processes is direly needed. However, a link must be maintained with conventional culture methods and production systems for the purposes of validation and scale up studies. Further, the device must be highly versatile to facilitate evaluation and optimisation of the wide range of process variables that affect stem cell fate. In particular, this versatility must extend beyond the control of the physico-chemical properties of the culture medium to include auto/paracrine factors, growth surfaces, shear forces, and cell seeding densities. Finally, rapid online assessment of the product (cells) is crucial.

We have developed a microfabricated cell culture device ('micro bioreactor') where cell culture growth is monitored online. Using human and mouse embryonic stem cell expansion protocols as a model system, we investigated microfluidic design concepts to enable controlled and low-shear perfusion culture of stem cells that maintain the expression of pluripotency markers. Characterization of our microfabricated cell culture device included fluid flow modeling, device reliability testing, verification of autoclavation as a means of sterilization for our device, and also the careful establishment of protocols to prime these devices in a bubble-free fashion. We showed that with our device, we adhere closely to protocols typically employed for stem cell maintenance in laboratory scale vessels, which will facilitate comparison of results, and creates a link with traditional small-scale culture devices for validation and scale-up studies. To understand both the device and the culture data, an image-processing algorithm rapidly detects cell culture confluency (as an indirect method of cell culture growth). The algorithm is also capable to detect changes in cellular morphology, mean fluorescence expression per confluency (for example to establish time course data of differentiation marker expression), and an estimate of cell density. Our image-processing algorithm furthermore provides heat-map like visualization of temporal and spatial cell growth patterns during microfluidic culture. Bulk and peri-cellular oxygen concentrations are monitored in real time, time course data for 6-day cultures of mouse embryonic stem cells (mESCs) is presented and oxygen profiles analysed, and we have also started with efforts to characterize a multiplexed culture device platform.

References:

Jaccard, N., Griffin, L. D., Keser, A., Macown, R. J., Super, A., Veraitch, F. S., Szita, N. (2013). Automated method for the rapid and precise estimation of adherent cell culture characteristics from phase contrast microscopy images. *Biotechnol Bioeng* doi:10.1002/bit.25115.

Kirk, T. V., Szita, N. (2013). Oxygen transfer characteristics of miniaturized bioreactor systems. *Biotechnol Bioeng* 110(4), 1005-1019 doi:10.1002/bit.24824.

Reichen, M., Veraitch, F. S., Szita, N. (2013). Development of a Multiplexed Microfluidic Platform for the Automated Cultivation of Embryonic Stem Cells. *J Lab Autom* 18(6), 519-529, doi:10.1177/2211068213499917.

Reichen, M., Macown, R. J., Jaccard, N., Super, A., Ruban, L., Griffin, L., Veraitch, F. S., Szita, N. (2012). Microfabricated Modular Scale-Down Device for Regenerative Medicine Process Development. 7(12), e52246 doi:10.1371/journal.pone.0052246.

Differentiating Intestinal Stem Cells in a 3D Niche

March, J. C., Cornell University
Costello, C.,
Originally presented on: 3/17/2014 17:56:00 - 17:58:00

Differentiating Intestinal Stem Cells in a 3D Niche

Cait M. Costello and John C. March

Purpose: In-vitro intestinal models are useful tools for studies into small intestinal function, including cellular growth and proliferation mechanisms, drug absorption capabilities and host-microbial interactions. In the past, these models have typically been limited to simple cultures on 2-D scaffolds or transwell inserts, but we have shown previously that epithelial cells cultured in 3-D environments exhibit different phenotypes that are more reflective of native tissue. It is widely accepted that different microbial species will preferentially adhere to select locations along the intestinal villi. Some of this niche determination is based on topography, but some is based on epithelial cell function. Our focus was to develop a porous, synthetic 3-D tissue scaffold with villous features that could support the co-culture of different epithelial cell types derived from a single stem cell precursor that we could co-culture with select bacterial populations. Our end goal was to establish microbial niches along the crypt-villus axis in order to mimic the natural microenvironment of the small intestine, which could potentially provide new insights into microbe-induced intestinal disorders, as well as enabling targeted probiotic therapies.

Methods and Results: We recreated the surface topography of the small intestine by fabricating a biodegradable and biocompatible villous scaffold using poly-lactic-glycolic-acid (PLGA), made porous through particulate leaching and phase separation of the solvent (Fig 1). By initially seeding the scaffolds with a simple co-culture of Caco-2 and HT29-MTX, we demonstrated that our scaffolds enable cellular differentiation along the crypt-villus axis in a similar manner to native intestines, with greater epithelial cell polarization and brush-border enzyme expression near the tips of the villi (Fig 2) compared to the crypt region, as well as statistically significant supporting TEER values. In addition, we are also able to simulate the intestinal mesenchyme by either feeding growth factors directly, or by co-culturing with a basolateral fibroblast feeder layer. This enables us to increase the complexity of our epithelial layer by allowing the growth and differentiation of intestinal stem cells, including the established stem cell line rat IEC-6, and through primary cultures of mouse crypts. Intestinal stem cells can differentiate into secretory epithelial cell types, as well as enterocytes, and these have been shown to affect the spatial location of bacteria along the crypt-villus axis. Using RT-PCR and immunofluorescence, we have shown expression of markers for goblet, paneth, absorptive and enteroendocrine cell types with similar numbers and location along the crypt-villus axis to real small intestine epithelium (Fig 3). In addition, we have shown through preliminary experimentation with intestinal pathogens that on our scaffolds we are able to mimic the adhesion and invasion profiles of both salmonella and pseudomonas, when co-cultured with intestinal epithelial and probiotics (Fig 4).

Conclusions: We have demonstrated the feasibility of developing an in-vitro artificial intestine from a biocompatible polymer, which can be molded into villous shapes to mimic the topography of the small intestine, providing a platform for the differentiation of intestinal stem cells into epithelial cell types, and the subsequent adhesion/invasion of pathogenic bacteria.

Physical and Genetic Preconditioning of Adipose-Derived Stem Cells to Survive Ischemia

Chung, E.,
Leal, S., The University of Texas at Austin
Merchant, A.,
Santiesteban, D.,
Sun, I. C.,
Suggs, L.,
Originally presented on: 3/17/2014 17:58:00 - 18:00:00

The survival rate of stem cells following implantation can determine the regenerative potential of stem cells. After cells are exposed to the severe stress environment of injured tissues which are beyond their cyto-protective capacity, the majority of stem cells undergo apoptosis within one week. According to numerous prior studies, the post-stress damage of cells can be abrogated by inducing an elevated level of stress proteins (*e.g.* heat shock protein). However, the detailed interaction between stress proteins and stem cell activities are not well-described. In the current study, we seek to develop a robust stem cell source with strong apoptotic resistance for tissue engineering applications. We are interested in whether or not up-regulation of stress proteins, as a result of stem cell preconditioning, can improve the survival and subsequent regenerative potential of stem cells. We successfully up-regulated heat shock protein 70 (HSP70) in adipose-derived stem cells (ASCs) without significant cytotoxicity by either exogenous delivery of a HSP70 cDNA plasmid or physical stress treatment (*e.g.* mild heating or oxidative stress). Current studies are focused on comparing the effect of physical stresses and gene manipulation strategies on the behaviors (viability growth, protein production, morphology) of HSP70-upregulated ASCs under an *in vitro* ischemic environment. The result of this work will identify an effective stem cell preconditioning protocol to produce a stem cell source more resistant to severe stress environments.

Culture of Human Embryonic Stem Cells (hESC) on Matrigel Matrix without Conditional Media

Allahbakhshian-Farsani, M., Department of Hematology and Blood Banking, Faculty of Allied Medicine, Shahid Beheshti University of Medical Sciences
Ghasemi-Dehkordi, P., Cellular and Molecular Research Center, Shahrekord University of Medical Sciences
Abdian, N., Cellular and Molecular Research Center, Shahrekord University of Medical Sciences
Mirzaeian, A., Cellular and Molecular Research Center, Shahrekord University of Medical Sciences, Shahrekord, Iran
Hashemzadeh-Chaleshtori, M., Cellular and Molecular Research Center, Shahrekord University of Medical Sciences
Originally presented on: 3/17/2014 18:00:00 - 18:02:00

Human embryonic stem cell (hESC) is thought to have great potential in regenerative medicine. In order to maintain the hESCs in undifferentiated status, researchers use feeder cells to provide both a suitable attachment substrate and essential soluble factors. Although mouse embryonic fibroblasts (MEFs) support hESC growth but feeder cells from animal sources expose hESC to many possible infectious agents and unknown factors. On the other hand, it is difficult to interpret the results obtained from hESC-MEF coculture during any intervention. Matrigel, a complex protein mixture, is very interesting substitute for optimal growth of hESC in feeder free system for long culture. In the present study, hES cells were cultured on BD Matrigel matrix without conditional media. hESCs were cultured on BD Matrigel matrix with ESC media for 25 passages. During this period, cytogenetic and phenotypic stability of hESC colonies were analyzed. Our results from karyotype test and morphology evaluation showed that the hESC colonies were cytogenetically stable and showed criteria of undifferentiated colonies. These findings indicate that BD Matrigel matrix is an effective surface that supports feeder-free expansion and maintenance of cytogenetically stable undifferentiated hES cells.

Keywords: Human embryonic stem cells (hESC), BD Matrigel matrix, Conditional media

Revivals the Mouse Liver Damage Using Induced Pluripotent Stem Cells (iPSC)

Ghasemi-Dehkordi, P., Cellular and Molecular Research Center, Shahrekord University of Medical Sciences
Allahbakhshian-Farsani, M., Department of Hematology and Blood Banking, Faculty of Allied Medicine, Shahid Beheshti University of Medical Sciences
Abdian, N., Cellular and Molecular Research Center, Shahrekord University of Medical Sciences
Hashemzadeh-Chaleshtori, M., Cellular and Molecular Research Center, Shahrekord University of Medical Sciences
Amiri, S. B.,
Originally presented on: 3/17/2014 18:02:00 - 18:04:00

Induced pluripotent stem cells (iPSCs) are a type of pluripotent stem cell and could be generate from reprogramming of adult somatic cells. The goal of present study was to culture of iPS cells on feeder free layer and injection of iPS cells to mouse with liver failure for evaluation of effect of these cells. In this study two groups include liver failure mouse and control (with no harm) were used for experiment. IPS cells were cultured on BD Matrigel matrix and sufficient colonies were obtained. Then, mice were anesthetized and mechanically were injured in their liver and the iPS cells were injected via to the tail vein. The healthy of mice after injection were evaluated by biochemical tests. In this period liver dysfunction include variety of symptoms related to blood sugar problems, digestive problems, and immune disorders were not detected. Since 14 days, they were killed and revival effects of iPS cells on liver tissue were examined via pathological investigation. The results showed that mice with injured liver had a same appearance with livers of controls and physical signs of tissue damage were not observed. These findings indicate that iPS cells can revival the liver tissue and recovery the damages of hepatocytes cells.

Keywords: iPSCs, BD Matrigel matrix, Liver damage

Alginate Microencapsulation Parameters Modulate Embryonic Stem Cell Aggregate Expansion and Phenotype

Wilson, J. L., Georgia Institute of Technology and Emory University
Najia, M. A., Georgia Institute of Technology and Emory University
Saeed, R., Georgia Institute of Technology and Emory University
McDevitt, T. C., Georgia Institute of Technology
Originally presented on: 3/17/2014 18:06:00 - 18:08:00

Pluripotent embryonic stem cells (ESCs) are of great interest in regenerative medicine due to their capacity for expansion and differentiation into all somatic cell types; however, most current processes are incapable of manufacturing sufficient quantities of homogeneous stem cell populations. Microencapsulation of ESCs within alginate capsules is a potential platform for scalable biomanufacturing, as the capsules shield cells from hydrodynamic forces found in suspension bioreactors while providing a controlled environment for differentiation. Despite the advantages of an encapsulation-based system, there has been a deficiency in examining the impact of encapsulation parameters on ESC behavior. Alginate is a common material used for encapsulation, but the composition of the polymer varies based on the source, which has led to variability in reported results. Therefore, the objective of this study was to systematically examine the impact of alginate composition, specifically the ratio of guluronic (G) and mannuronic (M) acid, on microencapsulated ESC expansion and phenotype.

Murine ESCs (D3 line) were formed into 500 cell aggregates using forced centrifugation prior to encapsulation in 1.5% alginate (10^4 aggregates/mL) using a Nisco electrostatic bead generator. Two alginate compositions were examined: one with a higher content of G residues (>60%, termed High G) and one with a higher content of M residues (>50%, termed High M). Aggregates were cultured in serum-free N2B27 media for up to 14 days. Bead mechanical properties were determined using a microscale compression system, cell viability was determined using a LIVE/DEAD assay, cell number was assessed using a CyQuant assay, and gene expression was quantified via RT-PCR. Histological analysis was performed on cryosections for immunostaining. Conditioned media was collected from the cultures and analyzed using ELISAs for vascular endothelial growth factor (VEGF) and bone morphogenic protein-4 (BMP-4).

The elastic modulus of High G alginate beads was higher than High M alginate (8.2 kPa and 6.2 kPa, respectively; $p = 0.011$), indicating that the capsule types had divergent bulk mechanical properties. Aggregates in the High G alginate remained round whereas aggregates in the High M alginate exhibited an elongated shape. Though cell viability remained high over 14 days (>80%), decreased proliferation was observed in the encapsulated groups, chiefly in the stiffer High G alginate, compared to unencapsulated controls. Gene expression of the pluripotency markers *Oct4* and *Nanog* decreased over time, although the encapsulated cells, particularly those in the stiffer High G alginate, exhibited a slower decay of *Oct4* expression when compared with unencapsulated controls. The cells within the High M conditions exhibited increased gene and protein expression of the endoderm marker alpha-fetoprotein (AFP) and the cardiac markers myosin light chain (MLC-2v) and alpha-smooth muscle actin (α-SMA). Additionally, cells within High M alginate produced higher levels of VEGF and BMP-4 on a per cell basis than unencapsulated cells or cells within High G alginate.

Together, these findings suggest that alginate material properties can be used to control stem cell fate and aid in the development of robust and scalable stem cell bioprocessing strategies.

Longitudinal Study of Neurosphere Metabolism in Bioreactor Culture for Scalable Production of Human Embryonic Stem Cell Derived Neurons

Åstrand, C., KTH - Royal Institute of Technology
Wiklund, L., Karolinska Institutet
Hagrot, E., KTH - Royal Institute of Technology
Valluru, M., KTH - Royal Institute of Technology
Marutle, A., Karolinska Institutet
Chotteau, V., KTH - Royal Institute of Technology
Gertow-Gillner, K., KTH - Royal Institute of Technology
Originally presented on: 3/17/2014 18:08:00 - 18:10:00

Human embryonic stem cells (hESCs) offer great potential for multiple tissue engineering applications including drug screening and future clinical applications such as treatment of neurological disease. Today, state-of-the-art technology lack the capacity to support systematic larger scale applications. The challenge resides not only in enlarging the cultivation size but also in controlling the quality Herein, long-term expansion of hESC derived free-floating neurospheres in stirred bioreactor systems enhanced proliferation and viability and also improved the differentiation potential into various neuronal lineages. We examined in detail variations in cellular metabolism during expansion under hypoxic and normoxic conditions in static versus stirred cultivation systems, orbitally or impeller-mixed over an extended time by analyzing nutrients and metabolites including amino acid content in growth medium. We found that an orbitally stirred bioreactor system promotes larger sphere size and higher cell yield. Furthermore, the bioreactor system promoted a higher ammonium release, mainly a by-product of oxidative phosphorylation. This was seen also during hypoxia and was accompanied by changes in glutamine and glutamate consumption, indicating a switch towards increased efficiency of metabolic pathways generally activated at atmospheric oxygen. This metabolic pattern observed in the stirred bioreactor was accompanied with enhanced consumption and/or production of many non-essential amino acids mimicking a high oxygen condition. Furthermore, glutamic acid, aspartic acid and serine could be partially or fully depleted thus indicating that these amino acids may be a limiting factor during large-scale production of neurospheres. Together, these data give new insight in the specific needs when designing platforms for large-scale propagation of neuronal progenitors and differentiated neuronal subtypes.

A New Economical Cord Blood Storage Method

Hasse, M.,
Youmans, A.,
Mamott, D.,
Gigliotti, N.,
Hassemer, E., Milwaukee School of Engineering
Afshan, G., Milwaukee School of Engineering
Originally presented on: 3/17/2014 18:10:00 - 18:12:00

Currently, umbilical cord blood (UCB) is cryopreserved for medical applications by private and public banks with initial processing costs averaging $1950 with $125 of annual storage fees. This can be a significant burden on expecting guardians. Therefore, a method to overcome this cost is currently under development which utilizes lyophilization of UCB cells. These nucleated cells, found in the buffy coat of gradient centrifuged UCB, contain the highly desired therapeutic hematopoietic and mesenchymal stem cells. Previous research has shown that lyophilization of mammalian cells yields low cell viability upon rehydration. This approach seeks to overcome this problem through the implementation of a highly specified environment for lyophilization in a millifluidic device. The device will serve as both a sample chamber and long term storage vessel; therefore, it needs to be able to withstand the extreme conditions that occur during lyophilization and autoclaving. It appears to be especially promising because this device will introduce control to lyophilization as well as rehydration. Prior research has discovered that the cell recovery process is directly related to the cell viability. Through this process, it is anticipated that the rehydrated cells will yield higher cell viability than thawed cells, since thawing has been shown to be harsh on cells. This technology is being developed to increase the economic availability of UCB banking to middle and lower class caregivers by eliminating the annual fee for liquid nitrogen, maintenance, and storage space. The small millifluidic device would allow for storage at the guardian's residence in a refrigerator or at room temperature. This project offers an innovative, marketable solution for utilizing umbilical cord blood stem cells in a growing number of therapeutic uses.

Type I Collagen Matrix Activates the Hedgehog Pathway in CD34+ Cells and Enhances Their Function

Ostojic, A., University of Ottawa
McNeill, B., University of Ottawa Heart Institute
Ruel, M., University of Ottawa Heart Institute
Suuronen, E., University of Ottawa Heart Institute
Originally presented on: 3/17/2014 18:12:00 - 18:14:00

Background. Acute myocardial infarction (MI) is a leading cause of death in the western world. Cell therapy using the $CD34^+$ fraction of circulating endothelial progenitor cells (EPCs) is one approach under investigation for restoring blood flow and function to the ischemic heart. However, the number of $CD34^+$ circulating cells is inversely proportional to the severity of cardiovascular disease and age. Therefore, there is a need to increase their numbers and/or function for therapy. One possibility is to deliver the therapeutic cells in combination with a biomaterial. We have previously demonstrated that the therapeutic potency of EPCs is improved by their culture on a type I collagen-based matrix. The aim of this study was to investigate the ability and mechanism through which the collagen matrix may enhance $CD34^+$ cell function.

Methods/Results. Human blood from healthy donors was used to generate EPCs by 4-day culture on fibronectin. $CD34^+$ cells were sorted from the heterogeneous EPC population and re-seeded onto fibronectin or collagen matrix for up to another 4 days and then used for *in vitro* experiments. $CD34^+$ cells from matrix culture demonstrated increased migration towards VEGF (by 3.2-fold) in a Boyden chamber assay versus fibronectin-cultured cells (p=0.03). Also, the number of $CD34^+$ cells derived from matrix that incorporated into capillary-like structures in an *in vitro* angiogenic assay was 3.1-fold greater than cells from fibronectin cultures. Furthermore, survival upon exposure to serum deprivation and hypoxia for 48h was enhanced by approximately 15% in matrix-derived $CD34^+$ cells compared to fibronectin (p=0.02). In contrast, the adhesion potential of $CD34^+$ cells was not affected by matrix culture, as the ability to adhere to fibronectin or collagen matrix substrates was not different between matrix- and fibronectin-derived cells. Other studies have identified the importance of the Hedgehog (Hh) signaling pathway in the therapeutic potency of $CD34^+$ cells. Therefore, we assessed the expression of known Hh pathway members by performing qRT-PCR. $CD34^+$ cells within unsorted EPCs cultured on matrix for up to 96h had increased mRNA expression of Gli1 (28.9-fold), CyclinD1 (10.4-fold) and Hes1 (6.2-fold), compared to fibronectin-cultured cells ($p < 0.05$ for each).

Conclusion. In the present study, we demonstrated that the interaction of $CD34^+$ progenitor cells with our collagen matrix enhances their migratory potential, angiogenic properties and survival. Our data also indicates the possible involvement of the Hh pathway in these functional changes as key downstream targets are up-regulated in matrix-derived $CD34^+$ cells. This work provides insight into novel mechanisms for the observed matrix-enhanced function of therapeutic $CD34^+$ cells.

Islation of Neural Crest-Derived Pluripotent Stem Cells in Adult Dental Tissue

Cheung, H. S.,
Pelaez, D.,
Huang, C.,
Originally presented on: 3/17/2014 18:16:00 - 18:18:00

Isolation of Neural Crest-derived pluripotent stem cells in adult dental tissues

D. Pelaez, C.Y. CH Huang, HS Cheung (Miami VA Healthcare System, Miami, FL)

Cellular potency, or the potential for a cell to differentiate down various tissue lineages, is the defining characteristic as to the viability of the cell source for a given application. With the establishment of ESC, the idea of utilizing pluripotent cell sources for regenerative medical applications was introduced as a promising alternative to traditional medical practices. However, there are ethical dilemmas as well as immunological concerns arising from the use of ESC in clinical settings. The advent of induced pluripotent stem cells (iPSCs) attempted to circumvent both of these issues and brought the concept of patient-specific pluripotent stem cell attainment to the forefront of the field. This iPSC technology is still in its early stages, the scientific community is trying to decipher the mechanisms underlying the induction of pluripotency and overcome the aberrant epigenetic changes arising from the implementation of the technique. Nevertheless, the potential of obtaining pluripotent stem cells from adult tissues is of significant importance and remains as the ultimate goal in overcoming the limitations presently found with other post-natal stem cell populations.

Identification and isolation of pluripotent stem cells in adult tissues represents an important advancement in the fields of stem cell biology and regenerative medicine. Work has been performed on the identification of biomarkers that can isolate stem cells residing in neural crest-derived adult tissues. The neural crest is an excellent model in stem cell biology as cells from it migrate extensively and contribute to the formation of diverse tissues in the body during organogenesis. Migration of these cells is modulated in part by gap junction communication amongst the cell sheets. Here we present a study in which selection of connexin 43 (Cx43) expressing cells from human adult periodontal ligament yields a novel pluripotent stem cell population. $Cx43^+$ periodontal ligament stem cells express pluripotency-associated transcription factors OCT4, Nanog, and Sox2, as well as neural crest-specific markers Sox10, p75, and Nestin. When injected *in-vivo* into an immunodeficient mouse model, these cells were capable of generating teratomas with tissues from the three embryological germ layers: endoderm, mesoderm, and ectoderm. Furthermore, the cells formed mature structures of tissues normally arising from the neural crest during embryogenesis such as eccrine sweat glands of the human skin, muscle, neuronal tissues, cartilage, and bone. Immunohistochemical analysis confirmed the human origin of the neoplastic cells as well as the ectodermal and endodermal nature of some of the structures found in the tumors. These results suggest that Cx43 may be used as a biomarker to select and isolate remnant neural crest pluripotent stem cells from adult human tissues arising from this embryological structure.

The isolation of these cells through routine medical procedures such as wisdom teeth extraction further enhances their applicability to the regenerative medicine field.

Defining the Surface Chemistry for Pluripotent Stem Cell Culture on 2D and 3D Porous Substrates

Zonca, M. R. Jr., College of Nanoscale Science and Engineering, University at Albany, State University of New York
Imbrogno, J., Rensselaer Polytechnic Institute
Unser, A. M.,
Belfort, G., Rensselaer Polytechnic Institute
Xie, Y., College of Nanoscale Science and Engineering, University at Albany, State University of New York
Originally presented on: 3/17/2014 18:18:00 - 18:20:00

Surface chemistry is critical for growing pluripotent stem cells in an undifferentiated state and expanding them in large number of cells. The challenge is to identify the optimal surface chemistry of the substrata for pluripotent stem cell attachment and maintenance. Using a high-throughput polymerization and screening platform, a chemically defined, synthetic polymer grafted coating that supports strong attachment and high expansion capacity of pluripotent stem cells has been discovered using mouse embryonic stem (ES) cells as a model system. This optimal substrate, N-[3-(Dimethylamino)propyl] methacrylamide (DMAPMA), sustains long-term self-renewal of ES cells. DMAPMA supports cell attachment of ES cells through integrin β1 in a RGD-independent manner and is similar to another recently reported polymer surface. The results obtained on a 2-D porous flat sheet membrane were then transferred to 3D by grafting DMAPMA to synthetic poly(ether sulfone) (PES) fibrous matrices through both photo-induced and plasma-induced polymerization. These 3D modified fibers exhibited higher cell proliferation and greater expression of pluripotency markers of mouse ES cells than 2D PES porous membranes. Our results indicated that desirable surfaces in 2D can be scaled to 3D and that both surface chemistry and structural dimension strongly influence the growth and differentiation of pluripotent stem cells.

Understanding the Mathematical and Geometric Language of Endothelial Progenitor Cells

Garcia, J., Universidad de Monterrey
Rivera-Silva, G., Universidad de Monterrey
Originally presented on: 3/17/2014 18:20:00 - 18:22:00

Stem cell activities also depend of mathematical principles and physical laws and tissue engineering has originated new methods to understand cell behaviors. Using mathematical and computational modeling facilitates the understanding of tissue growth process. This study focuses on the reactions of endothelial progenitor cells to fluid frictional stress by utilizing a mathematical and morphological analysis. A flow system was built where cultures of endothelial progenitor cells were exposed. We identified that the establishment of lamellipodia networks was related with a 103° angle, which could be the mathematical expression of Rac proteins. The cell morphology of these cells in culture submitted to fluid frictional stress was determined by the flow strength. An original vectorial approach to identify cell migration direction was used. We concluded that endothelial progenitor cells under flow conditions adopt different morphologies to optimize their hydrodynamic capacities. Geometric factors and surface dynamics conduct are significant features of endothelial coverage and angiogenesis.

Enhancing hMSC Attachment to Fibrin Microthreads

Hansen, K.,
Cunha, A.,
Gaudette, G.,

Originally presented on: 3/17/2014 18:22:00 - 18:24:00

Purpose: A limiting factor to the success of cellular therapies for cardiac regeneration is the lack of an efficient and targeted delivery system. Our lab has developed a fibrin suture based approach that has a delivery efficiency of over 60%. However, the method used to seed cells onto the fibrin sutures remains inefficient; this study seeks to improve cell-loading efficiency. **Methods**: 2cm long fibrin sutures, comprised of 24 microthreads twisted together around the eye of a needle to form a suture, were seeded with 100ul of a 2.5×10^5, 5.0×10^5, and 1×10^6 human mesenchymal stem cells/1ml suspension for 6, 12, and 24 hours. Cell attachment on the suture was visualized through phalloidin (F-actin) and Hoechst staining. Cell quantity, obtained through a Cyquant assay, and seeding efficiency were calculated. **Results**: No statistical significance was found in cell attachment or seeding efficiency between the different incubation times, indicating that cells seed equally well at all time points. There were no significant differences in cell attachment for the 5.0×10^5 and 1×10^6 groups; the 2.5×10^5 group was significantly different than the previous groups . However, there was a significant difference in percent efficiency between the 5.0×10^5 and 1×10^6 groups, indicating that seeding fewer cells yields a more efficient cell attachment. **Conclusions**: These data indicate that incubating fibrin sutures with fewer cells compared to previously reported studies yield higher loading efficiency and shorter loading times. This improved cell loading efficiency results, together with improved delivery efficiency, reduces the number of cells that need to be grown for implantation in the heart.

Differential Expression of Neuron-Glial Antigen 2 (NG2) Selects for Cell Survival during High-Throughput Enrichment of Marrow-Derived Mesenchymal Stem Cells

OConnor, K., Tulane University
Russell, K., Tulane University
Tucker, A., Tulane University
Bunnell, B., Tulane University
Andreeff, M., The University of Texas MD Anderson Cancer Center
Schober, W., The University of Texas MD Anderson Cancer Center
Lacey, M., Tulane University
Originally presented on: 3/17/2014 18:24:00 - 18:26:00

Cellular heterogeneity of mesenchymal stem cells (MSCs) impedes their use in regenerative medicine. The objective of this research is to identify potential biomarkers for the high-throughput enrichment of progenitors from heterogeneous MSC cultures. To this end, the present study examines variation in expression of neuron-glial antigen 2 (NG2) and melanoma cell adhesion molecule (CD146) on the surface of MSCs derived from human bone marrow in response to culture conditions and among cell populations. Multipotent cells isolated from heterogeneous MSC cultures exhibit a >3-fold increase in surface expression for NG2 and >2-fold increase for CD146 as compared with parental and lineage-committed MSCs. For both antigens, surface expression is downregulated by ≥6-fold when MSCs become confluent. During serial passage, maximum surface expression of NG2 and CD146 is associated with minimum doubling time. Upregulation of NG2 and CD146 during loss of adipogenic potential at early passage suggests some limits to their utility as potency markers. A potential relationship between proliferation and antigen expression was explored by sorting heterogeneous MSCs into rapidly and slowly dividing groups. Fluorescence-activated cell sorting revealed that rapidly dividing MSCs display lower scatter and 50% higher NG2 surface expression than slowly dividing cells, but CD146 expression is comparable in both groups. Heterogeneous MSCs were sorted based on scatter properties and surface expression of NG2 and CD146 into high (HI) and low (LO) groups. $Sc^{LO}NG2^{HI}$ and $Sc^{LO}NG2^{HI}CD146^{HI}$ MSCs have the highest proliferative potential of the sorted groups, with colony-forming efficiencies that are 1.5-2.2 times the value for the parental controls. The Sc^{LO} gate enriches for rapidly dividing cells. Addition of the $NG2^{HI}$ gate selects for enhanced cell survival. Further addition of the $CD146^{HI}$ gate does not significantly improve cell division or survival. This is the first report of a role for NG2 in the survival of normal diploid cells. The relationship between NG2 and MSC survival is particularly significant because cell survival is a basic prerequisite for any MSC therapy to be effective. As such, there are numerous basic research and clinical applications for the $Sc^{LO}NG2^{HI}$ sort strategy. This research was funded by the National Science Foundation.

An Animal-Free Recombinant Lectin Molecule, Vpu-102, Supports Adult Stem Cells Ex Vivo Expansion and Differentiation Under Serum-Free Condition

Ngo, A.,
Satomaa, T.,
Ni, J. H. T.,
Originally presented on: 3/17/2014 18:26:00 - 18:28:00

The uses of human adult tissue-derived stem/progenitor cells for cell-based therapies have increased in the past few years. Many of these applications have started clinical trials. However, for majority of cell culture processes associated with these clinical applications still requires the use of animal-derived components, including serum. This dependency of animal-derived components in the stem cell cultures conveys the high level of safety risks, lot-to-lot inconsistencies and supply issues associated with the process. At Irvine Scientific (IS), we developed a portfolio of new serum-free (SF), ready-to-use, and adaptable media products for *ex vivo* expansion and differentiation of adult stem/progenitor cells. The coating substrate is often required to achieve optimal performance of IS SF media. Majority of coating substrates used for SF stem cell culture, such as fibronectin, laminin or vitronectin that can be cost-effectively used for large-scale productions is derived from animal-origins. In this study, we examined an animal-free recombinant Erythrina cristagalli lectin (agglutinin) (ECA) molecule, VPU-102, together with IS SF media for its application in supporting expansion and differentiation of human mesenchymal stem/stromal cells (MSCs) and neural progenitor cells (NPCs). ECA was initially identified for its ability to support human pluripotent stem cell cultures through its binding to N-acetyl lactosamine glycoconjuates commonly expressed structures by undifferentiated pluripotent stem cells. We will present data to demonstrate VPU-102, a plant-expressed recombinant ECA, can be used to replace human plasma-derived fibronectin to support optimal expansions of human MSCs and NPCs up to 3 passages while maintain their proper marker expressions and multipotency. We will also demonstrate how VPU-102 was validated for its uses in SF differentiation process. The availability of VPU-102 that can deliver fibronectin-comparable function in SF stem/progenitor cell cultures facilitates the potentials of human adult stem cells in any clinical application.

Three-Dimensional Microscaffolds for Enrichment and Transplantation of iPSC-Derived Neurons

Francis, N. L., Rutgers University
Bennett, N. K., Rutgers University
Carlson, A. L., Rutgers University
Halikere, A., Child Health Institute of New Jersey
Wang, X.,
Pang, Z.,
Moghe, P. V., Rutgers University
Originally presented on: 3/17/2014 18:28:00 - 18:30:00

Introduction: Human induced pluripotent stem cells (iPSCs) can be directly converted into neurons by ectopic expression of neuronal transcription factors, providing a cell source for *in vitro* disease modeling, drug screening, and regenerative therapies. Cell transplantation therapies present a strategy to treat the functional deficits of neurodegenerative diseases, such as Parkinson's disease, however the poor survival rates of transplanted cells *in vivo* limits the efficacy of this treatment. In contrast to the traditional delivery of dissociated cells, we hypothesize that using fibrous microscaffolds to transplant adherent, functional networks of neurons will enhance cell survival and engraftment *in vivo*. In this study, we evaluated the ability of fibrous 3-D microscaffolds to support the growth and maturation of iPSC-induced neurons (iNs) *in vitro*, and to enhance survival of iNs transplanted into an *ex vivo* mouse organotypic brain slice culture. This work serves as a preliminary study for the future *in vivo* transplantation of microscaffold-seeded dopaminergic neurons into a Parkinson's disease mouse model.

Materials and Methods: Fibrous microscaffolds were fabricated by electrospinning sheets of synthetic tyrosine-derived polycarbonate polymers, and sectioning the sheets into squares 100 μm in side length. iPSCs were transfected with the drug-inducible neuronal transcription factor NeuroD1 and were seeded onto microscaffolds shortly after neuronal induction. Maturation and functionality were assessed by immunocytochemistry and calcium imaging. Microscaffolds were used to transplant networks of iNs into *ex vivo* organotypic mouse brain slices for evaluation of survival and functional integration.

Results and Discussion: Seeding iNs within fibrous microscaffolds yielded an enriched, highly dense population of neurons. The scaffold geometry enabled cell infiltration and cell-cell contact, supporting neuronal maturation. iNs within the microscaffolds expressed βIII-tubulin and MAP2 within 8 days of neural induction, and responded to exogenous electrical stimulation by live cell calcium imaging. Microscaffold-seeded iNs extended significantly longer neurites in an *ex vivo* organotypic mouse brain slice culture compared to dissociated iNs.

Conclusions: 3-D microscaffolds can be used to enrich the neuronal population of NeuroD1-transfected iPSCs, and these scaffolds support the growth of mature, functional neurons. The use of these scaffolds to enhance neuronal integration into an *ex vivo* brain slice suggests that this platform will be effective as a transplantation vehicle for *in vivo* regenerative therapies.

Acknowledgments: Research reported in this publication was supported by NIH T32EB005583, NJCST Core grant, NIH EB000146 RESBIO, NJSCC10-3090-SCR-E-0 and graduate fellowship 11-2957-SCR-E-0, and NSF-IGERT grant 0801620. The authors gratefully acknowledge the RUCDR for the gift of the iPSC line used in this study and the NJCBM for providing the polymers used in this study.

Prolonged Induction of Human Mesenchymal Stem Cells Does Not Sustain the Osteogenic Phenotype

Hoch, A., UC Davis
Mittal, V., UC Davis
Leach, K., UC Davis

Originally presented on: 3/17/2014 18:30:00 - 18:32:00

Prior to transplantation, mesenchymal stromal/stem cells (MSC) can be osteogenically induced toward the osteoblast phenotype in vitro using a cocktail of osteogenic supplements. However, there is little evidence of induced MSC directly participating in bone formation in vivo, suggesting that MSC may return to an undifferentiated phenotype upon removal of the osteogenic stimulus. As MSC are commonly induced for only a few days in culture, the induction period may be insufficient to sustain the osteoblast phenotype. We hypothesized that increasing the duration of MSC osteoinduction would preserve the osteoblast phenotype after removal of the osteogenic stimulus. MSC derived from three male donors (20-30yrs old) were expanded under standard conditions: α-MEM supplemented with 10% FBS and 1% penicillin (10,000 U/mL) and streptomycin (10 mg/mL). Osteogenic media consisted of growth media supplemented with 50 μg/mL ascorbic acid, 10 mM β-glycerophosphate, and 10 nM dexamethasone. MSC were seeded at 30,000 cells/cm2 and cultured in growth media or osteogenic media for 2, 4, and 6 weeks. After each duration, induction was removed from osteogenically induced MSC, which were subsequently cultured in growth media for an additional 5 days (OM/GM). As a positive control, osteogenically induced MSC remained in osteogenic media (OM). MSC cultured in growth media for the full duration (GM) served as the negative control. Osteogenic markers were assessed at 0 (baseline), +1, and +5 days where 0d marks the day OM/GM was switched from osteogenic media to growth media. After 24h following induction removal, OM/GM exhibited a regression in both normalized ALP activity and calcium deposition compared to OM regardless of induction duration: 2 (Fig. 1A), 4 (Fig. 1B), and 6 weeks (Fig. 1C). The trend in calcium deposition for 6 week regression extended across three donors, wherein the bars span data points and the line signifies the mean (Fig. 1D). After 2 week differentiation, OM/GM decreased to a calcium deposition more similar to that of GM at +5d (Fig. 1A). Significant differences in normalized ALP activity and calcium deposition between OM/GM and OM were not attributed to significant differences in DNA. In 24h, induction withdrawal from 4 week differentiated MSC (OM/GM) diminished bone sialoprotein (IBSP) expression, thereby demonstrating the implications of induction withdrawal extend to the transcriptional level. Osteogenically induced MSC retained their ability to differentiate toward adipocytes despite mineral production and morphological transition from spindle to cuboidal-shaped. MSC induced in osteogenic media exhibited enhanced colony-forming efficiency (CFE) compared to MSC in growth media. Upon removal of the osteogenic stimulus, MSC secreted similar or greater concentrations of vascular endothelial growth factor (VEGF) compared to cells maintained in growth or osteogenic media for the entire culture period. These data demonstrate osteoinduction withdrawal abrogates the potency of the acquired osteogenic phenotype regardless of induction duration. Our findings underscore the clinical requirement to strike a balance between maximizing the therapeutic potential of MSC while minimizing costly, time-consuming, and potentially deleterious ex vivo manipulation. These data highlight the importance of understanding the merit of 2D soluble cue differentiation for bone tissue engineering applications.

Recapitulating Early Mammalian Morphogenesis in Microfluidic Culture

Razian, G.,
Ungrin, M., University of Calgary
Originally presented on: 3/17/2014 18:32:00 - 18:34:00

The progression from a fertilized oocyte to a fully developed newborn animal involves multiple cell divisions and developmental transitions, as initially pluripotent stem cells gradually commit to distinct cell lineages. A key developmental stage is the blastocyst stage â?? a time when embryonic (epiblast) and extra-embryonic (hypoblast) lineages are specified and segregate into distinct tissues. The epiblast, which gives rise to the embryo proper, is a structured pluripotent epithelium - a micro-scale tissue with tremendous developmental potential. The epiblast undergoes gastrulation to give rise to the three canonical germ layers of definitive endoderm, ectoderm, and mesoderm, which in turn generate all the tissues and organs in the developing embryo. Understanding this stage of development is thus of great importance to both developmental biology and tissue engineering.
The majority of our present understanding of mammalian embryology has been obtained in mouse, much of it using in-vivo-isolated blastocysts. This is a very powerful system, but it can also be labour intensive and expensive, with single experiments requiring several hundred embryos in some cases. There is therefore a significant need for in vitro model systems that will allow precise recapitulation of the earliest stages of embryogenesis.

To mimic early developmental stages, investigators have used embryoid bodies (EBs), which are aggregates of embryonic stem cells (ESCs) that form three dimensional structures that can be induced to differentiate into different cell lineages. While EBs have differences in geometry from the embryonic bilaminar disc, they have been widely and successfully used as a model system for this stage of development. While this work has primarily focused on murine cells, we have previously demonstrated that human ESCs exhibit similar potential, extending the applicability of this approach into other species where developmental studies have previously been limited for both ethical and practical reasons.
Subsequent induction of tissue-level
gastrulation behaviour has proved more challenging. This challenge is not unexpected, due to the well defined and oriented gradients of signalling molecules that are known to play a major role in embryonic axis formation and cell fate specification. There is thus a significant need for culture systems that permit recapitulation of the microenvironment of graded signals that is native to the bilaminar disc. I will present progress towards the development of microfluidic devices that will allow us to culture and manipulate individual colonies of pluripotent stem cells and provide a model system for the study of peri-implantation morphogenesis.

Nanog Reverses the Effects of Senescence on Proliferation and Myogenic Differentiation of Human Mesenchymal Stem Cells

Mistriotis, P., University at Buffalo, State University of New York
Liang, M. S., State University of New York at Buffalo
Han, J., University at Buffalo, the State University of New York
Andreadis, S. T., University at Buffalo, The State University of New York
Originally presented on: 3/17/2014 18:34:00 - 18:36:00

Nanog reverses the effects of senescence on proliferation and myogenic differentiation of human mesenchymal stem cells

Panagiotis Mistriotis[1], Maoshih Liang[1], Juhee Han[1], Stelios T. Andreadis[1, 2]

1. Bioengineering Laboratory, 908 Furnas Hall, Department of Chemical and Biological Engineering, University at Buffalo, State University of New York, Amherst, NY 14260-4200, USA

2. Center of Excellence in Bioinformatics and Life Sciences, Buffalo, NY 14203, USA

Introduction: Human Mesenchymal Stem Cells (hMSCs) show great promise as an autologous smooth muscle cell source for vascular tissue engineering and regenerative medicine. However culture senescence and organismal aging reduce the proliferative and myogenic differentiation capacity of MSCs. To address this issue, we recently reported that ectopic expression of Nanog can restore the lost contractile phenotype due to aging and enhance the proliferative potential of ovine MSCs (Han et al., 2012, Stem Cells). In the current study we aim in identifying the molecular mechanism that mediates the rejuvenating effects of Nanog in MSCs.

Materials and Methods: MSCs isolated from both hair follicle or bone marrow and transduced with a tetracycline regulatable vector that carries the Nanog gene. This system enables Nanog expression upon Doxycycline (Dox) treatment. After serial passaging, MSCs were induced to senescence and Dox was added to the culture medium. Subsequently, the effect of Nanog on proliferative and myogenic capacity of MSCs were evaluated and compared with early passage (EP) MSCs.

Results and Discussion: Ectopic expression of Nanog in late passage (LP) MSCs significantly increases the proliferation (fig. 1A) and decreases the percentage of senescent cells as indicated by SA-β-GAL staining (fig. 1B). Furthermore, we show that Nanog controls the expression of key regulators of senescence such as cyclin inhibitors and DNA methyltransferases. Next we studied the myogenic differentiation upon Nanog expression. SMC-specific markers such as ACTA2 and Calponin were highly upregulated in Nanog^{+}LP MSCs resembling EP MSCs (fig. 1C). In agreement, the contractility in response to vasoactive agonists was also enhanced (fig.1D). Interestingly, we also report that Nanog reverses the loss of the expression of key myogenic transcription such as SRF, MRTF-b, SMAD2 and MYOCD due to senescence (fig. 1E,F). Notably, we identified that Nanog co-immunoprecipitatets with SRF and enhances the transcriptionaly activity of the SRF DNA binding site (CArG box) (fig. 1G). At this framework, SRF knockdown abolished the effect of Nanog on myogenic markers and contractility (fig. 1H). On the other hand, over-expression of SRF in LP MSCs enhances the myogenic phenotype resembling the effect of Nanog.

Conclusion: Taken together these data suggest that Nanog reverses the effects of senescence on MSC proliferation and myogenic differentiation capacity, thereby increasing the potential of MSC from aged donors for cellular therapy and tissue regeneration.

Expansion and Secretory Profiles of Human Platelet Lysate Conditioned BM-, UC- and ESC-MSCs for Inflammatory and Immune Therapies

Griffiths, S.,
Mandrycky, C.,
McDevitt, T., Georgia Institute of Technology
Nerem, R. M., Georgia Institute of Technology
Originally presented on: 3/17/2014 18:36:00 - 18:38:00

Inflammatory and autoimmune diseases can occur for a number of reasons including the dysregulation of the normal immune response and aggressive inflammation as a result of infection which can affect multiple tissue types, resulting in damage and increased risk of morbidity and/or mortality. Current drug treatment regimens are typically not sufficient to treat moderate to severe cases of inflammatory and autoimmune diseases. Mesenchymal stem cells (MSCs) secrete a cadre of potent immunomodulatory factors and are generally considered to be immunotolerant, thus the ability of MSCs to be manufactured and stored long-term, present a unique opportunity to produce an allogeneic "off-the-shelf" cell therapy for immunomodulation. However, bone marrow (BM)-MSCs are a heterogeneous cell population that have limited expansion capabilities *in vitro*, additionally, several concerns remain regarding the immunogenicity of xenogeneic cultured MSCs. Recently, culture media supplementation with allogeneic human platelet lysate (hPL) has been demonstrated to promote BM-MSC growth whilst maintaining differentiation potential. However, MSCs derived from earlier developmental stages, including umbilical cord (UC) or embryonic stem cells (ESC), with highly replicative phenotypes may be far superior for generating large numbers of clinical-grade allogeneic MSCs for inflammatory and immune disease treatment.

MSC from BM, UC and ESC were cultured in hPL (10%) or FBS (10%) and doubling time calculated. To determine paracrine factor secretion MSC conditioned media was collected after 24 h under serum free conditions and immunomodulatory factors quantified via ELISA. UC-MSCs secreted the highest levels of immunomodulatory factors PGE2, MCP-1, HGF and IL-6 when compared with ESC-MSCs and BM-MSCs. Notably, UC-MSCs secreted greater than a 2-fold amount of all four factors when compared with ESC-MSCs. Xeno-free hPL culture promoted the expansion of human of both UC-MSCs and ESC-MSCs. The lowest doubling time was achieved with ESC-MSC under hPL conditions (35.4 h $\pm$ 1.4), compared to FBS conditions (38.2 $\pm$ 1.3). BM-MSCs cultured under FBS conditions demonstrated the highest doubling time (74.7 $\pm$ 9.5). BM-MSCs under hPL culture conditions secreted significantly higher levels of cytokines including the immunomodulatory cytokines IL-6 and MCP-1 when compared to FBS culture. Results demonstrate increased proliferation of all MSC types under hPL conditions and potent paracrine secretion from UC-MSCs compared with the other MSC sources. However, greater understanding of MSC source and culture environment on immunomodulatory properties is needed to generate therapeutic allogeneic MSCs for inflammatory and autoimmune diseases. Equally, studying the ability of immature MSCs to suppress the proliferation and cytokine production from specific subsets of immune cells, involved in inflammation and immune disorders, can help to indicate target treatments.

Engineering 3D Cardiac Microtissues from Human Pluripotent Stem Cells

Hookway, T.,
Nguyen, D.,
Wagner, M.,
Xu, C.,
McDevitt, T. C., Georgia Institute of Technology
Originally presented on: 3/17/2014 18:38:00 - 18:40:00

There has been limited progress toward the development of *in vitro* cardiac tissue models and cellular cardiac therapies due to the lack of an easily obtainable and expandable cell source. Human pluripotent stem cells (hPSCs) represent a potentially unlimited source of cardiomyocytes for the study of cardiovascular development, disease, and tissue homeostasis, as well as regenerative cardiac therapies. An enhanced understanding of interactions between cardiomyocytes and their surrounding microenvironment is required in order to generate appropriate *in vitro* models of cardiac diseases. In this study, we model cardiac microenvironments by generating 3D cardiospheres of hPSC-derived cardiomyocytes using controlled formation and culture conditions to modulate cardiomyocyte phenotype and physiological function.

Differentiation of cardiomyocytes from hPSCs was achieved within 14 days with both growth factor- and small molecule-directed protocols on 2D monolayer cultures using a serum-free medium. Following 2D differentiation, cultures were trypsinized and seeded into 400 μm diameter microwells (Aggrewell™) at a density of 500-1500 cells/microwell. After 24 hours of formation, 3D cardiospheres were transferred to suspension culture and maintained for 7 days. Spheroid size and the percentage of beating spheroids were monitored over time and cardiac marker expression was analyzed after 7 additional days.

The differentiation protocols yielded weakly beating cells by day 10 of differentiation and strongly beating regions across the cultures by day 14. Regardless of the initial cardiomyocyte efficiency, cells aggregated into 3D cardiospheres with nearly 100% exhibiting spontaneous contractile beating, synchronous intracellular calcium transients, and expression of cardiac markers α-actinin and connexin 43. Within a given experiment, cardiospheres maintained their size for the duration of culture, however, the relative sizes of cardiospheres was modulated by controlling the initial cell seeding density (500-1500 cells/sphere). The cell population within the 3D aggregates expressed higher levels of cardiac specific genes, and cells dissociated from cardiospheres exhibited an enrichment of cardiomyocytes from a starting population of 10-40% to a final population of 80-99% cardiomyocytes when cultured in 3D cardiospheres as opposed to the purity of parallel 2D monolayer cultures, which remained similar to the starting population.

These studies demonstrate that the combination of forced aggregation and suspension culture enables reliable generation of 3D cardiospheres with nearly 100% spontaneously beating constructs independent of the starting purity, providing a straight-forward method for enrichment of cardiomyocytes. Additionally, microparticles can be incorporated within both hPSC aggregates and cardiospheres, enabling the delivery of factors to direct differentiation or promote cardiomyocyte maturation and survival within the 3D constructs. Ongoing studies aim to understand the complex physical and biochemical elements of 3D cardiac environments and how aggregated cardiospheres can be utilized both as *in vitro* models to study cardiac health and disease, as well as serve as transplantable constructs for regenerative cardiac therapies.

Induced Pluripotent Stem Cells Generated from Multiple Somatic Cell Types Via Feeder-Free Reprogramming in TeSR™-E7™ Medium

Chang, W. Y., STEMCELL Technologies Inc
Hunter, A.,
Ng, A.,
Wong, M.,
De Jong, S.,
Duronio, C.,
Yu, I.,
Wognum, B.,
Wagey, R.,
Peters, C.,
McQueen, K.,
Fairhurst, M.,
Hadley, E.,
Antonchuk, J., STEMCELL Technologies Inc
Thomas, T. E., STEMCELL Technologies Inc
Eaves, A. C., British Columbia Cancer Agency
Louis, S. A., STEMCELL Technologies Inc
Originally presented on: 3/17/2014 18:40:00 - 18:42:00

There is a general movement towards generating induced pluripotent stem (iPS) cells in more defined, feeder-free, and clinically relevant systems, as we move closer to realizing the use of iPS-derived cells in cellular therapies. Moreover, there is growing need to reprogram cells from easily accessible tissues such as skin (fibroblasts) and blood, to develop disease models and patient-specific therapeutics. TeSR™-E7™medium, based on a formulation by Chen et al. (2011), is a low-protein, xeno-free, and defined medium for feeder-free reprogramming. Here, we demonstrate the use of TeSR™-E7™ in reprogramming of human fibroblast cells into fully characterized iPS cells, and furthermore extend its utility to include reprogramming of other somatic cell types such as bone marrow-derived mesenchymal stromal cells (MSCs), CD34+ hematopoietic cells, and culture expanded erythroid cells. Fibroblast suspensions from different sources were transfected with episomal vectors containing Oct4, Sox2, Klf4, L-Myc, and then plated onto Matrigel®-coated plates containing either TeSR™-E7™, TeSR™-E8™, or KnockOut Serum Replacement (KOSR)-containing human embryonic stem (hES) cell maintenance medium (KOSR-hES). Cells were cultured in each medium with daily changes and iPS cell colonies exhibiting human pluripotent stem cell (hPSC)-like colony morphology were enumerated after 21 - 28 days. A range of reprogramming efficiencies were observed in TeSR™-E7™ from human fibroblasts of adult ($0.125 \pm 0.05\%$, n=3), neonatal ($0.016 \pm 0.003\%$, n=2) and fetal (0.015%, n=1) origin. Notably, reprogramming efficiencies in TeSR™-E7™ were 2-fold higher compared to KOSR-hES for neonatal fibroblasts (n=6, $p<0.05$) but equivalent for adult fibroblasts (n=6). Reprogramming efficiencies were on average 9.4- and 6.1-fold higher in TeSR™ E7™ compared to TeSR™-E8™ for both neonatal and adult fibroblasts (n=6, $p<0.05$). TeSR™-E7™-derived iPS cell colonies were then picked and sub-cultured in mTeSR™1 or TeSR™-E8™ maintenance media on Matrigel or Vitronectin-XF™ matrices for further characterization. iPS cell lines generated in TeSR™-E7™ exhibited characteristic hPSC-like morphology, >95% positivity for Oct4, SSEA-3, Tra-1-81 and were karyotypically normal. In addition, iPS cells derived in TeSR™-E7™ exhibited the capacity to differentiate efficiently to cells of the three germ-layers using STEMdiff™ Neural Induction Medium, STEMdiff™ Definitive Endoderm kit, or STEMdiff™ APEL™ Medium supplemented with cytokines. Finally, we show reprogramming efficiencies for multiple somatic cell types, including MSCs (0.001%, n=2), CD34+ hematopoietic cells (0.015%, n=1), and culture expanded erythroid cells (0.007%, n=1) with successful establishment of iPS cell lines from these cell sources. All iPS cell lines derived from these cell sources exhibited hPSC morphology and are now being further characterized for pluripotency. Overall, our studies demonstrate the robustness and versatility of TeSR™-E7™ for reprogramming multiple human somatic cell types.

Controlled BMP4 Delivery within ESC Aggregates Via Pnipmam Microparticles

Sullivan, D. D., Georgia Institute of Technology & Emory University
Saxena, S.,
Gaulding, J. C., Georgia Institute of Technology
Hettiaratchi, M. H., Georgia Institute of Technology and Emory University
Lyon, L. A., The Parker H. Petit Institute for Bioengineering and Bioscience
McDevitt, T. C., Georgia Institute of Technology
Originally presented on: 3/17/2014 18:42:00 - 18:44:00

Strategies to direct differentiation of embryonic stem cell (ESC) aggregates, or embryoid bodies (EBs), often employ an â??outside-inâ?□ approach by addition of soluble factors to culture medium. However, diffusion into EBs is limited. One method to direct differentiation and potentially overcome diffusion limitations is incorporation of engineered microparticles (MPs), which have been widely used to control molecular delivery locally, within 3D multicellular aggregates. Poly(n- isopropylmethacrylamide) (pNIPMAm) is a thermoresponsive polymer that exhibits a volume phase transition temperature at
43Â°C, allowing the polymer to maintain a swollen hydrated state at physiological temperatures to permit passive diffusion of entrapped soluble factors. Coupling of pNIPMAm microgels to core particles combines the hydrophilic and morphogen-releasing properties of hydrogels with a dense core material that enhances incorporation within EBs. The objective of this study was to characterize pNIPMAm MPs for controlled delivery of bioactive BMP4 within EBs to direct ESC differentiation.
Core-shell MPs were constructed with the shell consisting of microgels composed of 68% pNIPMAm, 2% *N,Nâ??* methylene- bisacrylamide, 30% acrylic acid and synthesized by precipitation polymerization. Core carboxyl-functionalized polystyrene MPs (~4.5 Î¼m) were conjugated to 4-amino-benzophenone using EDC chemistry. Microgels were coupled to functionalized cores by UV excitation. The BMP4 binding capacity was examined by loading MPs with a range of BMP4 doses (10 â?? 300 ng/mg MPs) over 18 hours at 4Â°C. An ELISA was used to quantify passive release of BMP4 from MPs over 12 days. Bioactivity of BMP4-laden MPs was evaluated using an *in vitro* alkaline phosphatase (ALP) assay to quantify ALP activity of skeletal myoblasts (C2C12s), following treatment with soluble BMP4, â??emptyâ?□ MPs, or BMP4-laden MPs. MPs were incorporated within EBs by forced aggregation in micro-well inserts at MP:cell seeding ratios of 1:10, 1:3, 1:1 and 3:1. BMP4 was delivered solubly and via MPs to ESC aggregates to evaluate gene expression of pluripotency marker, *Oct-4*, and mesoderm marker, *Brachyury-T*.
MP formation produced approximately 14.5 million MPs/mg of polymer with an average diameter of 4.6 Â± 1.0 Î¼m. BMP4- laden MPs released less than 25% of BMP4 for all loading amounts. At higher loading amounts (200 and 300 ng/mg MPs),
less than 5% of BMP4 was released from MPs. Results demonstrate that MPs maintain BMP4 bioactivity as treatment with BMP4-laden MPs induced ALP activity similar to delivery of soluble BMP4. However, treatment with â??emptyâ?□ MPs and soluble BMP4 induced less ALP activity as compared to soluble BMP4 alone, suggesting that MPs actively sequester free BMP4 and prevent interactions with cells. Maximum MP

incorporation within EBs was achieved with ~80 MPs per EB at a MP to cell seeding ratio of 1:1. *Oct-4* expression decreased over the course of differentiation in all groups. BMP4-laden MPs demonstrated comparable expression of *Brachyury-T* to soluble BMP4, despite the delivery of 20-fold less total protein from the MPs. These results suggest that pNIPMAm MPs can be used to control the release of bioactive BMP4 and induce ESC mesoderm differentiation.

Controlled Presentation of Cytokines within 3D Mesenchymal Stem Cell Constructs Enhances Immunomodulatory Activity

Zimmermann, J., Georgia Institute of Technology and Emory University
McDevitt, T. C., Georgia Institute of Technology
Originally presented on: 3/17/2014 18:44:00 - 18:46:00

Mesenchymal stem cells (MSCs) offer a potent cell therapy for the treatment of inflammatory and immune disorders due to their ability to modulate components of innate and adaptive immunity. Although MSCs have been investigated as cell therapies for diseases including graft-versus-host, type I diabetes, and inflammatory bowel disease (IBD), experimental results thus far have been inconsistent. Recent studies have suggested that MSC immunomodulation can be greatly influenced by the cellular microenvironment and the concentration of inflammatory cytokines, including IFNγ and TNFα, which stimulate MSC immunomodulation. The low levels of cytokines observed in chronic inflammation may not be sufficient to induce MSC immunomodulation, limiting the ability of MSCs to regulate inflammation in these disease states. Engineering the local microenvironment within MSC constructs via biomaterial-based presentation of inflammatory cytokines may serve as a novel means of regulating MSC immunomodulatory paracrine secretion *in vivo*. Therefore, the objective of this study was to regulate human MSC immunomodulation via the incorporation of immobilized-cytokine microparticles within 3D MSC spheroids.

TNFα was immobilized on the surface of microparticles by incubating streptavidin-coated polystyrene microparticles (~3μm diameter) with 100 ng biotinylated TNFα per $9x10^5$ particles overnight based on the number of streptavidins on the surface of the polystyrene microparticles ($1.2x10^6$ streptavidin-molecules per microparticle). After 18 hours of incubation, the supernatant was analyzed via ELISA for remaining TNFα to quantify binding to the particles. Microparticles with immobilized TNFα were subsequently suspended in PBS and incubated for 9 days to determine the stability of cytokine immobilization. Additionally, microparticles with or without immobilized-TNFα were incorporated into 3D MSC spheroids (6:1 microparticles-to-MSCs) by addition of particles and cells into Aggrewell? microwells to form 500-cell MSC spheroids. After 18 hours of formation, microparticle-incorporated MSC spheroids were transferred into rotary orbital suspension culture (65 rpm) and cultured for 4 days. A subset of spheroids without microparticles were also treated with soluble TNFα (10ng/mL). Immunomodulatory factor release (PGE2, TGFβ1, IL6) from MSC spheroids was quantified by ELISAs and normalized to cell numbers. Finally, MSC spheroids were co-cultured with IFNγ- and lipopolysaccharide-activated macrophages in 0.4 μm pore transwells for 16 hours and conditioned media was collected to determine macrophage expression of TNFα.

The loading efficiency of streptavidin-microparticles was determined to be approximately 47% ($1.7x10^{-4}$ ng TNFα per microparticle). Furthermore, less than 3% of the microparticle bound TNFα was released from the microparticles over 9 days of culture. Incorporation of microparticles with immobilized cytokines within MSC spheroids increased the secretion of the immunomodulatory factors PGE2 (2.2-fold, $p < 0.001$) and

IL-6 (3.6-fold, $p < 0.001$) compared to non-treated controls. This increase was comparable to treatment with soluble TNFα, demonstrating the cytokine remained bioactive after immobilization on the surface of the microparticles. Finally, greater suppression of the inflammatory response of IFNγ and LPS-activated macrophages was observed in co-cultures of MSC spheroids incorporated with immobilized-cytokine microparticles (71% reduction of macrophage TNFα secretion) compared to untreated spheroids (42% reduction). These results demonstrate the ability to immobilize bioactive cytokines to the surface of microparticles that can be stably incorporated in MSC aggregates to enhance immunomodulation.

Simple and Versatile Crispr Based Genome Editing Tool and Its Promise in Large Scale Genome and Cell Engineering Applications

Ravinder, N., Life Technologies

Originally presented on: 3/17/2014 18:46:00 - 18:48:00

Clustered Regularly Interspaced short palindromic repeat (CRISPR), a bacterial adaptive immune system from *Streptococcus pyogenes* has been recently engineered and demonstrated to function as a genome editing tool that can rapidly generate engineered cell lines, model organisms and perform large scale gene modifications in a wide variety of hosts. Due to its simple design including Cas9 nuclease and a non-coding guide RNA (gRNA) with target specificity defined by only a short 19-21 base nucleotide region, this system is an attractive tool for large scale genome engineering. Described here are different CRISPR delivery formats that address both small and large scale genome editing needs in a wide range of host and cell types including stem cells. Tools and methods tested through this work include a combination of ready to transfect formats that circumvent the need for host specific promoters and reduce the hands on time needed for generating target specific CRISPR system. It also allows the user to edit multiple loci or alleles simultaneously. The simple and easy to design tools described here holds great promise in large scale genome and cell engineering applications.

Novel Cell Surface Marker for the Identification of iPSC in Somatic Cell Reprogramming

Quintanilla, R. Jr., Life Technologies
MacArthur, C., Life Technologies
Lakshmipathy, U., Life Technologies
Originally presented on: 3/17/2014 18:48:00 - 18:50:00

The creation of human induced pluripotent stem cells (iPSCs) and their downstream applications demonstrate great promise as valuable tools for disease research, therapeutic applications and biological engineering. As different somatic tissues are reprogrammed into a pluripotent state using different technologies and under varying culture conditions, the critical step of establishing iPSC lines requires the early identification of true iPSC clones. It is important to identify and pick the right colonies for further expansion and characterization. Commonly, surface antibodies against pluripotent specific markers such as SSEA4, Tra-1-60 and Tra-1-81 are used but this method is expensive and sterility is a concern. We have utilized a global transcriptome comparison of human parental fibroblasts, and partially and fully reprogrammed clones to identify differentially expressed markers. One such marker is the cell adhesion surface molecule, CD44, which is highly expressed in parental fibroblasts and partially reprogrammed cells, but not in fully reprogrammed iPSCs.

CD44 has been used in combination with pluripotent markers to show distinct expression patterns between fully reprogrammed pluripotent colonies and partially or non-reprogrammed colonies during reprogramming. In this study, we utilized antibodies against CD44, in conjunction with other known pluripotent surface markers for elucidating reprogramming kinetics and for the rapid enrichment of pluripotent stem cells from reprogrammed somatic tissues. This combined approach of positive and negative markers for identification of true iPSCs during the reprogramming process permits the selection of bonafide clones during cell line creation.

Multiplexed Single-Cell Proteomic and Transcriptomic Analysis of Embryonic Stem Cells

Wang, J., California Insitute of Technology
Originally presented on: 3/17/2014 18:50:00 - 18:52:00

While immunofluorescent labeling of ESC surface receptors can suggest a signal dependence on the cognate ligand for those receptors, it provides limited insight as to whether this pathway is autocrine or paracrine in nature. Bulk analysis of media conditioned by ESCs and feeders have been profiled with proteomic studies, but the distinct molecules within a heterogenous mixture cannot be traced back to the secretory cell of origin. We use a single-cell platform uniquely enables the high thoroughput, quantitative profiling of candidate signaling proteins secreted by a single ESC or ESC-derivative. The ability to simultaneously assay a multiplex of secreted proteins can define and classify subpopulations of cells with distinct secretomes, several of which may be characteristic of pluripotent cells. In addition, this novel diagnostic tool will also be useful in the field of regenerative medicine, particularly in studies related to the generation and application of induced pluripotent stem cell (iPSC) technology. The reprogramming of a somatic cell to a pluripotent cell is highly inefficient, and the mechanism underlying this process remains largely unknown. Probing the signaling aspects of a single cell undergoing the reprogramming process will yield novel insights into molecular mechanisms and thereby enhance the efficiency of the process.

: Investigating the Effect of microRNAs on Brain Progenitor Cells Population and Myelin Repair in an Experimental Model of Demyelination

Ghasemi Kasman, M.,

Originally presented on: 3/17/2014 18:52:00 - 18:54:00

Targeting endogenous stem cells via factors that affect their proliferation, migration and differentiation has been a priority in regenerative medicine. Beside various benefits, the most important bottleneck in this strategy is the limitation of endogenous stem cells resources. In recent years, the production of induced pluripotent stem cells (iPSCs) has led to more hopes to treat diseases. Application of Micro-RNAs and small molecules has been in the focus of attention over past few years to introduce safer methods in production of iPSCs. Recent reports showed the potential of miRs in generation of iPSCs. Here we report the efficacy of microRNAS in converting neural cells into neural progenitors, in vivo. Viral particles was prepared for inducible expression of miRs cluster and injected i.c.v, and then induced for expression in presence of valproic acid for 7 days. Quantitative RT-PCR against neural progenitor markers and histochemical analysis were used to study the effect of miRs on the expression of neural progenitor markers. For assessing of myelin repair, we used Y-maze test and luxol fast blue staining. qRT-PCR showed a significant increase in the expression level of Sox1, Oct4, c-Myc, Nestin but not in those of Sox2 and Nanog. Histochemical evaluation showed the expression of some neural progenitor markers in the rims of brain ventricle.Y- maze data revealed that this microRNAs could improve cuprizone induced demyelination. These findings suggest that microRNAs may increase progenitor cells by reprogramming of endogenous neural cells and provide new hope for increasing brain repair potential.

Hydrodynamics in Stirred Suspension Bioreactors from CFD Models for Expansion of Induced Pluripotent Stem Cells

Le, A., University of Calgary
Kallos, M. S., Pharmaceutical Production Research Facility (PPRF), Schulich School of Engineering, University of Calgary
Rancourt, D. E., University of Calgary
Gates, I. D., University of Calgary
Originally presented on: 3/17/2014 18:54:00 - 18:56:00

Induced pluripotent stem cells (iPSCs) have the ability to proliferate in culture and can differentiate into many other tissue types and as they can be derived from somatic cells, their use would overcome existing ethical issues associated with embryonic stem cells in future stem cell-based applications. Most of the current iPSC research focuses on improving the safety and efficiency of reprograming somatic cells to iPSCs. However, for iPSCs to be used in clinical applications, establishing a robust cell manufacturing process is an essential step. The challenge of cell manufacturing is to produce large quantities of cells while maintaining the desired phenotype. Stirred suspension bioreactors have been used by our group and many others for the expansion of a range of stem cell types. This culture system provides a well-mixed environment and for aggregate-forming cells, such as human iPSCs, the culture environment is able to control the size of the aggregates formed – thus generating more standardized cell populations. We have previously shown that shear stress in stirred suspension bioreactors plays an important role in stem cell proliferation and differentiation due to possible mechanotransduction mechanisms. Understanding the hydrodynamic environment and in particular the shear stress distribution inside stirred suspension bioreactors is a fundamental step to develop cell manufacturing processes. Here, we used computational fluid dynamics (CFD) to understand the flow regime and shear stress distribution at different agitation rates, viscosities, and geometries of vessel and impeller. The shear stress-cell growth relationship and shear stress-cell aggregate distribution was also examined by comparing our model to established experimental data. These key relationships provide the basis for designing robust cell manufacturing processes that meet specific clinical applications.

Myeloid Progenitor Cells-Suspension Cell (K562 cells)
Richardson, D. R., PSangelus
Originally presented on: 3/17/2014 18:56:00 - 18:58:00

10 liters of myeloid progenitor suspension cells were grown in a Wave bioreactor. The cells were sterily collected with the UniFuge single use centrifuge, using the AC module. The viability of the grown cells measured 95%. The suspension cells were grown for 4 days, doubling each day. The diameter size of the cells are 15 -20 micron. The cells reached a final cell density 4.93 million cells/ml at time of harvest. To harvest the cells, the UniFuge centrifugation parameters were feed rate (LPM) and G force. The cells were layered on a PBS buffer; a manual prefill step was utilized by switching supply lines. The suspension cells were collected aseptically. After Harvest, the cells were frozen for 24 hours, thawed and viabilities measured.

Tissue Engineered Vascular Grafts Created from Human Induced Pluripotent Stem Cells

Sundaram, S., Yale University

Originally presented on: 3/17/2014 18:58:00 - 19:00:00

The utility of human induced pluripotent stem cells (hiPSCs) to create tissue engineered vascular grafts (TEVGs) was evaluated in this study. hiPSC lines were first induced into a mesenchymal lineage via a neural crest intermediate using a previously reported serum-free, chemically defined differentiation scheme. Derived cells exhibited commonly known mesenchymal markers (CD90, CD105, CD73; negative markers – CD45), and could differentiate into several mesenchymal lineages (osteogenic, chondrogenic, adipogenic). Functional vascular grafts were then engineered by culturing hiPSC-derived mesenchymal progenitor cells in a pulsatile bioreactor system over 8 weeks, to induce smooth muscle cell differentiation and collagenous matrix generation. Histological analyses confirmed layers of calponin-positive smooth muscle cells in a collagen rich matrix. Mechanical tests revealed that grafts had an average burst pressure of 700 mm Hg, which is approximately half of that of native veins. Overall, these results provide significant insight into the utility of hiPS cells for vascular graft generation. They pave the way for creating personalized, patient-specific vascular grafts for surgical applications, as well as for creating experimental models of vascular development and disease.

Development Of A Scalable Manufacturing Process For Bone-Marrow Derived hMSCs In A Low-Shear Single-Use Bioreactor System

Giroux, D.,
Lee, B.,
Wesselschmidt, R.,
Serra, M., ITQB-UNL
Alves, P. M., IBET/ITQB
Carrondo, M. J. T., FCT/UNL
Sousa, M. F. Q., IBET
Originally presented on: 3/17/2014 19:00:00 - 19:02:00

Bone-Marrow derived Mesenchymal Stem Cells (BM-MSCs) are multi-potent plastic adherent non-hematopoietic cells found in bone marrow and stroma. They have been widely studied for 40 years and although a rare cell population, only 0.01-0.001%, they are relatively easy to isolate, grow well in vitro, and possess both immunomodulatory and low immune reactivity properties. For these reasons MSCs are thought by many to be an ideal cell for the development of "off the shelf" allogeneic cellular therapeutics for the treatment of a variety of diseases such as graft-vs-host disease, Crohn's, stroke, and some cardiomyopathies. MSCs have also been isolated from adipose and other tissues, expanding their therapeutic potential. A recent search of www.clinicaltrials.gov returns a list of more than 300 trials using mesenchymal stem cells. However, in order to move from these small trials to commercialization, cells must be manufactured at large scale while remaining potent and safe.

To address this manufacturing bottleneck, we have developed a unique MSC manufacturing process where BM-MSCs are cultured on microcarriers with low shear mixing using the Air-Wheel Bioreactor. Microcarrier culture was chosen for its productivity and scalability advantage over 2D culture. Successful microcarrier culture, particularly in large vessels, requires a judicious choice of bioreactor and impeller. The PBS single-use bioreactors are able to gently suspend microcarrier cultures at a Kolmogorov scale of the order of 130 um at all vessel sizes (3, 15, 80 and 500 L.)

BM-MSCs (StemCell Technologies) were cultured in xeno-free Mesencult medium (StemCell Technologies) on Synthemax II microcarriers (Corning) in a PBS 3 Air-Wheel Bioreactor. Processes for microcarrier seeding and expansion via bead-to-bead transfer took place within the Bioreactor. Cell growth and metabolic rates were monitored, and a traditional stirred-tank bioreactor was run in parallel as control. In the Air-Wheel Bioreactor, the cells attached to the microcarriers faster and reached confluence more quickly than a stirred-tank bioreactor. MSCs cultured as described met the International Society of Cellular Therapy (ISCT) minimum criteria for BM-MSC (> 95% CD73, CD105, CD90 and < 2% CD34, CD45, CD14, CD19, HLA-DR) and tri-lineage differentiation. In this study we show that BM-MSCs can be expanded on microcarriers using a low-shear Air-Wheel Bioreactor with bead-to-bead transfer that is amenable to scale up to commercial manufacturing. When BM-MSCs are cultured using this novel expansion system, microcarrier loading is fast and uniform, the MSCs are highly proliferative, and they retain the characteristic markers of BM-MSC.

Rapid Generation of Induced Pluripotent Stem Cells from Lymphoblastoid Cell Lines Using an Episomal Plasmid Containing Multiple Reprogramming Factors in a Single Cassette (Matisse Reprogramming System)

Sams, G.,
Paul, S.,
Dansithong, W.,
Figueroa, K.,
Pulst, S.,
Originally presented on: 3/17/2014 19:02:00 - 19:04:00

Objective: Reprogramming of skin fibroblasts is increasingly used for modeling of human neurological diseases. Cell lines from many patients, however, have been generated using EBV-immortalized B-lymphocytes (EBV-B). In contrast to skin fibroblasts, blood lymphocytes and especially lymphoblastoid (LB) cells are difficult to reprogram and iPSC colonies are usually formed after weeks and with very low efficiency.

Background: B cells represent a larger fraction of the peripheral blood mononuclear cell population (~20%) and can be transformed in vitro by EBV to generate lymphoblastoid cell lines (LB), creating an unlimited proliferative source of cells for reprogramming trials. LB cells are a precious resource for immunologic, epidemiologic, and rare disease studies. Generating iPSCs from LB cells offers the advantage of working with minimal amounts of blood from living donors as well as frozen LB collections banked worldwide. Recently, iPSCs have been generated by delivering the reprogramming factors via oriP/EBNA-1-based plasmids in fibroblasts and peripheral blood CD34+cells.

Design/Methods: We have generated a novel episomal plasmid where RFs (c-MYC, KLF4, SOX2 and OCT3/4) are controlled by individually modified CMV promoters in a single cassette, designated pPuro(CMV_{mt1})cMKSO (Matisse reprogramming system), which allows coordinated expression of the four RFs. EBV-B cells were electroporated with plasmid, pPuro(CMV_{mt1})cMKSO and plated on matrigel coated plate with MEF-derived conditioned media (MEF-CM) containing basic fibroblast growth factor (bFGF).

Results: After 6-8 days of culture of electroporated EBV-B cells with MEF-CM, cells tend to attach and start to form colonies. EBV-B cell line-derived iPSCs showed ES cell morphology, stained with alkaline phosphatase (AP), and expressed pluripotent cell-specific genes. The ability to reprogram banked patient EBV-transformed cell lines efficiently will offer an unprecedented opportunity to rapidly generate genetic disease models and also provide a translational platform for therapeutic drug development.

Conclusions: We developed a rapid method to generate iPSCs from EBV-immortalized B-lymphocyte cell lines (EBV-B cells) using a novel 4-factor vector.

Derivation of Induced Neural Crest from Human Postnatal Fibroblasts

Lee, G., Department of Neurology

Originally presented on: 3/17/2014 19:04:00 - 19:06:00

Recent direct conversion studies have employed multiple transcription factors to force somatic cells into different cellular fates. Here we report the direct conversion of human fibroblasts into induced neural crest (iNC) with a single genetic factor (*SOX10*) and environmental cues, including WNT activation. The isolated human iNC population exhibits transcriptional similarity to their hESC-derived counterpart, extensive migration ability and differentiation potentials toward multiple lineages. Furthermore, we identified a cell surface marker to prospectively isolate the iNC population from postnatal fibroblasts, allowing purification of disease-specific iNCs from fibroblasts of familial dysautonomia (FD) patients. Transcriptional characterization of FD-iNCs strongly demonstrated defective migration and alternative splicing, providing new insights into FD disease mechanism. Our studies demonstrate the feasibility of rapid and efficient direct conversion of human postnatal fibroblasts into iNCs with cell intrinsic and extrinsic cues, which could have important implications for studies of neural crest related disorders.

Biologically Driven Assembly of Human Tissues in Synthetic Environments

Murphy, W. L., University of Wisconsin

Originally presented on: 3/18/2014 0:00:00 - 0:20:00

In contemporary drug discovery and toxin screening there is a need to develop human, *ex vivo* models that represent critical features of human organs. Human pluripotent stem cells offer unprecedented potential for developing "organotypic" models for a variety of applications, including drug and toxin screening. Furthermore, recent studies demonstrate that human pluripotent stem cells may be capable of forming functional organotypic culture models via a process we term "biologically driven assembly", in which cells are allowed to undergo morphogenesis processes without substantial external control. This talk will introduce the role of synthetic biomaterials for defining the process of tissue assembly ex vivo. Controllable parameters such as the surrounding stiffness, protease-lability, and geometry can strongly influence cellular assembly. The talk will also describe biologically driven assembly of vascular and neural tissues for screening applications. Our studies to date emphasize the potential for synthetic biomaterials to define conditions for tissue assembly, with the ultimate goal of creating robust, reproducible, human tissues in a dish.

Actin-Mediated Contractility in Three-Dimensional Aggregates of Human Mesenchymal Stem Cells

Tsai, A. C., Florida State University
Liu, Y.,
Ma, T., Florida State University
Originally presented on: 3/18/2014 0:20:00 - 0:40:00

Human mesenchymal stem cells (hMSCs) are primary candidates in cell therapy and tissue engineering and are being tested in clinical trials for a wide range of diseases. Originally isolated and expanded as plastic adherent cells, hMSCs have intriguing properties of in vitro self-assembly into three-dimensional (3D) aggregates reminiscent of skeletal condensation in vivo. Recent studies have shown that hMSC 3D aggregation improved a range of biological properties, including multilineage potential, secretion of therapeutic factors, and resistance against ischemic condition. Although studies have suggested that hMSC aggregation is mediated through cell-cell contacts and cell-ECM interactions, the adaptive changes of hMSC cytoskeleton during self-assembly into 3D aggregates also play an important role that influences aggregate formation, compaction, functional activation, and multilineage differentiation. In the current study, we investigated the actin-mediated contractility in regulating hMSC 3D aggregation, contraction, apoptosis, functional activation, spreading on adherent surface, and coalescing. Individual hMSC aggregates with controlled initial cell number were formed by seeding a known number of hMSCs (500, 2,000, 5,000, and 20,000/well) in multi-well plates of ultra-low adherent (ULA) surface to form 3D aggregates. Tracking of individual hMSC aggregates revealed size-dependent contraction and decreasing viability with larger aggregates having lower contraction and apoptosis compared to smaller aggregates. Induction of osteogenic differentiation improved cell viability in hMSC aggregates while expressions of osteogenic markers were size-dependent. To assess the influence of actin-mediated contractility on hMSC aggregation and properties, actin modulators (i.e., blebbistatin, cytochalasin D, jasplakinolide, lysophosphatidic acid (LPA), Y-27632) were added at different stages of aggregation and their impacts on hMSC aggregate size, contraction, apoptosis, and multilineage potential were monitored. Interruption of actin polymerization by cytochalasin D prevented aggregate contraction but has minimal impact on cell viability in the aggregates. In contrast, addition of ROCK inhibitor (Y-27632) and LPA prevented aggregate contraction with increased cell viability in the aggregates. To further assess the role of actin and microtubule in mediating aggregate spreading and re-adhesion, aggregates treated by the actin modulators and nocodazole, a microtubule modulator, were placed on (a) adherent surface or (b) in close proximity on ULA surface. The results show that interruption of actin polymerization by cytochalasin D significantly reduced cell motility with reduced spreading on adherent surface and aggregate coalescing. On the other hand, hMSC aggregates treated with nocodazole, ROCK inhibitor (Y-27632), and LPA maintained their mobility on adherent surface and readily coalesced when in contact. Together, these results suggest that reorganization of actin cortical network during cell aggregation plays a prominent role in mediating hMSC aggregate assembly, contraction, and motility. The impact of actin-mediated contractility on hMSC phenotype selection and functional activation will also be presented.

Human Mesenchymal Stem Cell Spheroids in Fibrin Hydrogels Exhibit Improved Cell Survival and Potential for Bone Healing

Leach, K., UC Davis
Murphy, K., UC Davis
Fang, S., UC Davis
Originally presented on: 3/18/2014 1:00:00 - 1:20:00

Mesenchymal stem cells (MSC) have great therapeutic potential for the repair of nonhealing bone defects due to their proliferative capacity, multilineage potential, trophic factor secretion, and lack of immunogenicity. However, a major barrier to the translation of cell-based therapies into clinical practice is ensuring their survival and function upon implantation into the defect site. We hypothesized that forming MSC into more physiologic 3-dimensional spheroids, rather than employing dissociated cells from 2-dimensional monolayer culture, would enhance their survival when exposed to a harsh microenvironment while maintaining their osteogenic potential. MSC spheroids were formed using the hanging drop method with increasing cell numbers. Compared to larger spheroids, the smallest spheroids which contained 15,000 cells exhibited increased metabolic activity, reduced apoptosis, and the most uniform distribution of proliferating cells. Spheroids were then entrapped in fibrin gels and cultured in serum-free media and 1% oxygen. Compared to identical numbers of dissociated MSC in fibrin gels, spheroids exhibited significantly reduced apoptosis and secreted up to 100-fold more VEGF. We also observed that fibrin gels containing spheroids and those containing an equivalent number of dissociated cells exhibited similar expression levels of early and late markers of osteogenic differentiation. These data demonstrate that MSC spheroids exhibit greater resistance to apoptosis and enhanced proangiogenic potential, while maintaining similar osteogenic potential to dissociated MSC when entrapped in a clinically relevant biomaterial, supporting the use of MSC spheroids in cell-based approaches to bone repair.

Deriving Complex Endodermal Organ Tissues from Human Pluripotent Stem Cells As Models for Development and Disease Research

Wells, J.,

Originally presented on: 3/18/2014 1:20:00 - 1:40:00

Kyle McCracken*, Jorge Munera*, Patrick McGrath*, Michael Workman*, Katie Sinagoga*, Christopher Mayhew*, Noah Shroyer+, Carey Watson^, Michael Helmrath^ and James Wells#*.

Divisions of Developmental Biology*, Gastroenterology+, Endocrinology#, Pediatric Surgery^

Cincinnati ChildrenÕs Hospital Medical Center, Cincinnati OH 45229

Successful efforts to direct the differentiation of human embryonic and induced pluripotent stem cells (PSCs) into specific organ cell types *in vitro* have largely been guided by studies in embryonic development. We have used signaling pathways that control early endoderm organ morphogenesis *in vivo* to generate complex, three-dimensional organ tissues with improved functionality from human PSCs *in vitro*. We identified that the combined activity of FGF and Wnt signaling pathways was sufficient to promote PSC-derived endoderm to undergo gut tube morphogenesis *in vitro*. The resulting three-dimensional gut tube tissues can be further directed into specific organ tissue types using embryonic patterning pathways. For example we have been able to use a temporal series of growth factor manipulations that mimic embryonic intestinal development to generate three-dimensional human intestinal organoids (HIOs). HIOs consisted of a polarized, columnar epithelium with villus-like structures and crypt-like proliferative zones with all of the major cell types of the adult gut. Cell types within HIOs can be further using genetic gain- and loss-of-function approaches, or by adding progenitors of other lineages during the process of in vitro development. We have used similar three-dimensional approaches to generate human foregut organoids capable of giving rise to lung, pancreas, and stomach tissues. We have directed differentiation of human foregut tissue specifically into stomach/gastric organoids, which we are using to model the pathogenesis induced by infection with *Helicobacter pylori*. We are also using genetically modified PSCs, or iPSCs derived from patients, to establish models of disease including cystic fibrosis and malabsorption syndromes.

Grant funding: NIH, 1U18NS080815, R01DK080823 and S1 (J.W.), 1R01DK092456 (J.W. and N.S) and a CTSA (U54 RR025216 01).

In Vitro Scalable Production of Mature, Glucose-Responsive β Cells from Human Embryonic Stem Cells for Therapy and Drug Screening

Millman, J. R., Harvard University
Pagliuca, F. W., Harvard University
Gurtler, M., Harvard University
Segel, M., Harvard University
Van Dervort, A., Harvard University
Ryu, J. H., Harvard University
Peterson, Q., Harvard University
Melton, D. A., Harvard University
Originally presented on: 3/18/2014 0:00:00 - 0:20:00

The generation of functional, insulin-secreting pancreatic β cells from stem cells *in vitro* would provide an unprecedented cell source for cell replacement therapy and drug screening in diabetes. However, insulin-producing cells generated *in vitro* from human embryonic stem cells (hESC) lack many characteristics of bona fide β cells, including glucose-stimulated insulin secretion, appropriate marker expression, and rapid function after transplantation. Furthermore, production at a scale for practical use is highly desirable and technically challenging.

We have developed an *in vitro*, suspension-based protocol that produces β cells from hESC (ES-β cells) in a 300-mL spinner flask using defined factors. ES-β cells increased intracellular [Ca^{2+}] and secreted insulin in response to multiple sequential high glucose challenges at amounts comparable to cadaveric β cells. Insulin was packaged into secretary granules in a manner that was indistinguishable from cadaveric β cells when visualized by electron microscopy. In addition to insulin, ES-β cells expressed the β cell markers PDX1, MAFA, and NKX6-1 and did not express other non-β cell markers, including glucagon, somatostatin, ghrelin, and pancreatic polypeptide. This scalable protocol produced approximately $3x10^8$ cells, of which 30-50% are ES-β cells.

Two weeks after transplantation into SCID/Beige mice, high amounts of human insulin was detected in the serum of mice that received ES-β cells, similar to that found in mice transplanted with cadaveric β cells. The amount of human insulin increased after intraperitoneal injection of glucose, demonstrating the cells were functional *in vivo*. No human insulin was detected in mice transplanted with hESC-derived pancreatic progenitors and non-functional insulin-producing endocrine, consistent with previous findings.

To further demonstrate the utility of these ES-β cells, we treated ES-β cells with different categories of known anti-diabetic drugs that increase insulin secretion and with prolactin, which increases β cell proliferation. Increasing β cell function and mass are highly sought after drug targets for treating diabetes. The categories of anti-diabetic drugs tested were GCK activators, sulfonylureas, GLP-1 agonists, and meglitinide, and, as expected, treatment increased insulin release from ES-β cells. In addition, treatment with prolactin increased proliferation of ES-β cells, as indicated by an increase in Ki67 staining.

We have developed a scalable process to generate functional ES-β cells *in vitro* that are indistinguishable from cadaveric β cells. This protocol is able to produce >10^8 ES-β cells per batch, meaning it is the first scalable human source of functional β cells for therapy and drug discovery. Given that the only alternative is the very limited and unreliable supply of cadaveric β cells, this discovery represents a major advance in stem cell engineering for treating diabetes.

A Synergistic Biomaterial/Pharmacologic Approach to Rejuvenate the Aged Muscle Stem Cell Population

Gilbert, P.,

Originally presented on: 3/18/2014 0:00:00 - 0:20:00

Impaired muscle strength and regeneration are major problems of aging. We found that skeletal muscle stem cells (MuSCs) from aged mice are two-thirds less effective in regenerating muscle than young MuSCs, even in a young microenvironment. Culture of aged MuSCs in hydrogel niches in conjunction with a pharmacological inhibitor of p38 mitogen-activated protein kinase, induced MuSC self-renewal in culture and decreased the proportion expressing p21 and $p16^{Ink4a}$. The capacity of this population to regenerate damaged muscles was similar to young MuSCs following transplantation into mice, evaluated by non-invasive imaging, stem-cell repopulation, and serial transplantation assays. The rejuvenation of the aged muscle stem cell population was attributable to expansion of a functional subpopulation, as revealed by time-lapse microscopy. Molecular analyses identified a mechanism whereby substrate mechanics contributes to the modulation of MuSC stemness. Notably, strength was restored when the rejuvenated population was delivered to damaged muscles of aged mice. These findings suggest that a localized autologous muscle stem cell therapy post-injury could benefit aged individuals.

Amniotic Fluid Stem Cell Derived Endothelial Cells as a Vascular Source for Tissue Engineering

Jeffrey G. Jacot, Ph.D.

Assistant Professor of Bioengineering
Rice University
Director, Cardiac Bioengineering Lab
Texas Children's Hospital

RICE Texas Children's Hospital BCM Baylor College of Medicine

Heart defects are the most common fatal birth defect.

14,000 children/year have heart failure, increasing by 25% every 3 years.

Most surgeries bridge to the next surgery. A heart transplant is the only cure.

The waiting list for transplants is increasing, available hearts have been steady for >10 years.

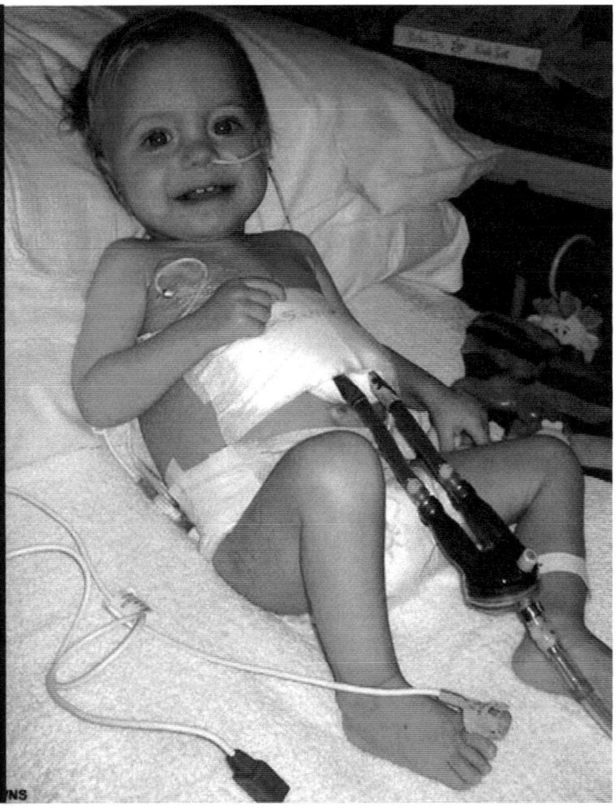

Amniotic Fluid Derived Stem Cells

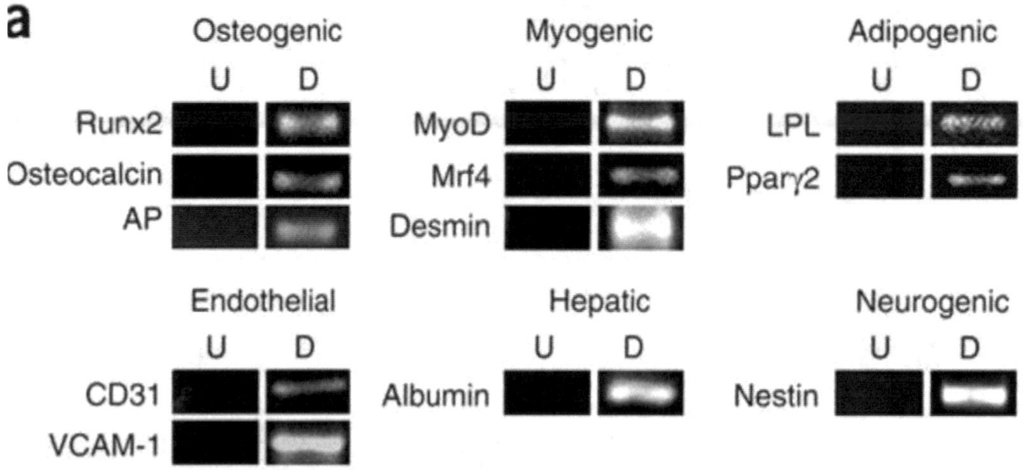

Di Cotta et al. 2007; Chiavegato et al. 2007

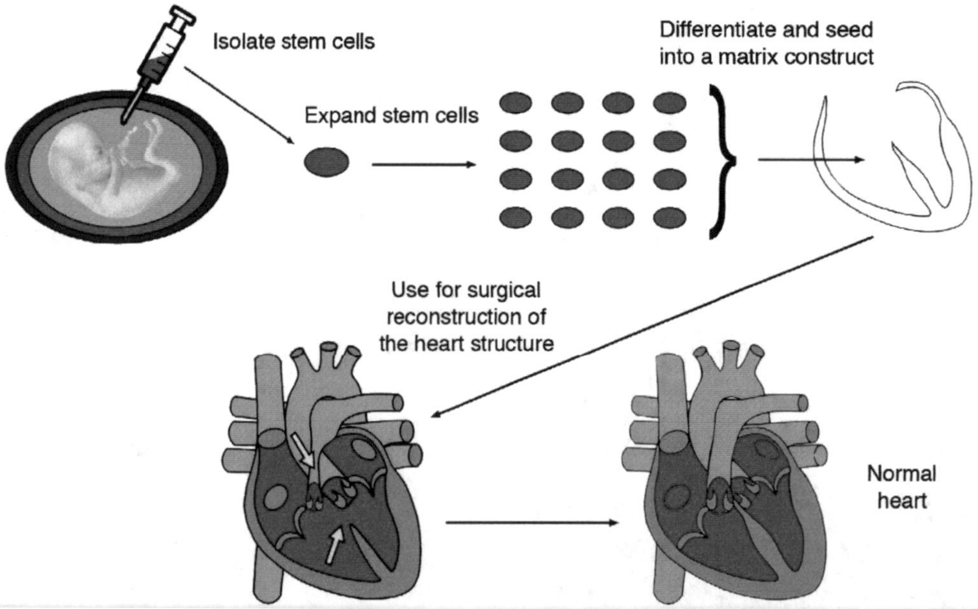

We have analyzed primary AFSC from amniotic fluid

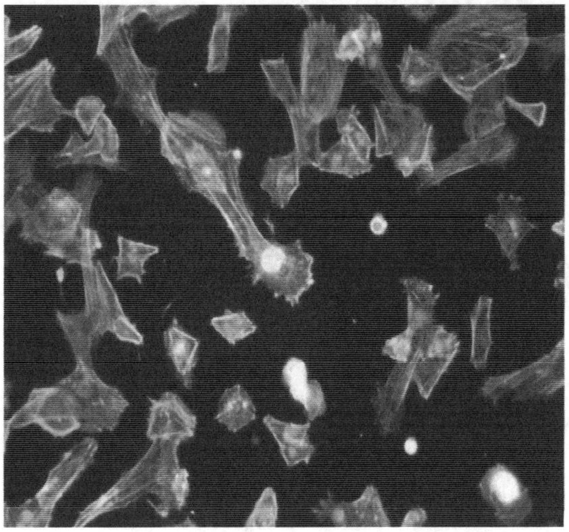

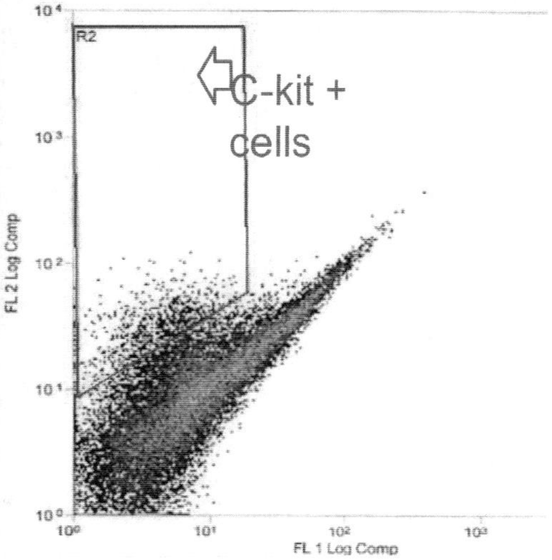

Cells are sorted for c-kit⁺ (~3%)

AFSC are karyotypically stable over 4 passages, and genetically matched to the fetus

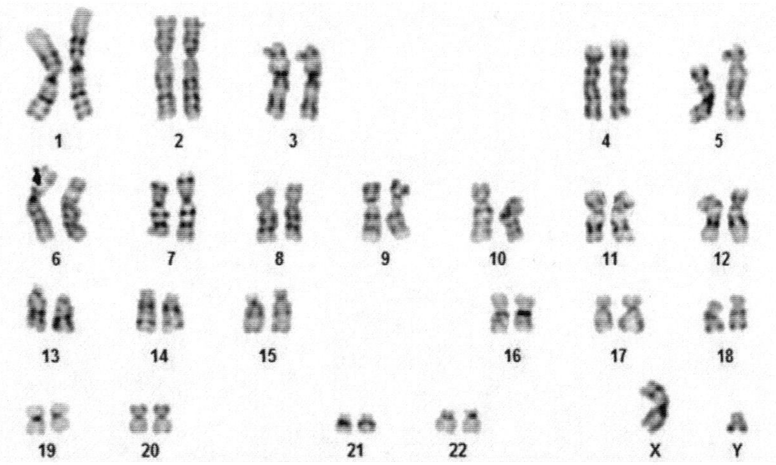

Benavides OM et al. *Tissue Engineering Part A*. 2012

Pre-Differentiation Analysis AFSC Sorted for c-kit$^+$ and passaged

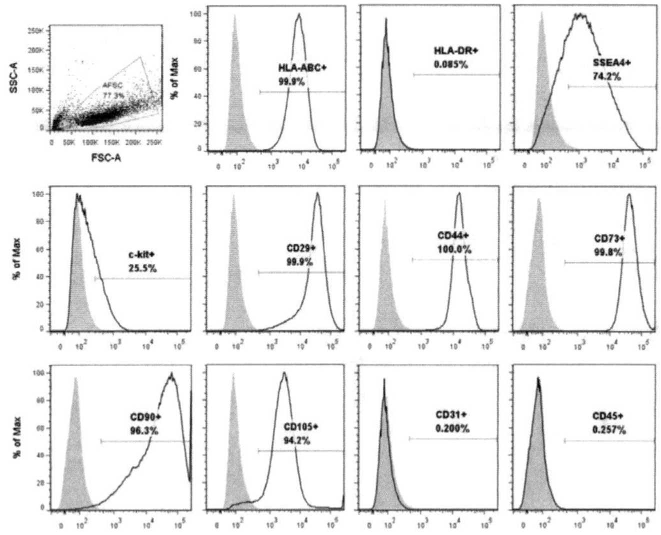

- Immunological Markers
  - HLA-ABC+
  - HLA-DR
- ESC Markers
  - SSEA4+
- HSC Markers
  - C-kit+
- MSC Markers
  - C-kit+
  - CD29+
  - CD44+
  - CD73+
  - CD90+
  - CD105+
- Differentiation Markers
  - CD31
  - CD45

Benavides OM et al. *Tissue Engineering Part A*. 2012

Pre-Differentiation Analysis AFSC Sorted for c-kit$^+$ and passaged

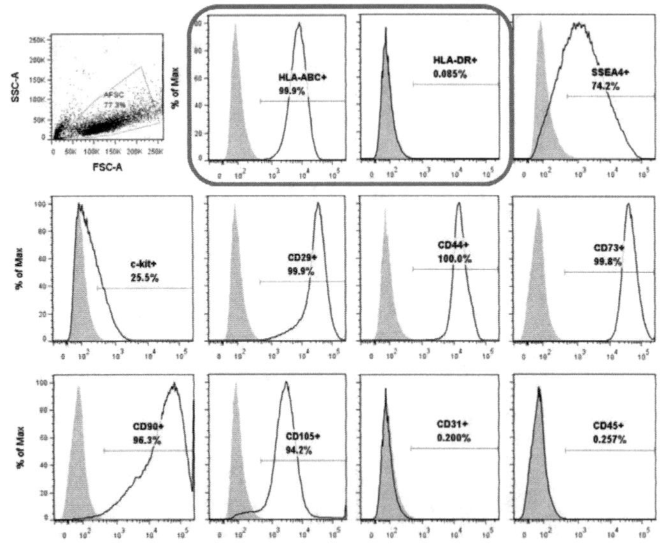

- Immunological Markers
  - HLA-ABC+
  - HLA-DR
- ESC Markers
  - SSEA4+
- HSC Markers
  - C-kit+
- MSC Markers
  - C-kit+
  - CD29+
  - CD44+
  - CD73+
  - CD90+
  - CD105+
- Differentiation Markers
  - CD31
  - CD45

Benavides OM et al. *Tissue Engineering Part A*. 2012

Pre-Differentiation Analysis AFSC
Sorted for c-kit$^+$ and passaged

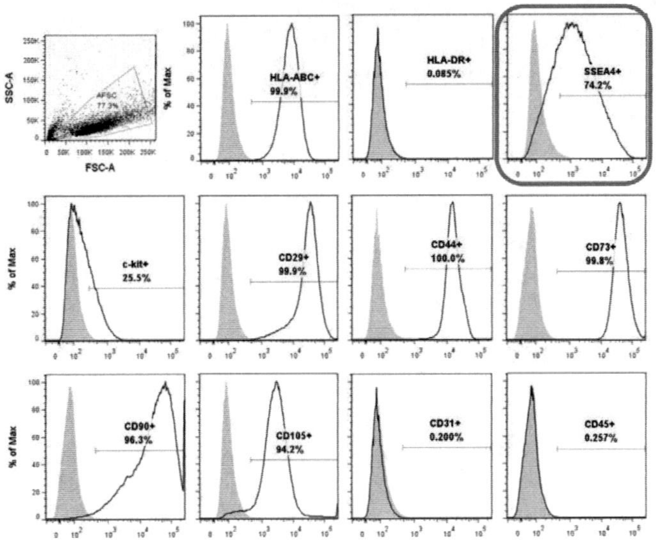

- Immunological Markers
  - HLA-ABC+
  - HLA-DR
- ESC Markers
  - SSEA4+
- HSC Markers
  - C-kit+
- MSC Markers
  - C-kit+
  - CD29+
  - CD44+
  - CD73+
  - CD90+
  - CD105+
- Differentiation Markers
  - CD31
  - CD45

Benavides OM et al. *Tissue Engineering Part A*. 2012

Pre-Differentiation Analysis AFSC
Sorted for c-kit$^+$ and passaged

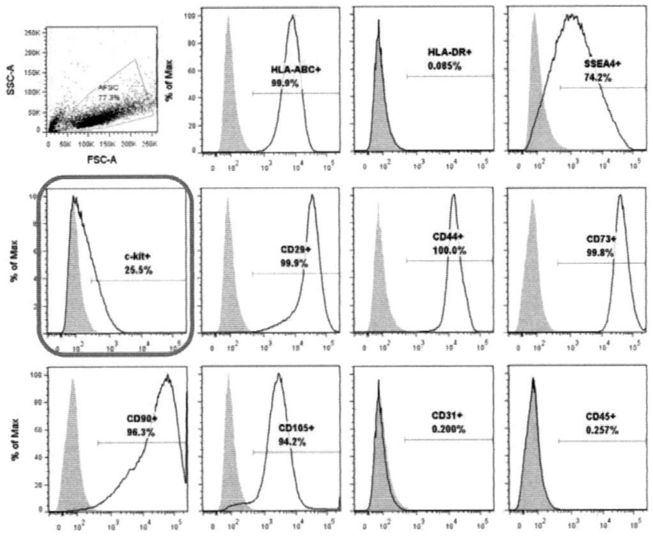

- Immunological Markers
  - HLA-ABC+
  - HLA-DR
- ESC Markers
  - SSEA4+
- HSC Markers
  - C-kit+
- MSC Markers
  - C-kit+
  - CD29+
  - CD44+
  - CD73+
  - CD90+
  - CD105+
- Differentiation Markers
  - CD31
  - CD45

Benavides OM et al. *Tissue Engineering Part A*. 2012

Pre-Differentiation Analysis AFSC Sorted for c-kit+ and passaged

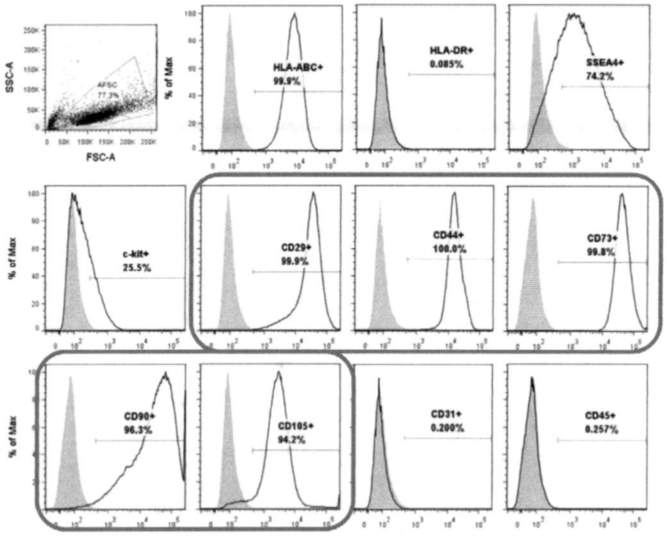

- Immunological Markers
  - HLA-ABC+
  - HLA-DR
- ESC Markers
  - SSEA4+
- HSC Markers
  - C-kit+
- MSC Markers
  - C-kit+
  - CD29+
  - CD44+
  - CD73+
  - CD90+
  - CD105+
- Differentiation Markers
  - CD31
  - CD45

Benavides OM et al. *Tissue Engineering Part A*. 2012

Pre-Differentiation Analysis AFSC Sorted for c-kit+ and passaged

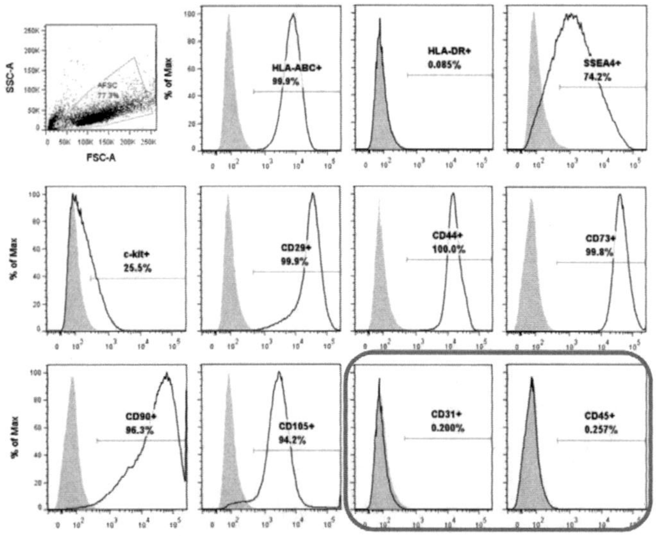

- Immunological Markers
  - HLA-ABC+
  - HLA-DR
- ESC Markers
  - SSEA4+
- HSC Markers
  - C-kit+
- MSC Markers
  - C-kit+
  - CD29+
  - CD44+
  - CD73+
  - CD90+
  - CD105+
- Differentiation Markers
  - CD31
  - CD45

Benavides OM et al. *Tissue Engineering Part A*. 2012

Pluripotency Markers in AFSC

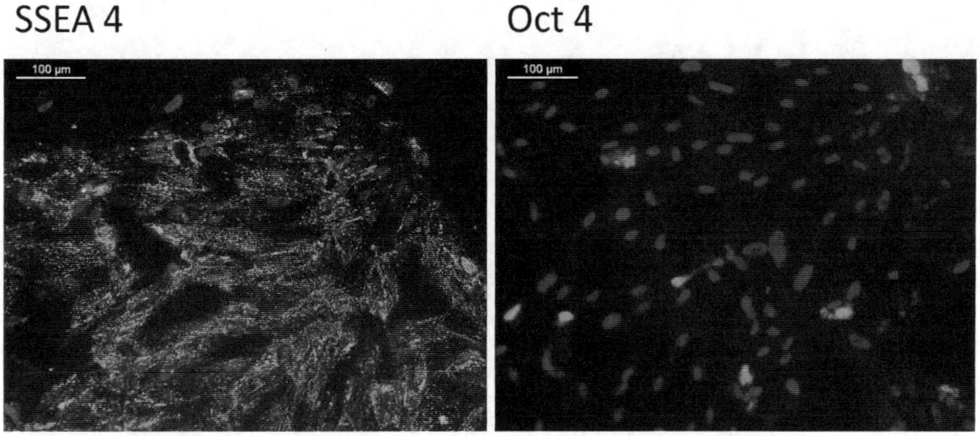

At passage 4, cells are 100% SSEA4 positive and 10% Oct4 positive

Benavides OM et al. *Tissue Engineering Part A*. 2012

Transmembrane coculture with rat cardiac myocytes leads to increased cell coupling

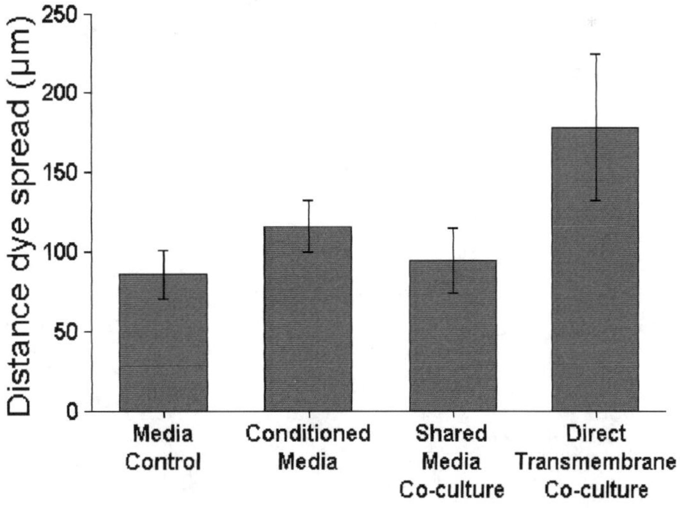

* p ≤ 0.001

Petsche Connell JJ et al. *Journal of Cellular and Molecular Medicine*. 2013

Human AFSC in contact with *human* myocytes express cardiac markers

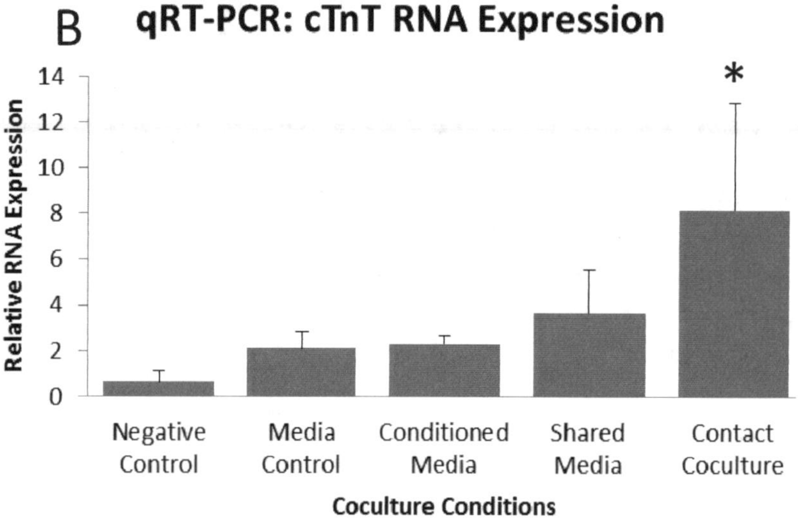

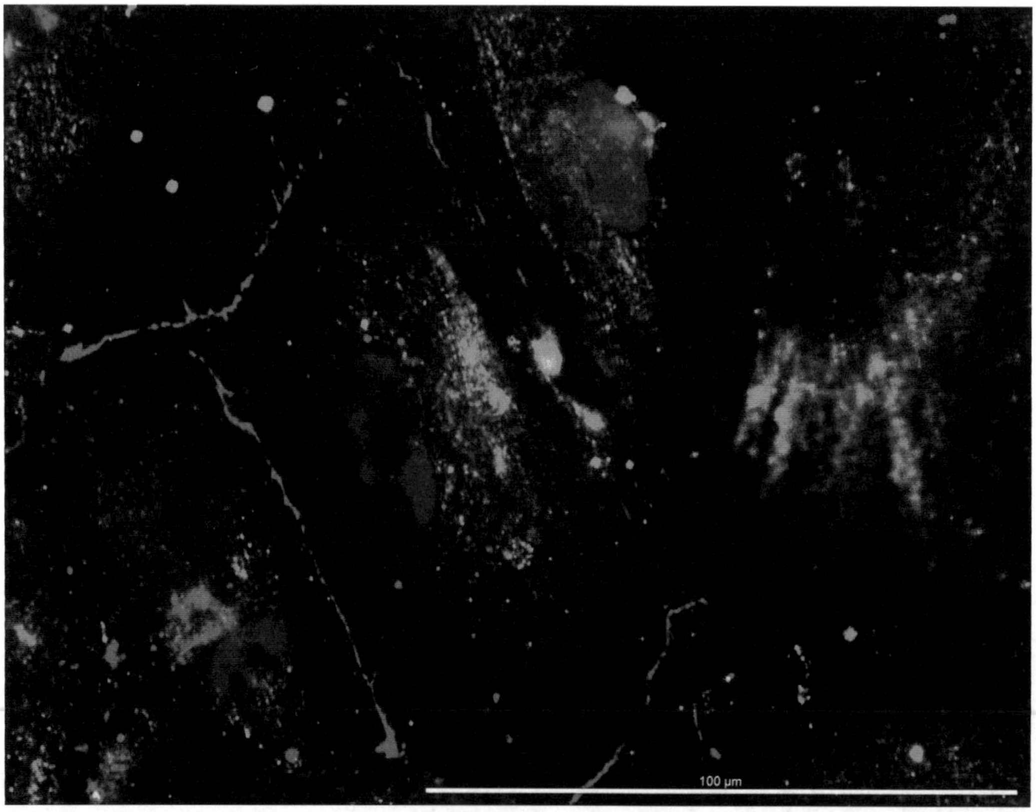

Endothelial differentiation of amniotic fluid stem cells

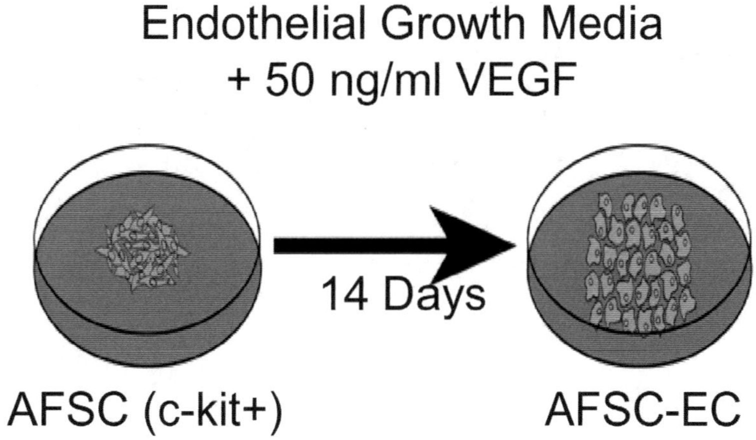

Benavides OM et al. *Tissue Engineering Part A*. 2012

Flow Cytometry for CD31/PECAM-1

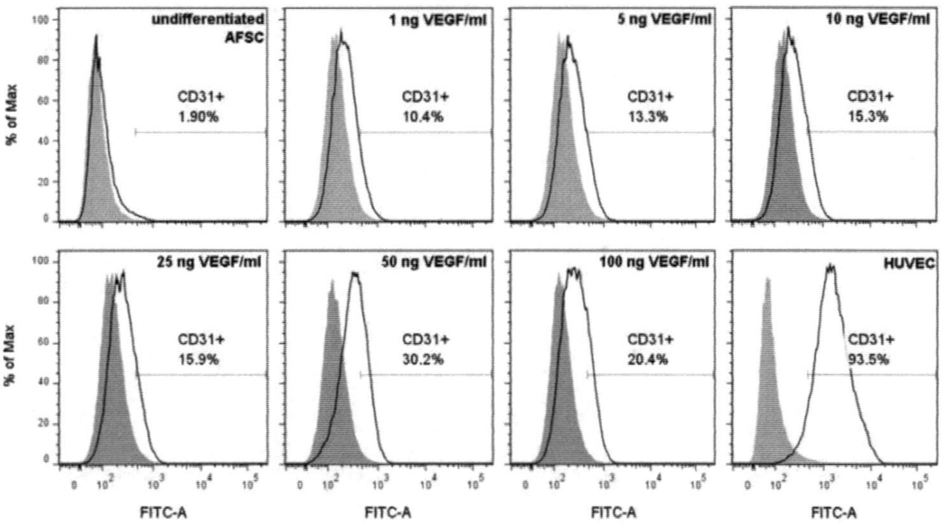

Benavides OM et al. *Tissue Engineering Part A*. 2012

AFSC-Endothelial Cells are triple positive for VE-Cadherin, VEGFR2 and CD31

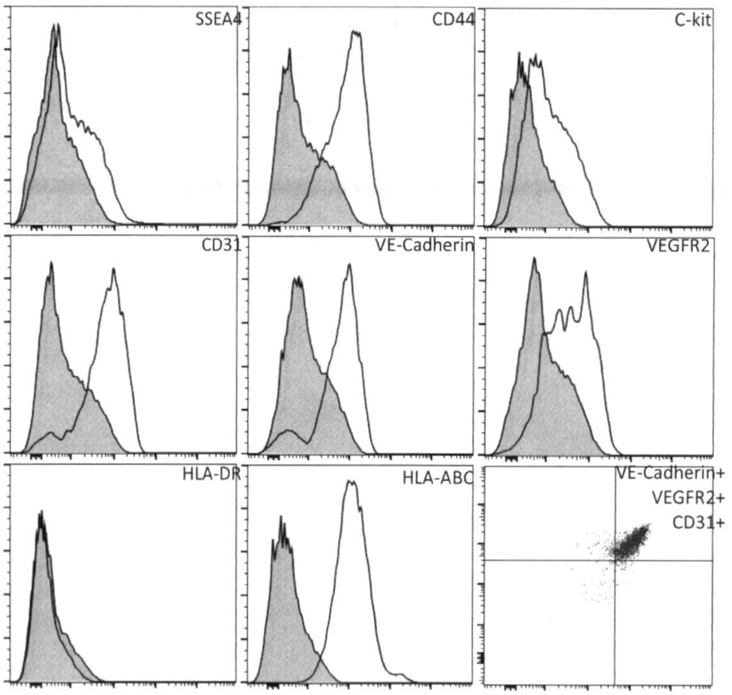

vWF/eNOS Immunofluorescence

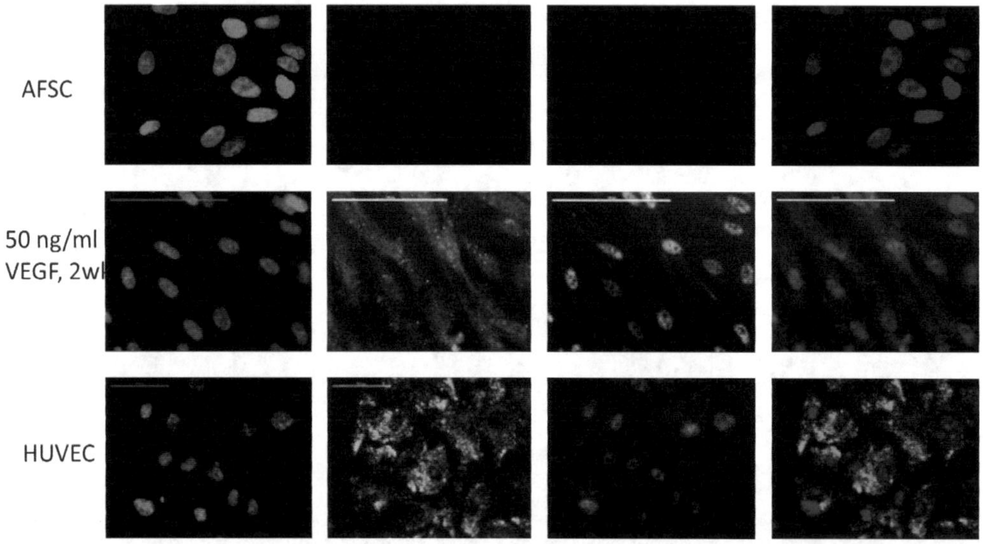

AFSC – negative control; HUVEC – positive control; vWF (green), eNOS (red), DAPI (blue). Scale bars are 100 um and 50 μm

Benavides OM et al. *Tissue Engineering Part A*. 2012

eNOS expression and activity

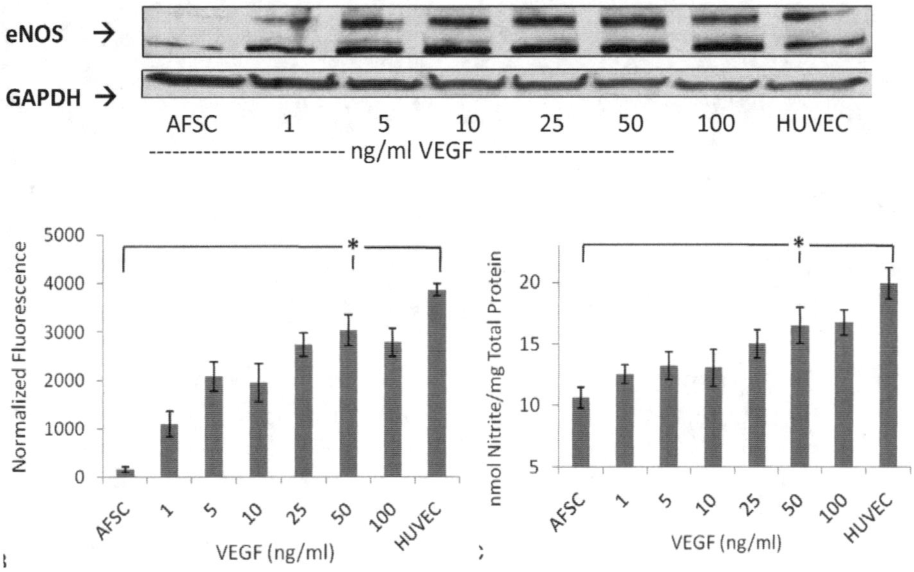

Benavides OM et al. *Tissue Engineering Part A*. 2012

Ac-LDL Uptake

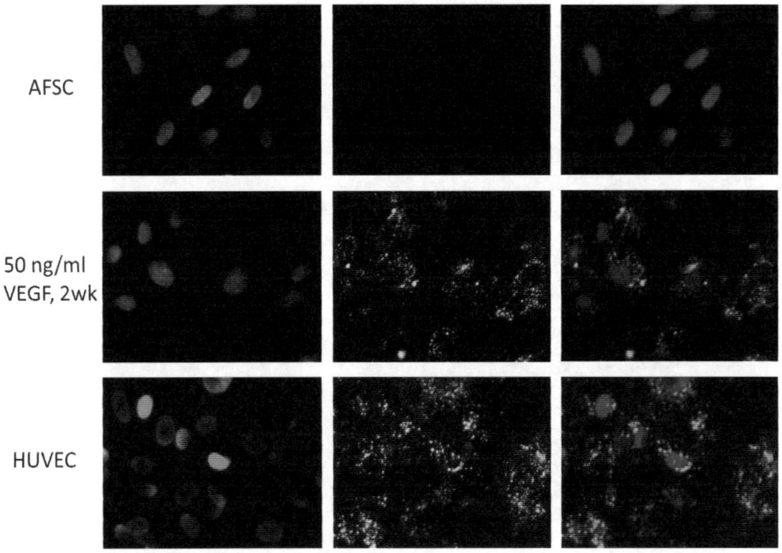

AFSC – negative control; HUVEC – positive control;
Ac-LDL(green/red), DAPI (blue).

Benavides OM et al. *Tissue Engineering Part A*. 2012

Network formation on Matrigel

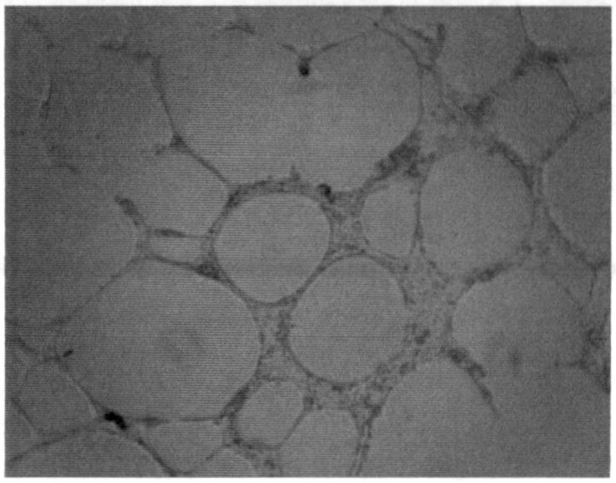

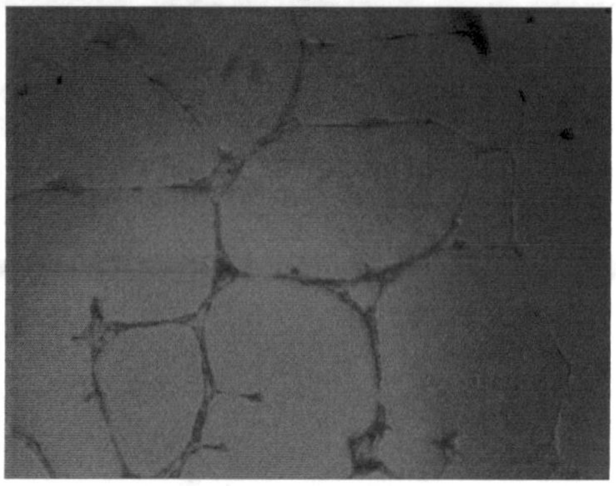

HUVEC

50 ng/ml VEGF, 2 wks

Benavides OM et al. *Tissue Engineering Part A*. 2012

Fibrin/PEG Gel

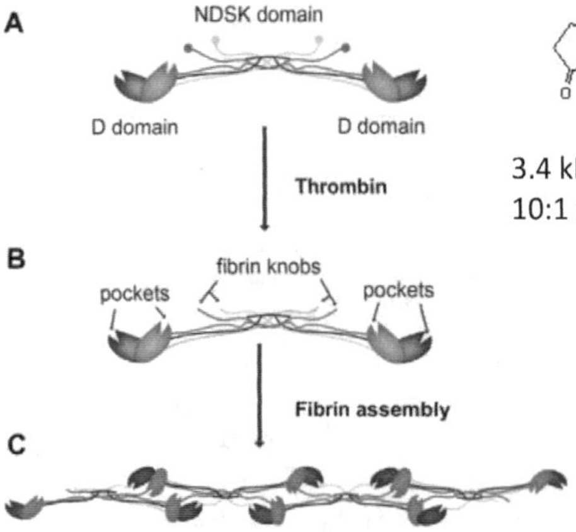

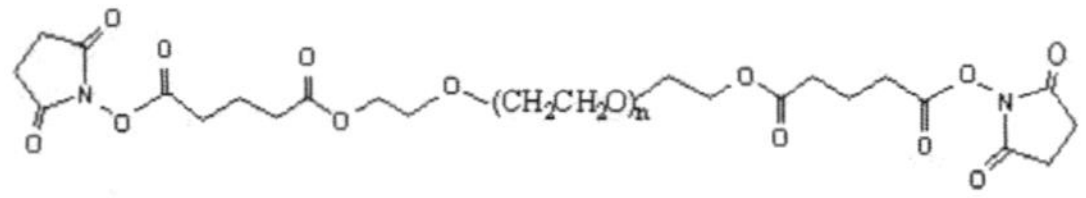

3.4 kDa poly(ethylene glycol)-succinimidyl glutarate
10:1 ratio of fibrinogen:PEG

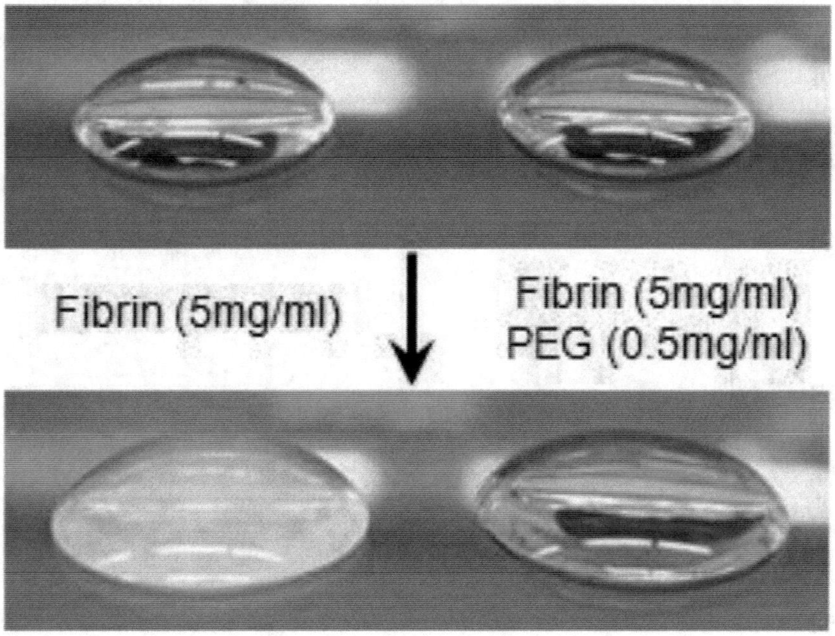

Gel Characterization

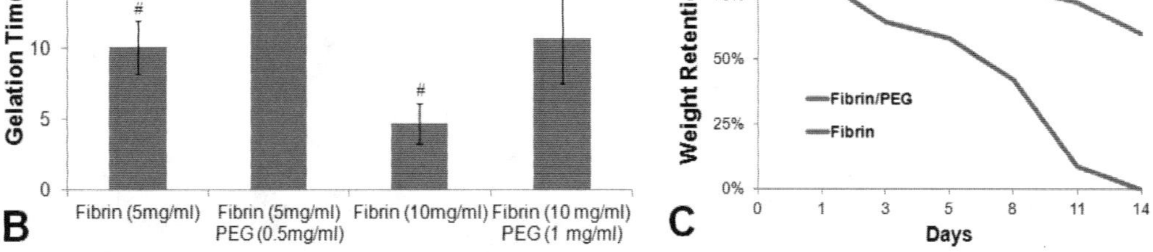

Gel Characterization

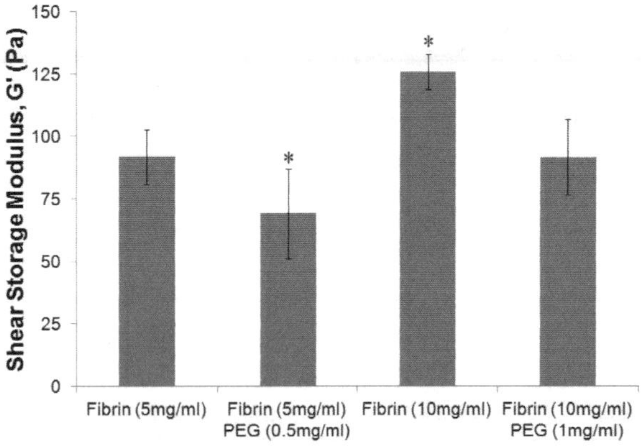

F

G

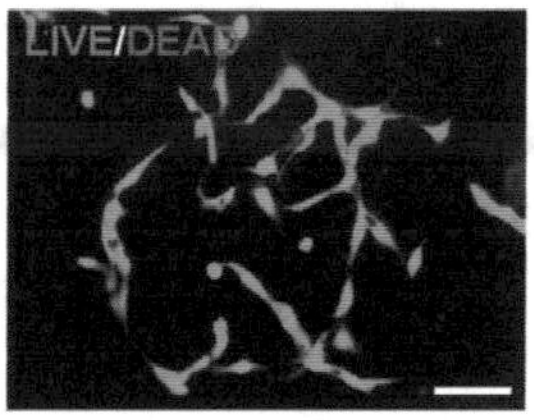

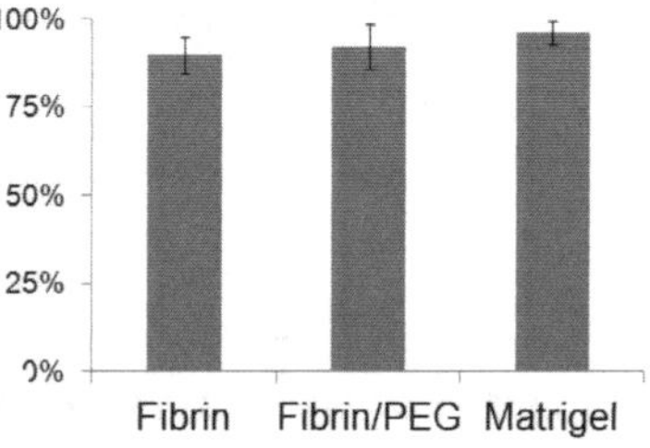

H

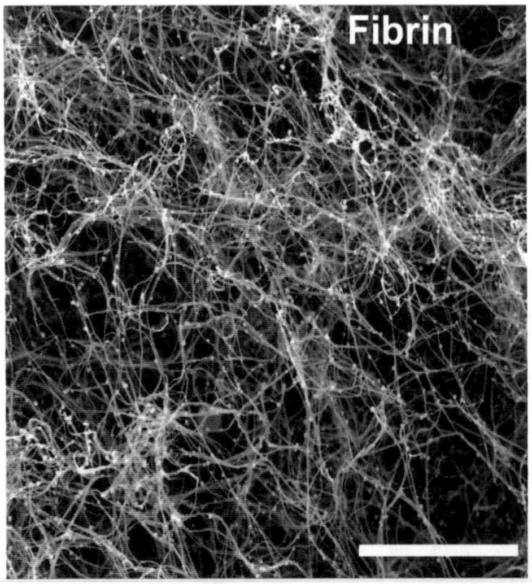

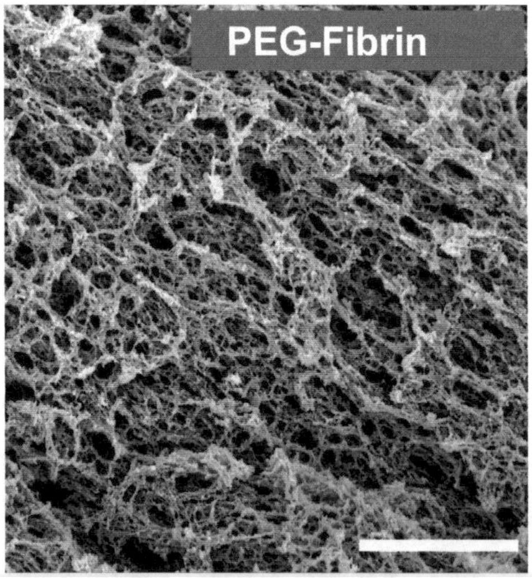

Co-Cultures on Fibrin/PEG

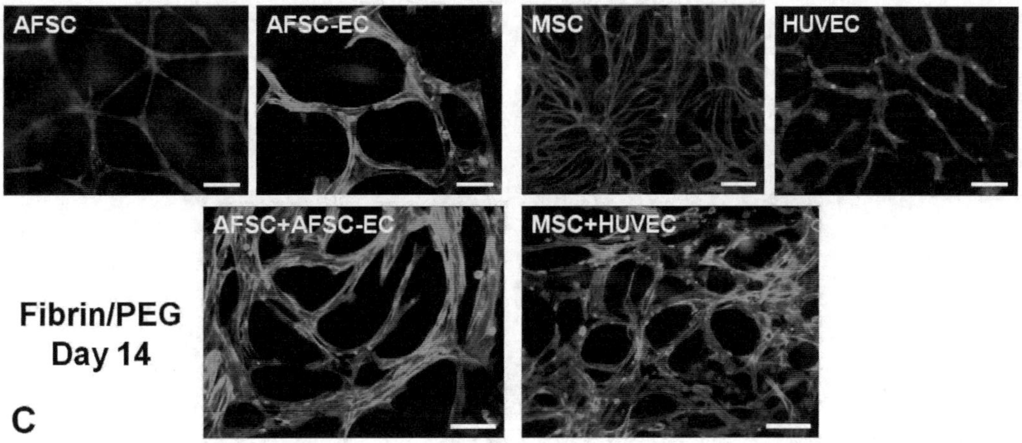

Network Analysis

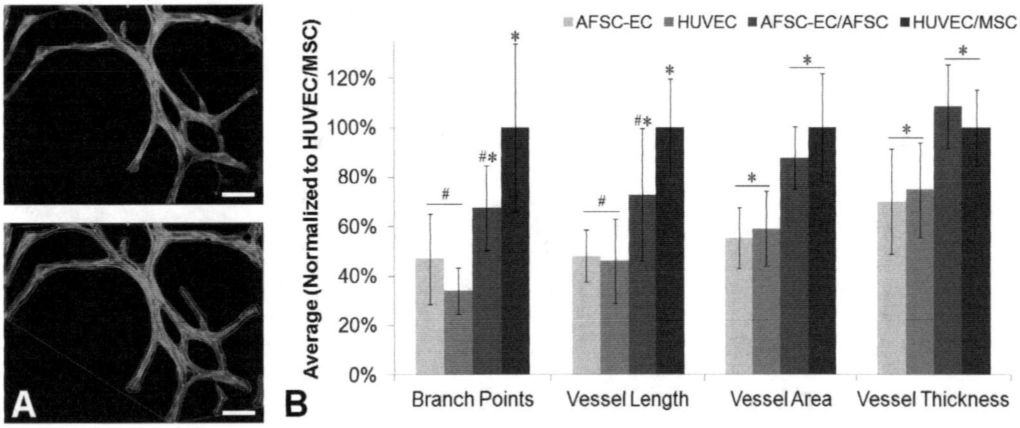

Histology of AFSC/AFSC-EC Gels at 14 days suggests open lumens

Right Ventricle Outflow Patch

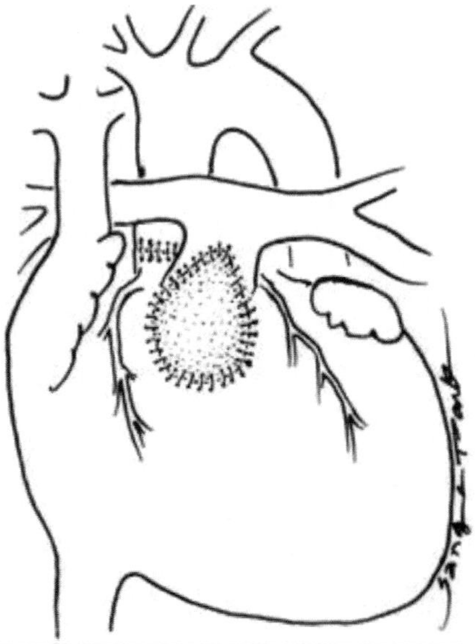

State of the art materials

Dacron

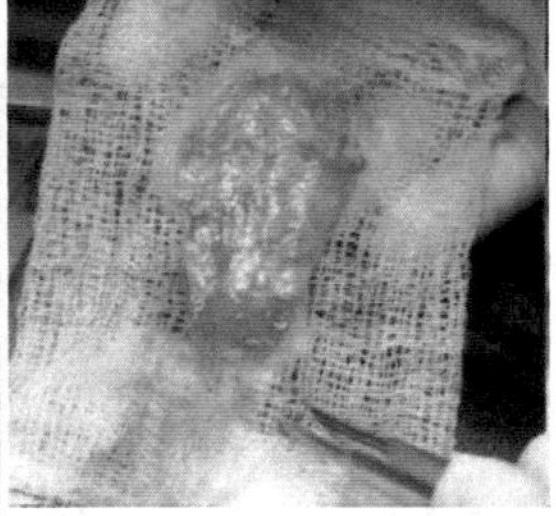

Fixed Pericardium

Both are nonfunctional tissues that do not act as native myocardium.

Heart Matrix Patch

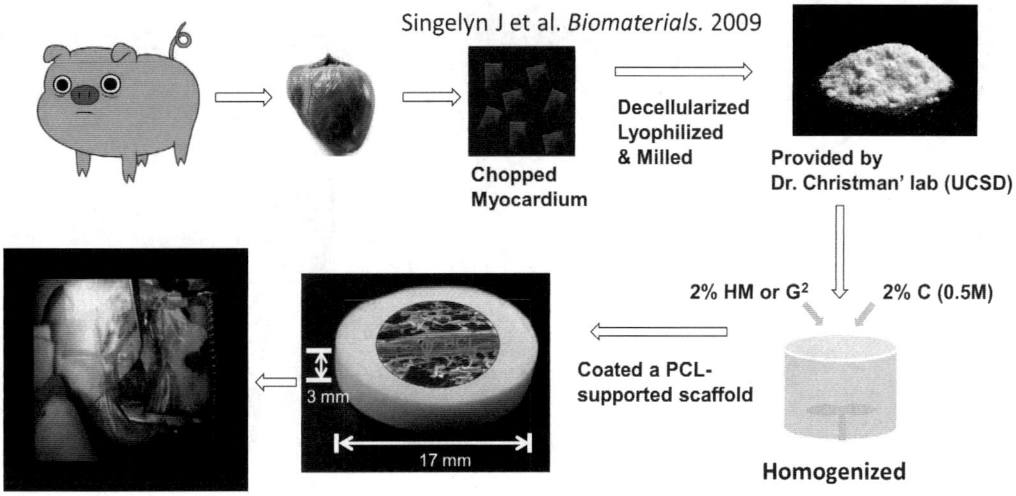

Pok SW et al. *Journal of Cardiovascular Translational Research.* 2011
Pok SW et al. *Acta Biomaterialia.* 2013
Pok SW et al. *Tissue Engineering.* 2014

Conduction velocity

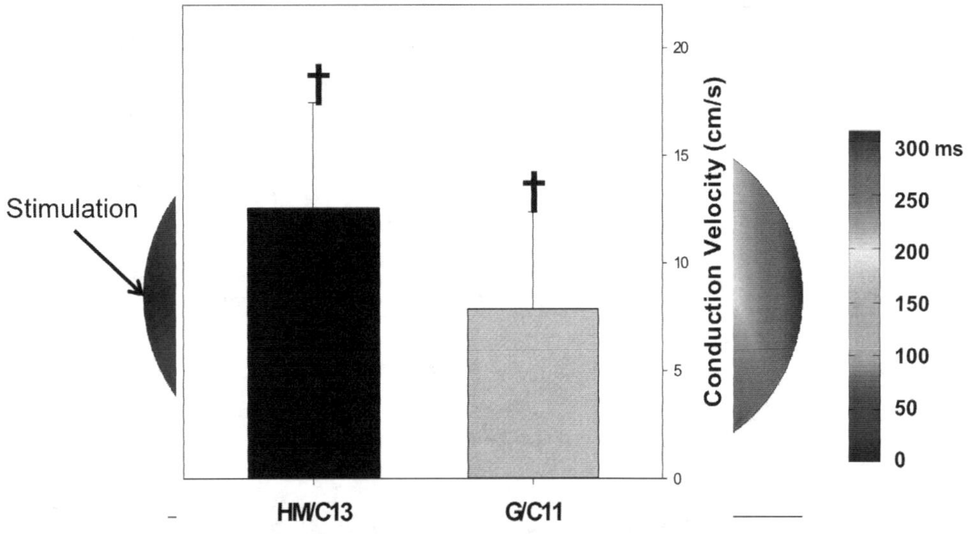

Pok SW et al. *Tissue Engineering Part A.* 2014

Contractile Force

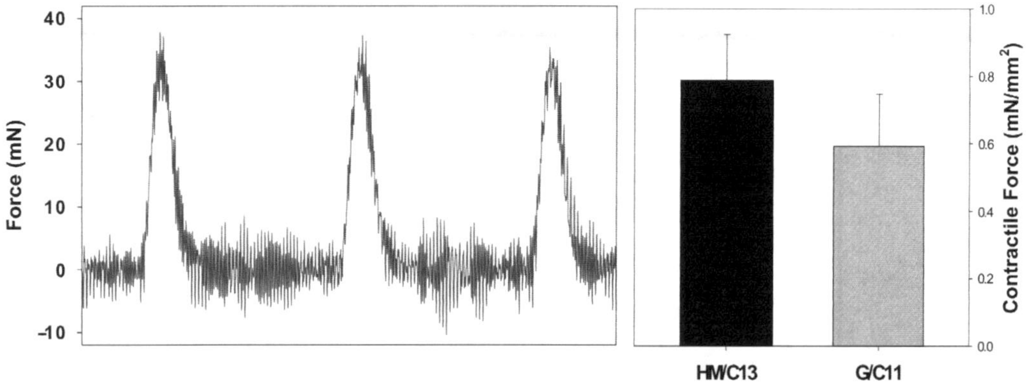

Pok SW et al. *Tissue Engineering Part A*. 2014

Pediatric Cardiac Bioengineering Lab at Texas Children's Hospital

Omar Benavides
Jen Petsche Connell
Yang Gao
Chris Tsao
Seokwon Pok
Jackson Myers
Emily Augustini
Sean Boothe
Patrick Hallal
Tomi Adetiba
Huiying Chen
A.J. Mao
Abraham Segura
Adriana Gamboa
Carolina White
Amy Liao
Sarah Mason

Texas Children's Hospital
Dr. Charles Fraser, Jr.
Dr. Rodrigo Ruano

Rice University
Dr. K. Jane Grande-Allen
Dr. Joel Moake
Dr. Antonios Mikos

Baylor College of Medicine
Dr. John Belmont
Dr. Dimiter Bissig

NSF
National Institutes of Health
RICE IBB Institute of Biosciences and Bioengineering
American Heart Association Learn and Live
Texas Children's Hospital
Collaborative Research Fund
Funded by the Virginia and L.E. Simmons Family Foundation

TÉCNICO LISBOA

INSTITUTE FOR BIOTECHNOLOGY AND BIOENGINEERING

PURIFICATION OF HUMAN INDUCED PLURIPOTENT STEM CELL-DERIVED NEURAL PROGENITORS FOR REGENERATIVE MEDICINE APPLICATIONS

Tiago Fernandes

Gonçalo Rodrigues, Andreia Matos, Maria Margarida Diogo and Joaquim M. S. Cabral

March 18th, 2014

Introduction
Objectives
Results
Conclusions

Human Induced Pluripotent Stem Cells (hiPSCs)

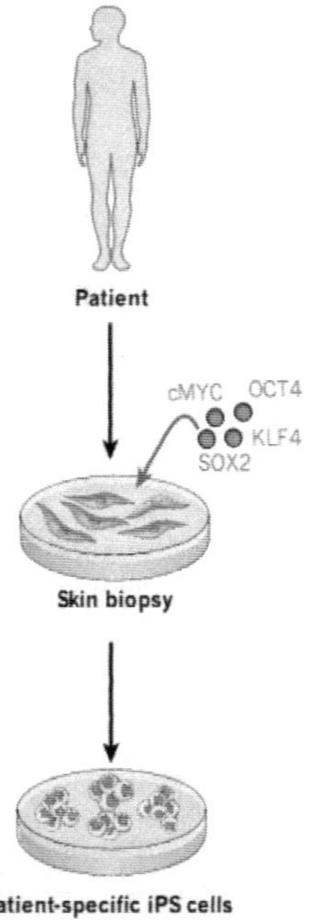

Source: Robinton, D. a & Daley, G. Q. *Nature* **481**, 295–305 (2012)

2

Introduction
Objectives
Results
Conclusions

Neural Commitment of hPSCs

Dual SMAD-inhibition Protocol – hESC-derived neural progenitor population, with more than 80 % of neural conversion, in monolayer and feeder-free cultures.

Dual SMAD-inhibition Protocol – an efficient and relatively fast neural induction protocol which adds the possibility for higher plating densities resulting in much higher absolute numbers

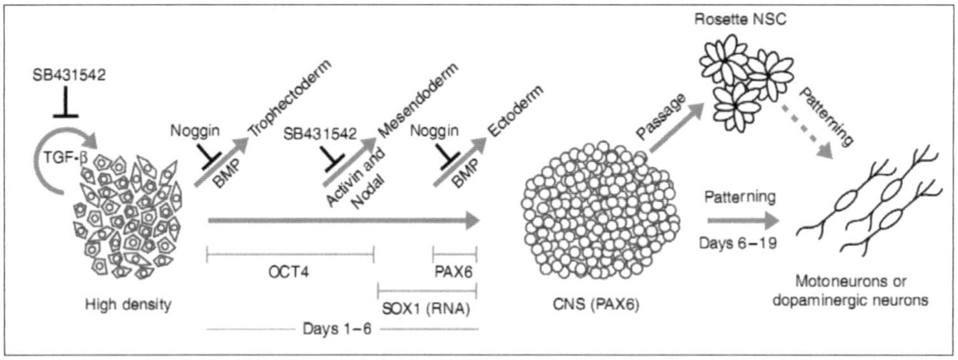

Source: Chambers, S. *et al.*
Nature biotechnology **27**, 275–280 (2009).

3

Introduction
Objectives
Results
Conclusions

hPSCs are Potentially Tumorigenic

➢ hPSCs have the ability to form teratomas after transplantation

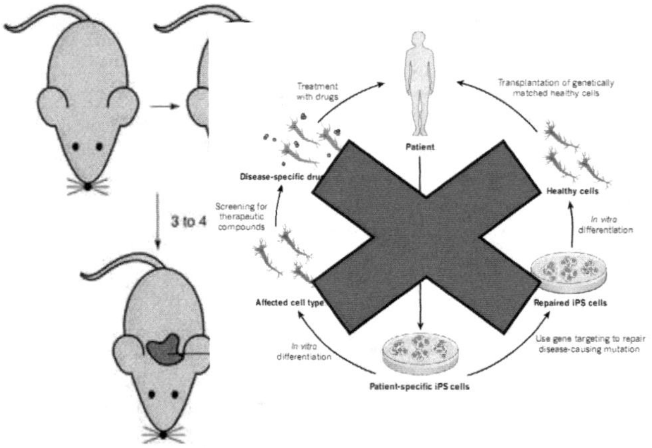

Source: Zhonghan Li, Tariq M. Rana. *Current Protocols in Stem Cell Biology* 4A.4 (2012).

4

Separation/Purification of hPSC-derived cells

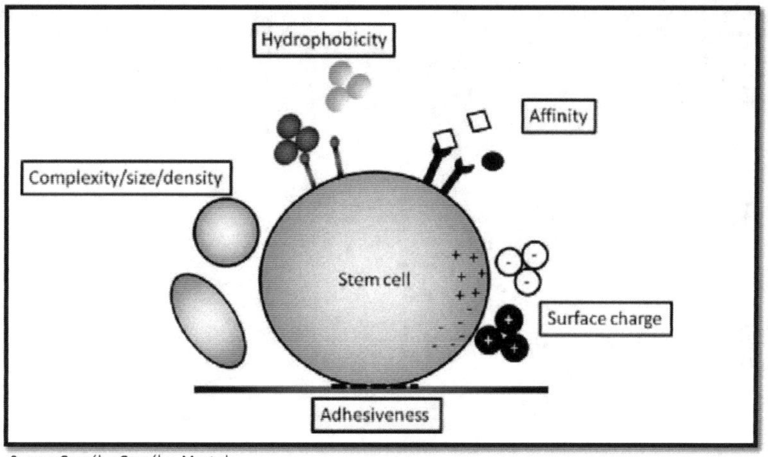

Source: González-González, M. et al.
J Chem Technol Biotechnol (2011).

➢ Separation techniques based on affinity properties are pointed as a possible solution for the complete separation of hPSCs

The 2 Objectives of this Project

The establishment of a robust and reproducible method for **neural differentiation of hPSCs** into neural progenitors, under chemically defined ... ing reco... ens/growth factors, a... **ed cultu...**

The development of a separation platform for the **preparative purifica...** differentiated cultures, by making the **depleti...** the remaining **hiPSCs**, present... cultures

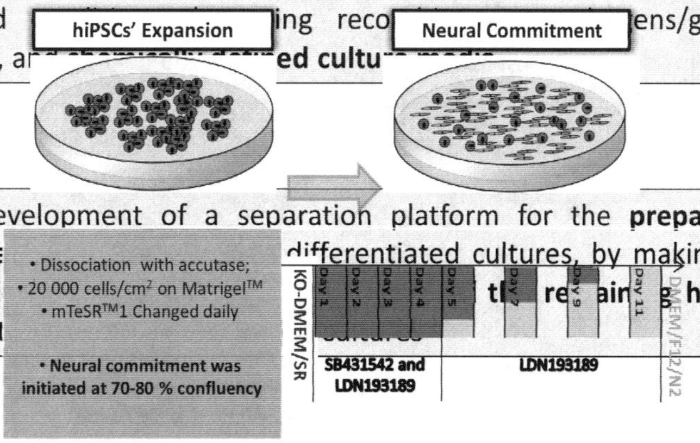

Introduction
Objectives
Results
Conclusions

Comparison of hiPSC Expansion and Neural Commitment

hiPSCs

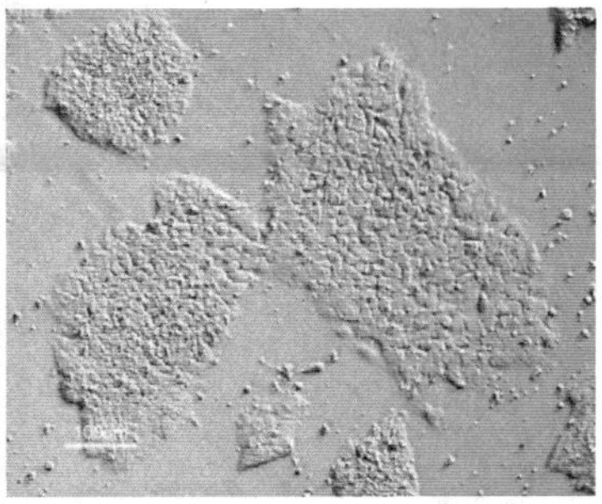

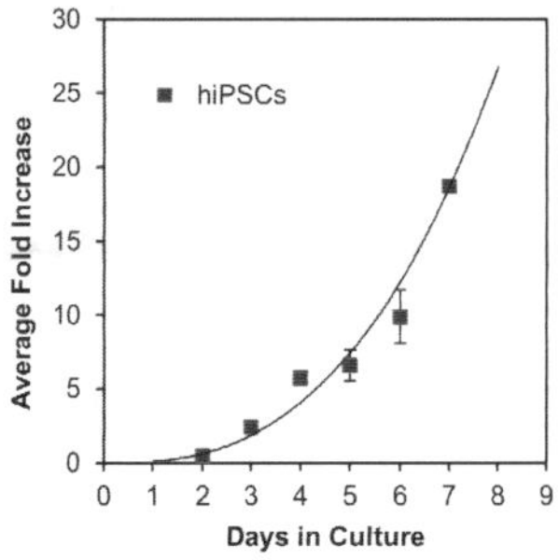

$\overline{\mu}$	$0.7 \pm 0.2\ day^{-1}$
$\overline{T_d}$	$23 \pm 4.7\ h$

hiPSC-derived Cells

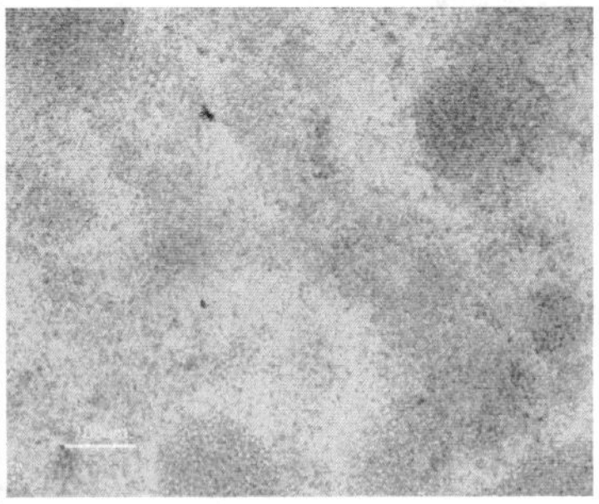

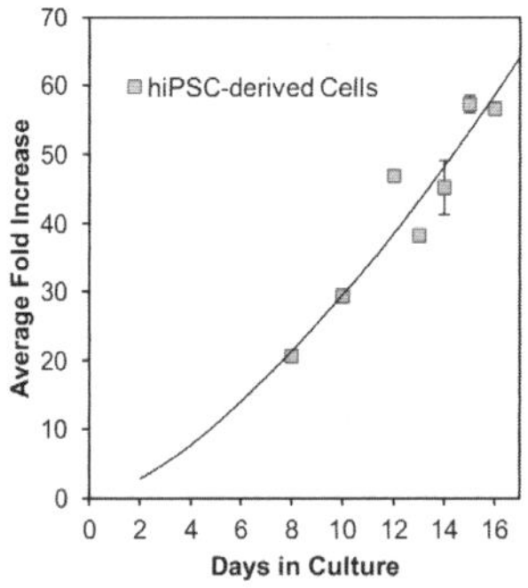

$\overline{\mu}$	$0.4 \pm 0.1\ day^{-1}$
$\overline{T_d}$	$48 \pm 6.1\ h$

Introduction
Objectives
Results
Conclusions

Comparison of hiPSC Expansion and Neural Commitment

hiPSCs

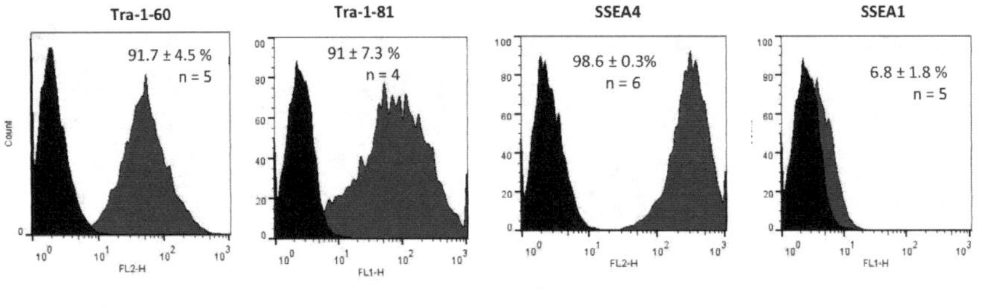

hiPSC-derived Cells

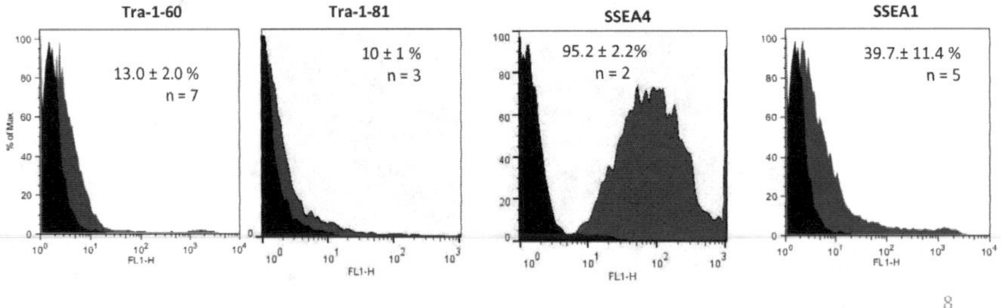

Introduction
Objectives
Results
Conclusions

Comparison of hiPSC Expansion and Neural Commitment

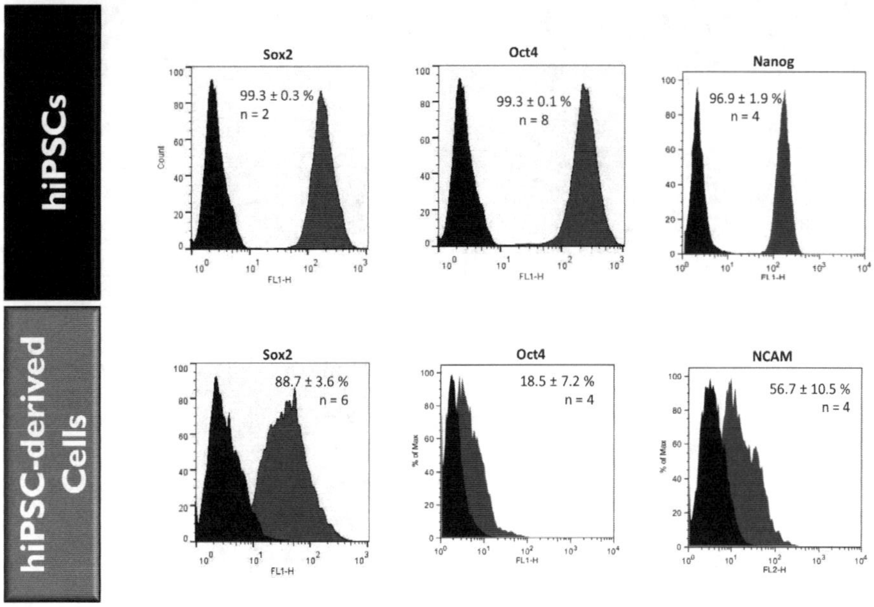

9

Introduction
Objectives
Results
Conclusions

Comparison of hiPSC Expansion and Neural Commitment

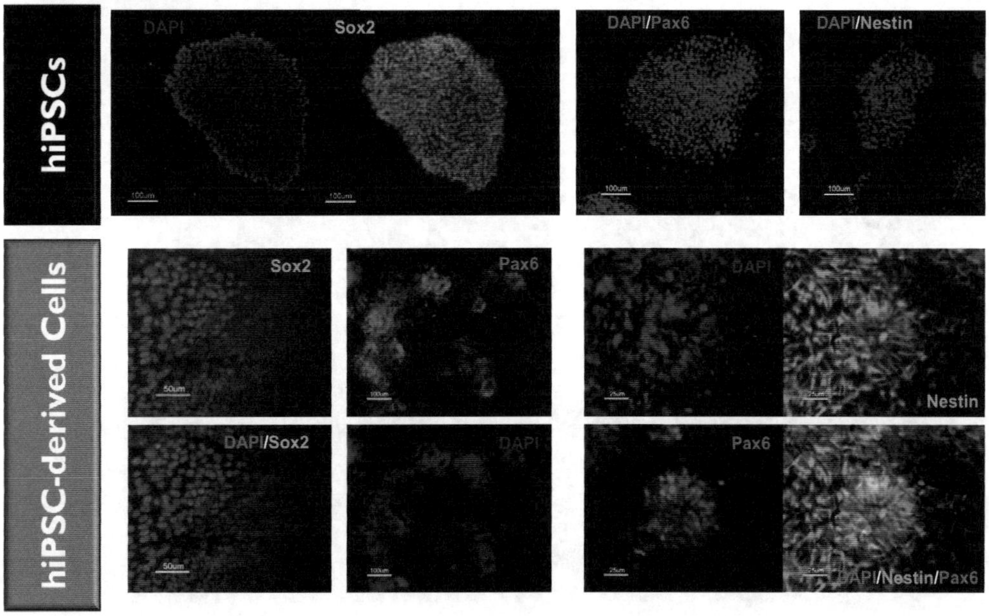

Introduction
Objectives
Results
Conclusions

Comparison of hESC Expansion and Neural Commitment

hESCs

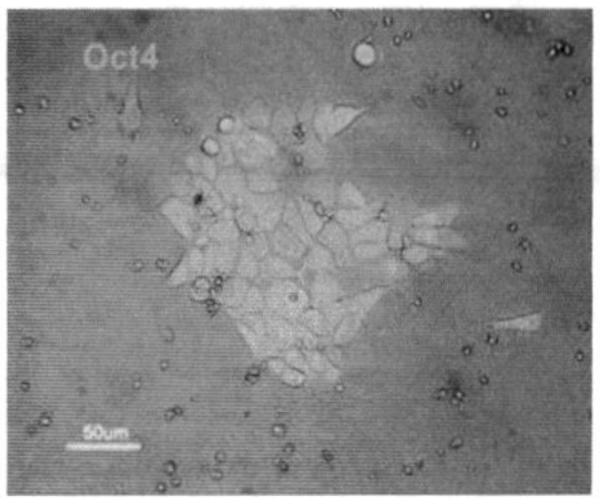

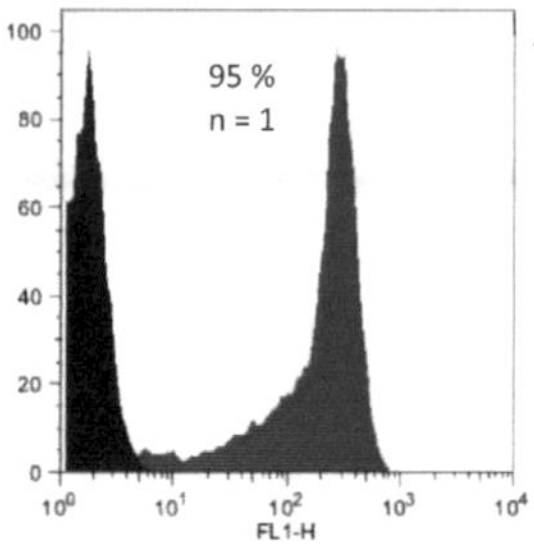

Oct4

hESC during Neural Commitment

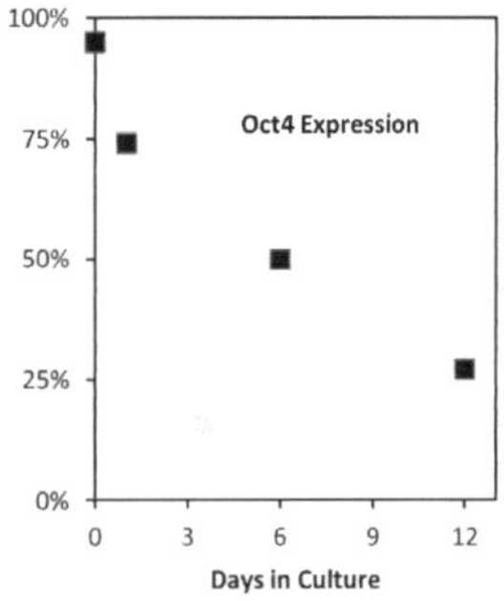

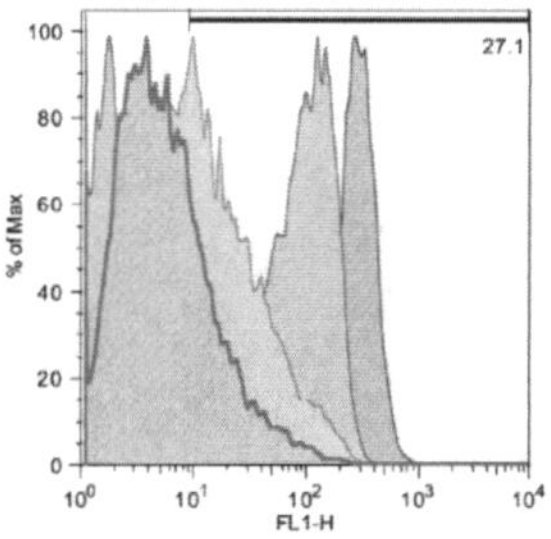

11

Introduction
Objectives
Results
Conclusions

Comparison of hiPSC Expansion and Neural Commitment

hiPSCs

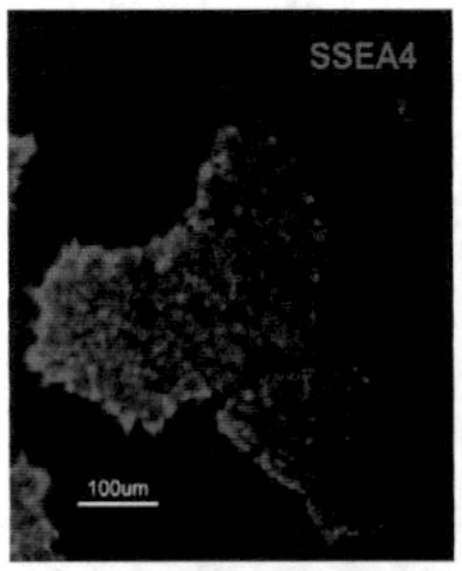

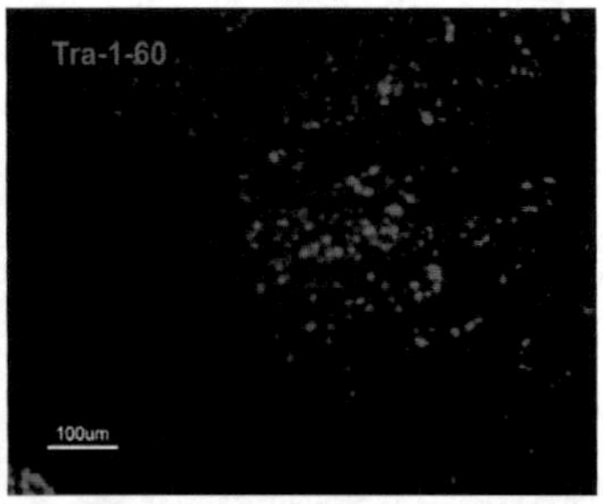

hiPSC-derived Cells

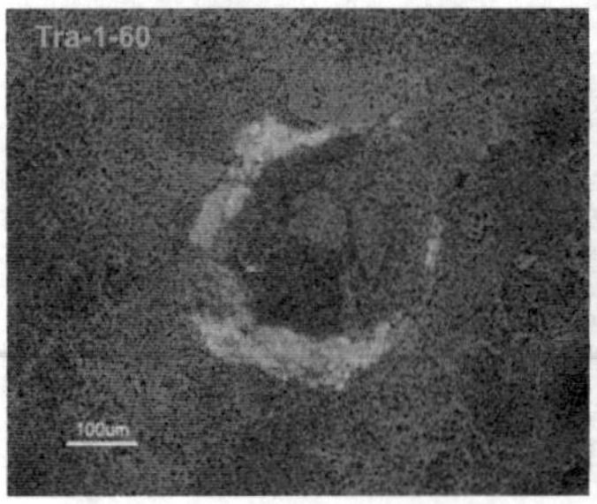

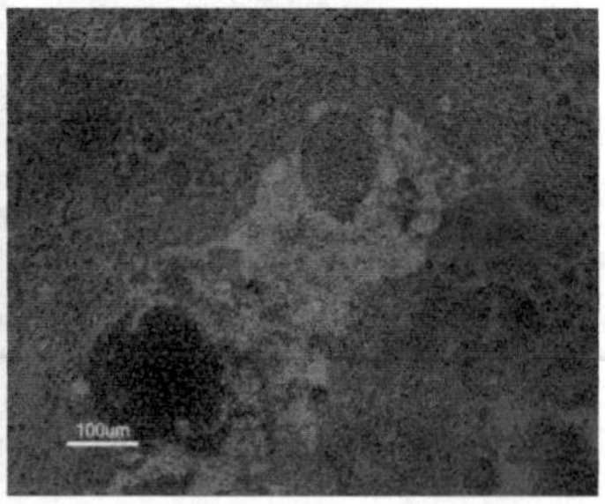

12

Comparison of hiPSC Expansion and Neural Commitment

hiPSCs

hiPSC-derived Cells

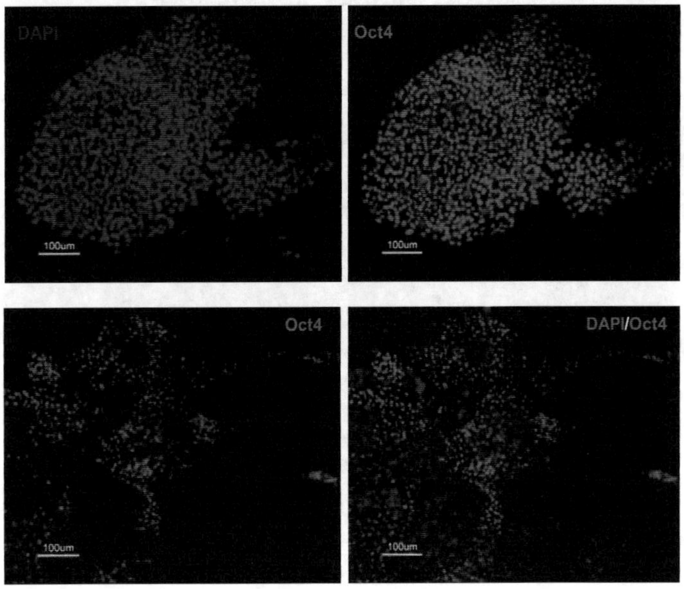

13

Separation Strategy Chosen

The **separation strategy projected** for the preparative purification step was based on a **negative selection method**

The pluripotency extracellular marker chosen was the **Tra-1-60** – commercially available, highly expressed in hiPSCs and was more expressed after the neural commitment than the Tra-1-81

14

Introduction
Objectives
Results
Conclusions

Separation of hiPSCs after the Neural Commitment Protocol

- **Magnetic Activated Cell Sorting (MACS)**

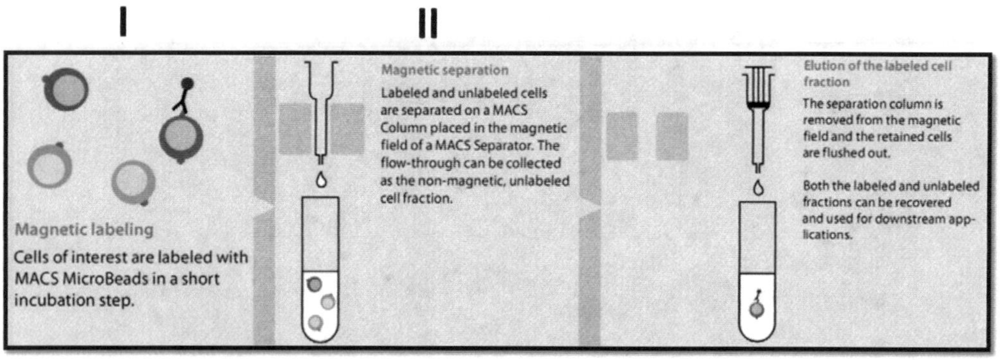

Adapted from MACS stem cell research brochure, Miltenyi Biotec

Introduction
Objectives
Results
Conclusions

MACS Mock Experiments

- **Deliberately mixed** populations of **hiPSCs** and **lt-NSCs** or hiPSC-derived cells after the neural commitment protocol were **purified** with the **anti-Tra-1-60** MACS kit

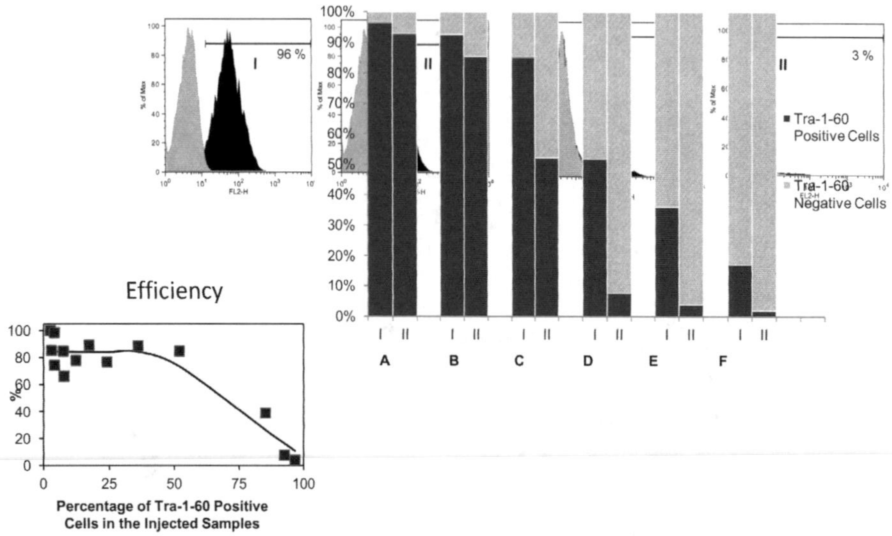

Introduction
Objectives
Results
Conclusions

MACS Experiments after the Neural Commitment Protocol

• hiPSC-derived cells after the neural commitment protocol were **purified** with the **anti-Tra-1-60** MACS kit

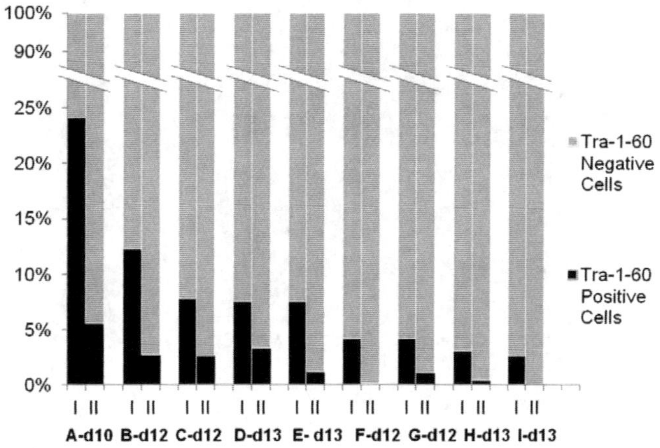

17

Introduction
Objectives
Results
Conclusions

Integrated BioProcess for Production of hPSC-derived Neural Progenitors (NPs)

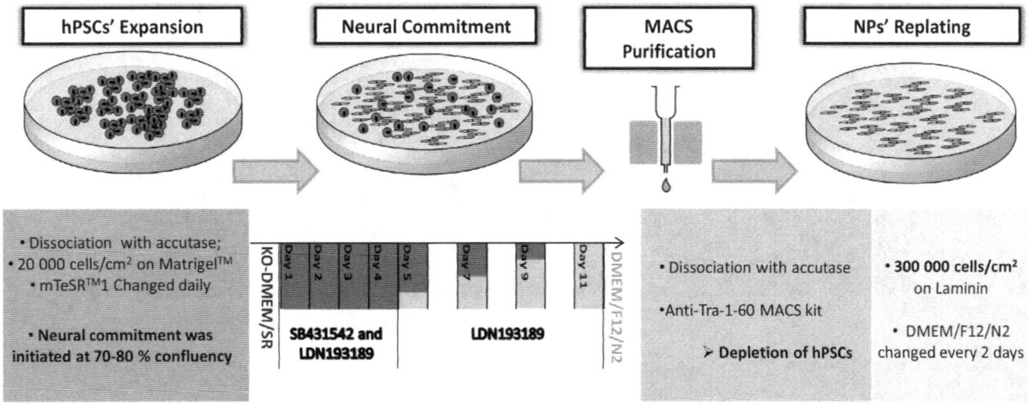

18

Introduction
Objectives
Results
Conclusions

Re-plating of the hiPSC-derived Cells

Dissociated and Re-Plated

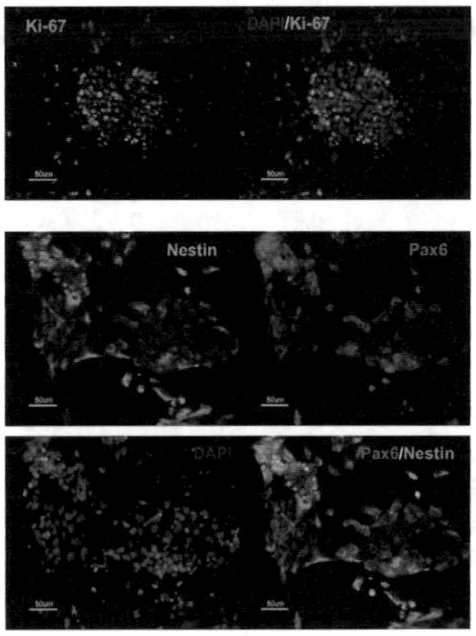

Dissociated, **Purified** and Re-Plated

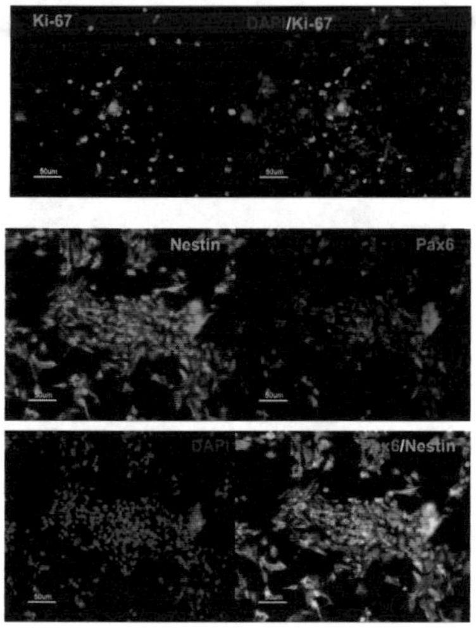

19

Introduction
Objectives
Results
Conclusions

Re-plating of the hiPSC-derived Cells

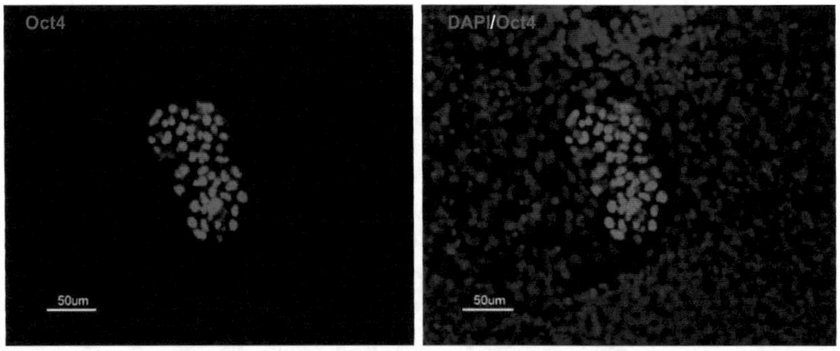

$\frac{Oct4^{+}\ Colonies\ formed}{10^{6}\ cells\ plated}$	Suspended	80 ± 4.8
	Separated	4.6 ± 1.3

20

Introduction
Objectives
Results
Conclusions

Conclusions

➢ A robust protocol for the **neural commitment** of hPSCs through the dual-SMAD inhibition protocol was established using two hPSC lines

➢ The results obtained with the **MACS anti-Tra-1-60** micro beads kit can suite a preparative-scale purification of hiPSC-derived neural progenitors, by negatively selecting the unwanted cells (hiPSC), if the initial cell-populations contain less than 30 % of "contaminant" hiPSC

➢ There were **94 % less Oct4 positive colonies** in the **Purified fraction** of hPSC-derived populations, when comparing with the Suspended fraction

21

Numbers

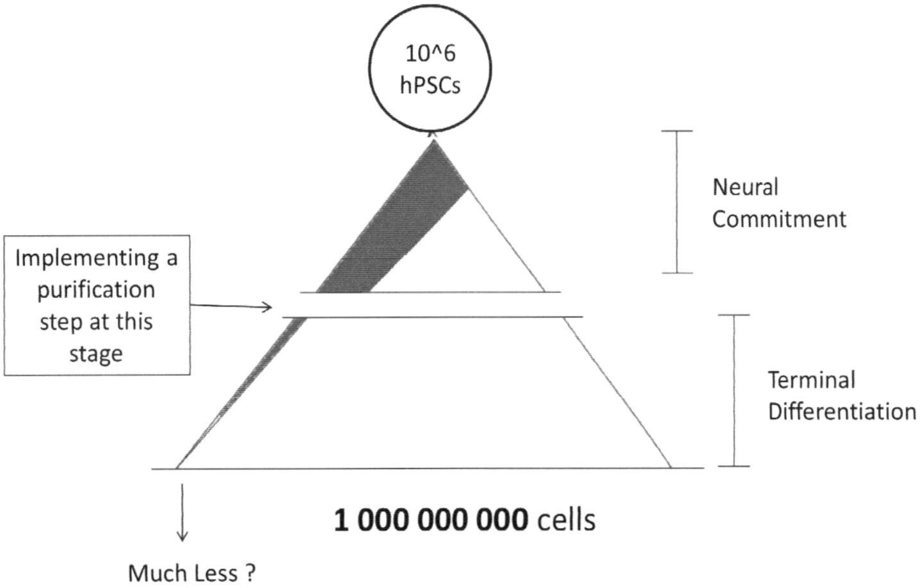

22

Acknowledgements

SCBL-RM | Stem Cell Bioengineering & Regenerative Medicine Lab

IBB Institute for Biotechnology and Bioengineering

FCT Fundação para a Ciência e a Tecnologia
MINISTÉRIO DA CIÊNCIA, TECNOLOGIA E ENSINO SUPERIOR

TÉCNICO LISBOA

universitätbonn

Supervisors:
Prof. Joaquim Sampaio Cabral
Dr. Maria Margarida Diogo
and co-workers

23

Numbers

1 000 000 000 of cells estimated for each patient

1 cell admitted per patient

2 000 is the minimum amount of hPSC detectable with qRT-PCR

1 000 000 is the minimum amount of hPSC detectable with FC

"Very little attention has been given to developing methods capable of detecting residual amounts of hiPSCs" – *Highly Sensitive In Vitro Methods for Detection of Residual Undifferentiated Cells in Retinal Pigment Epithelial Cells Derived from Human iPS Cells*, Kuroda, T. et al., 2012

1 000 – 1 000 000 cells are injected in teratoma assays

Is this enough?

24

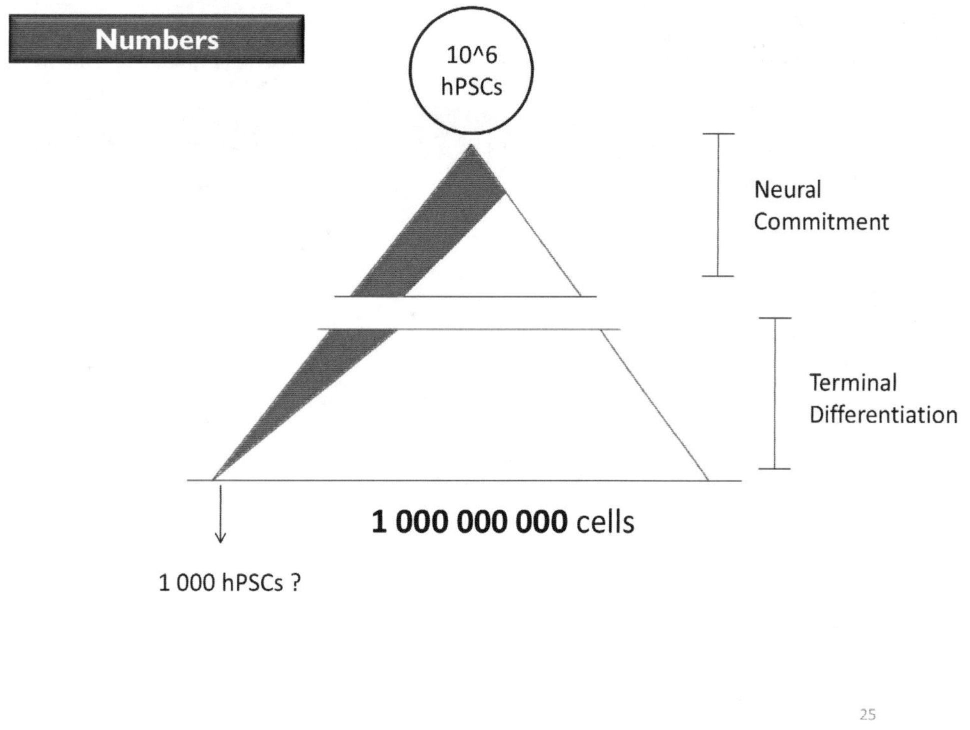
Numbers
10^6
hPSCs
Neural
Commitment
Terminal
Differentiation
1 000 000 000 cells
1 000 hPSCs ?
25

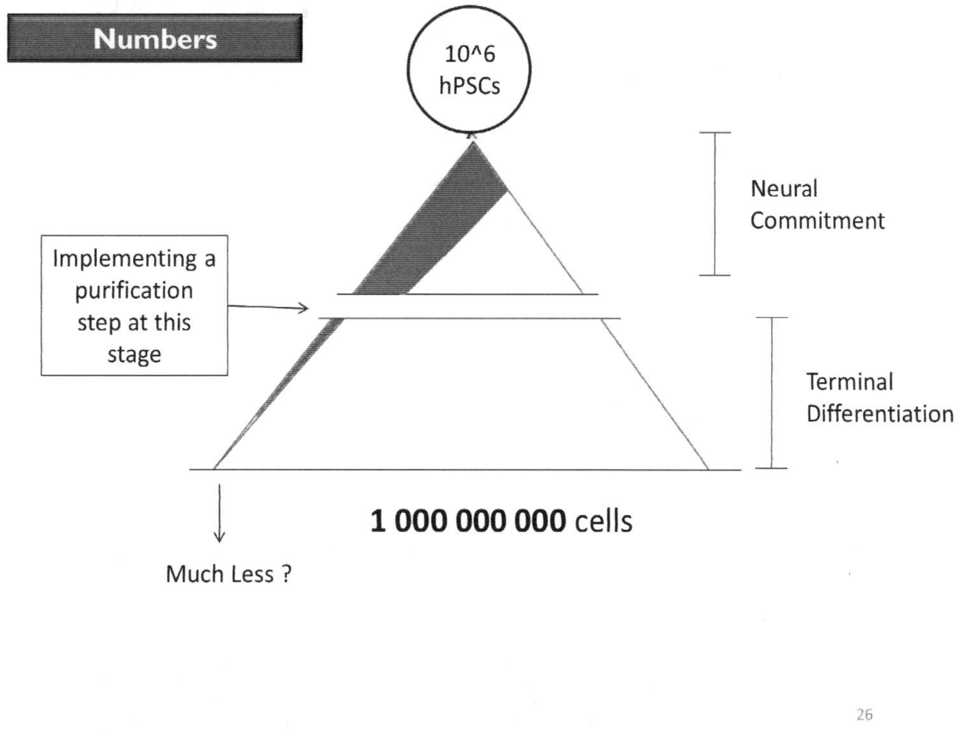
Numbers
10^6
hPSCs
Neural
Commitment
Implementing a purification step at this stage
Terminal
Differentiation
1 000 000 000 cells
Much Less ?
26

MACS Model

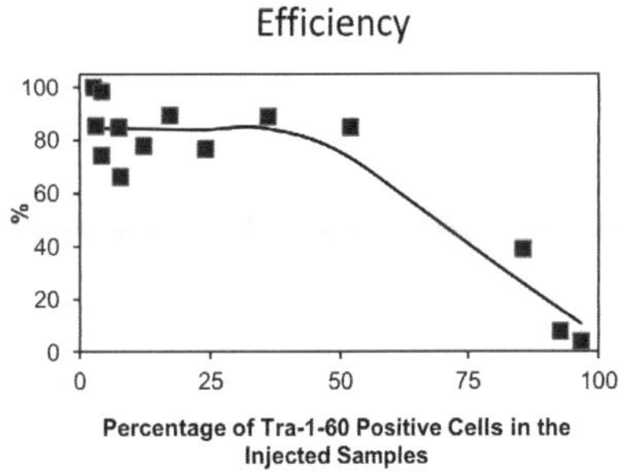

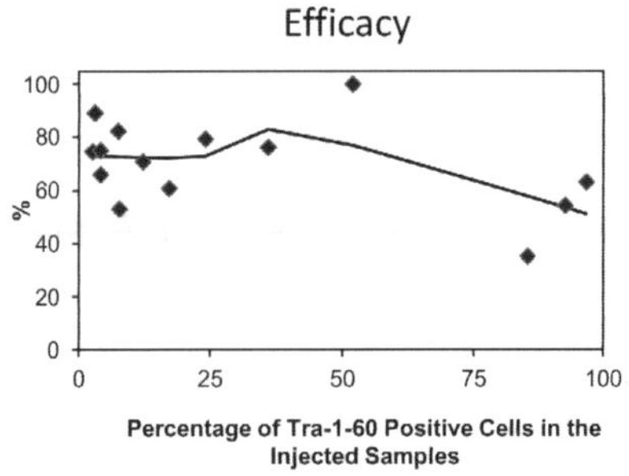

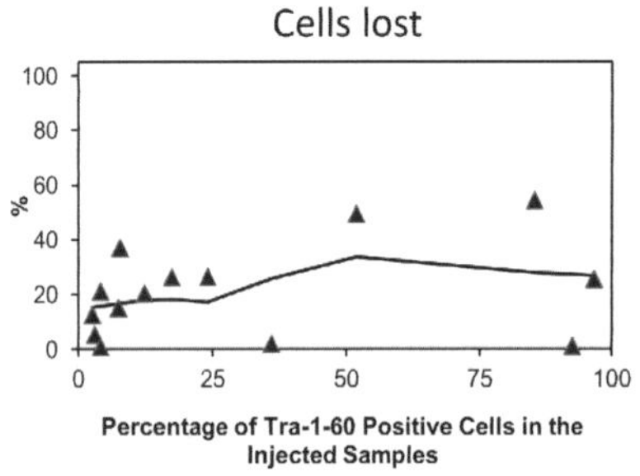

Relevance of Separation Processes in hPSC-based Bioprocesses

- The presence of **PSCs** within cultures **may alter the results** obtained in **transplantation** therapies as well as in ***in vitro*** **differentiation** protocols

- Most **neurological diseases** are **not life threatening** and the possibility of **teratoma** formation after a cell therapy is **unacceptable**

28

Stem Cell Bioprocessing of Skin-Derived Precursor Cells (SKPs) in Stirred Suspension Bioreactors

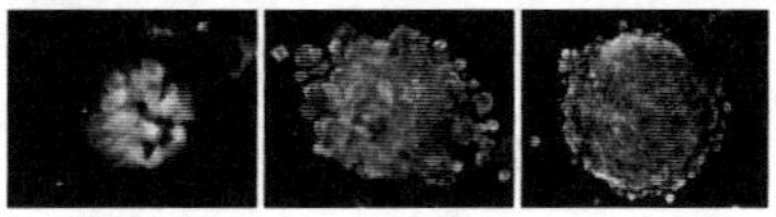

Kathryn V. Boon[1], Breanna Borys[1], Denver Surrao[1], Ranjan Kumar[2], Min Cheng[3], Eko Raharjo[3], Jeff Biernaskie[4], Michael S. Kallos[1]

(1) Pharmaceutical Production Research Facility, Schulich School of Engineering, University of Calgary, Calgary, AB, Canada,
(2) Hotchkiss Brain Institute, University of Calgary, Calgary, AB, Canada,
(3) Faculty of Veterinary Medicine, University of Calgary, Calgary, AB, Canada,
(4) Dept of Comparative Biology and Experimental Medicine, Faculty of Veterinary Medicine, University of Calgary, Calgary, AB, Canada,

AIChE SBE, 4th International Conference on Stem Cell Engineering, Coronado, CA, March 18, 2014

@kallosm mskallos@ucalgary.ca

Motivation / Background

- Clinical Need: Burn Patients
  - 195 000 deaths/year
  - Approximately 11 million people required hospitalization for burns in 2004 [1]
  - Direct costs for burn care for children in the United States > US$ 211 million in 2000 [1]
  - Current gold standard therapy for severe burns – Split Thickness Skin Graft

[1] World Health Organization, 2012

Split-Thickness Skin Graft (STSG)

- Autologous graft
- Skin harvested with dermatome
- Common donor sites:
  - Thigh
  - Buttock
  - Flanks
  - Back
  - Abdomen
  - Scalp
- Donor site heals
- Split-thickness: epidermis + part of dermis
- Graft can be meshed to cover more surface area (up to 9:1 ratio)

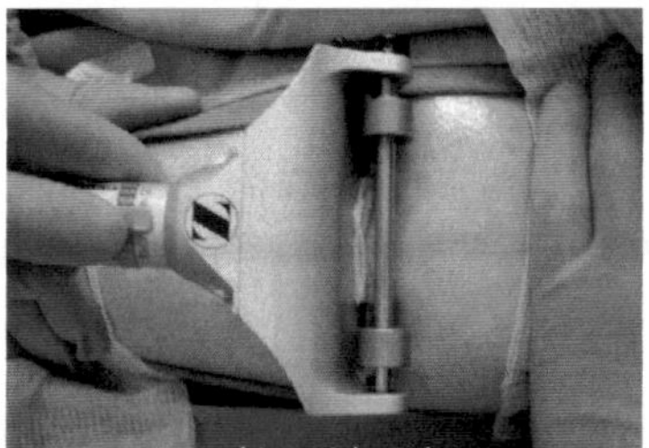
Harvesting skin with a dermatome

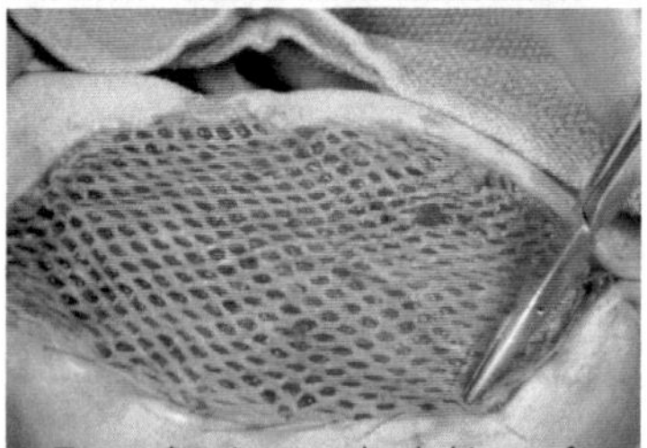
Transplanting meshed skin graft

[2] Miraliakbari et al., Operative Techniques in General Surgery, 2006

March 18, 2014 AIChE SBE Stem Cell 4

STSG Limitations

- Hypertrophic scarring
- Contracture
- Reduced sensory perception
- High incidence of Pruritus (Itching)

- 100% Function is not restored with STSG
- Possible Autologous Stem Cell Treatment - SKPs

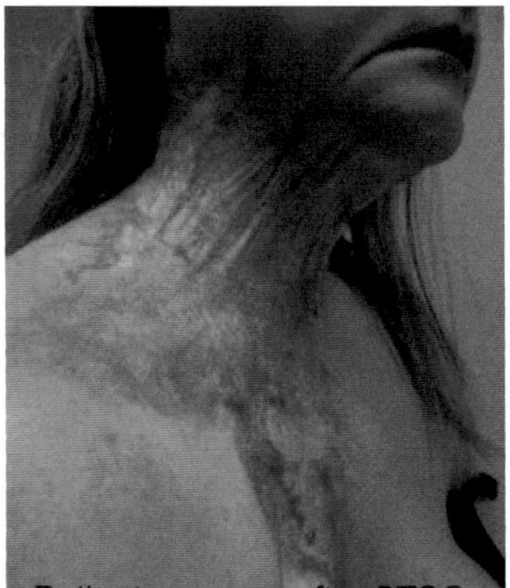
Patient one year after STSG

[3] Dr. Vincent Gabriel, Dr. Duncan Nickerson

March 18, 2014 AIChE SBE Stem Cell 5

Skin-derived Precursors (SKPs)

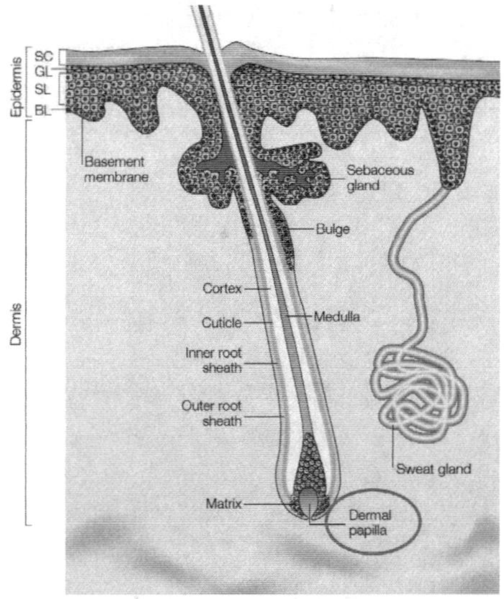

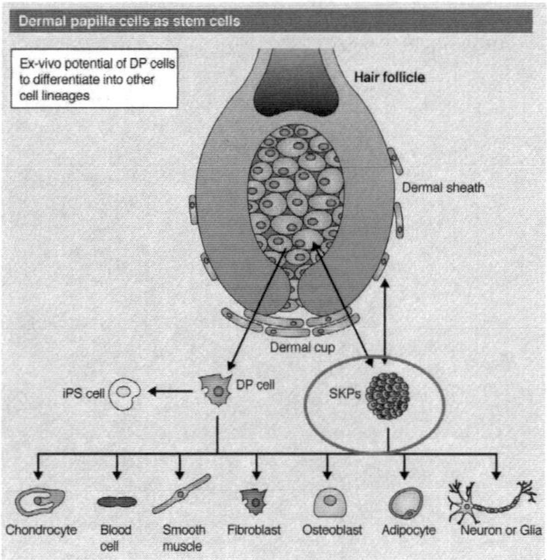

[4] Fuchs & Raghavan, Nature Reviews Genetics, 2002.
[5] Driskell et al., Cell Science, 2011.

Skin-derived Precursors (SKPs)

Skin-derived precursors (SKPs) can be isolated in vitro, self-renew and induce new hair follicle formation

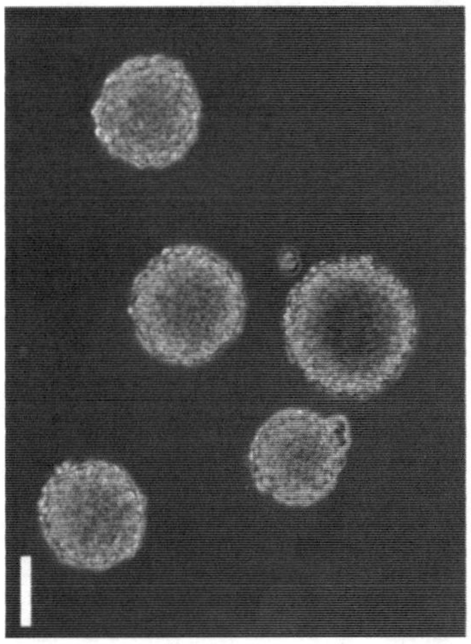

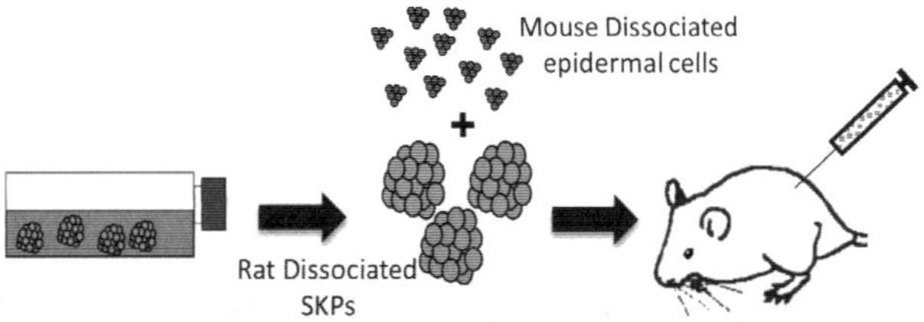

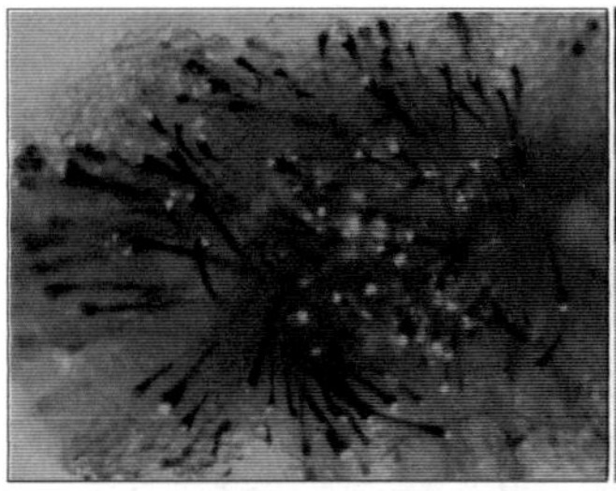

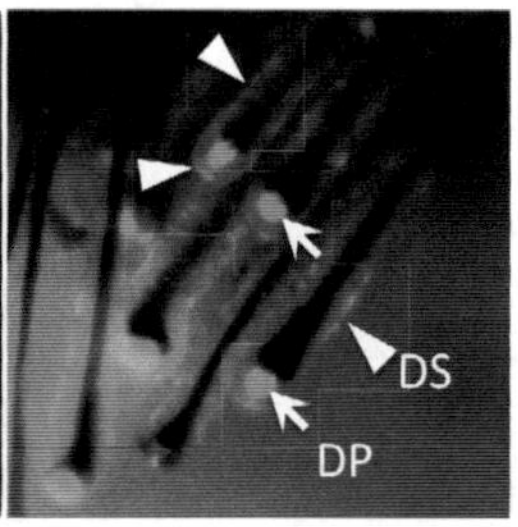

[6] Biernaskie et al., Cell Stem Cell, 2009.

Clonal Adult SKPs passage 6 Biernaskie

Skin-derived Precursors (SKPs)

- Isolated adult dermal stem cells transplanted directly into wounds generate new dermal tissue

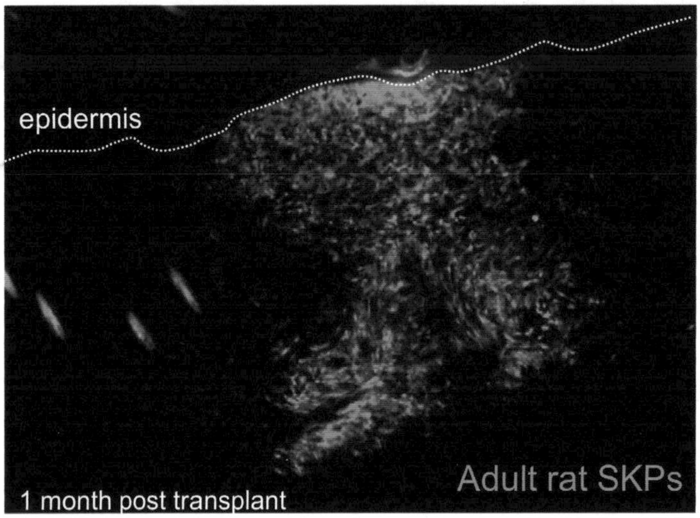

Biernaskie

Human SKPs – Therapeutic Potential

- SKPs can be isolated from numerous adult human donor sites
- SKPs can be expanded in vitro
- SKPs can form new dermal tissue and contribute to new hair follicle formation
- Need controlled bioprocess to expand cell numbers

Stem Cell Bioprocess Engineering

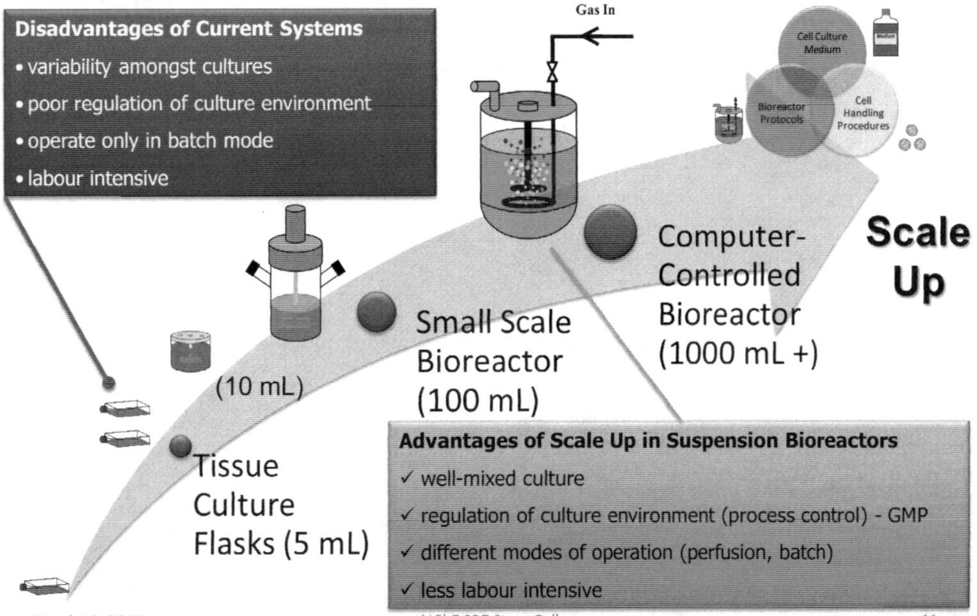

March 18, 2014 AIChE SBE Stem Cell 11

Objective

- To rapidly expand human SKPs in stirred-suspension bioreactors to produce relevant quantities of clinical quality human SKPs for use in autologous cell therapy
- Time limit for clinical relevance
- Limited starting population

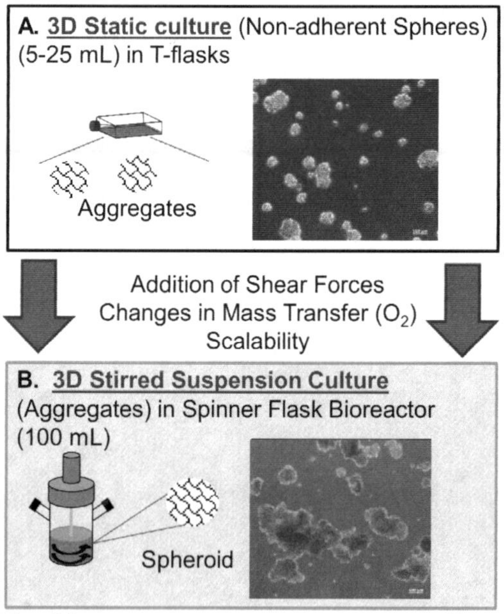

Methods

- Patient-derived human SKP cell lines
  - Various donors (age, gender, anatomical location)
- Culture vessels

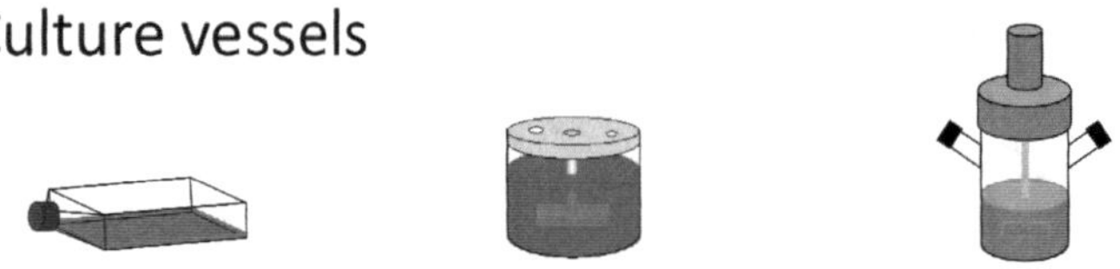

Static T-flask 10 ml bioreactor 100 ml bioreactor

- Cell Growth Kinetics
  - Cell counts
  - Cell cycle analysis
- Oxygen and metabolite consumption
- Phenotype – measured using immunocytochemistry

hSKPs Can be Expanded in Stirred Suspension Bioreactors (1/2)

- Batch Expansion of hSKPs (68yr female, thigh) - in static or 100 ml bioreactors
- Higher total viable cell density achieved with bioreactors
- Maximum total viable cell density achieved with agitation rate of 60 rpm

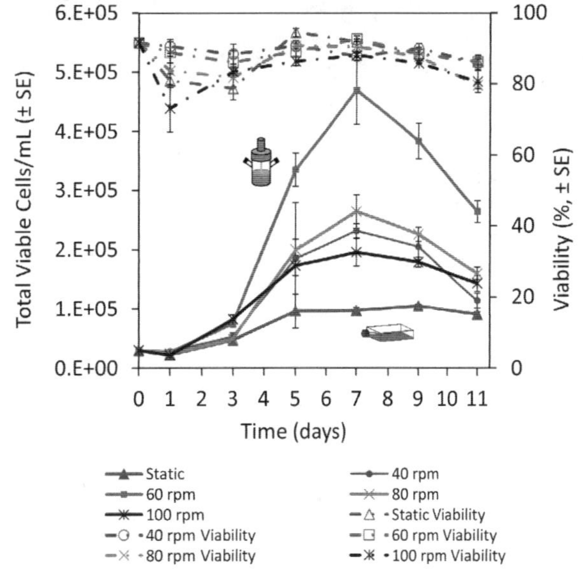

March 18, 2014 AIChE SBE Stem Cell 14

hSKPs Can be Expanded in Stirred Suspension Bioreactors (2/2)

- <u>Second Donor</u>
- Batch Expansion of hSKPs (28yr male, thigh) - in static or 100 ml bioreactors
- Higher total viable cell density achieved with bioreactors
- Maximum total viable cell density achieved with agitation rate of 60 rpm

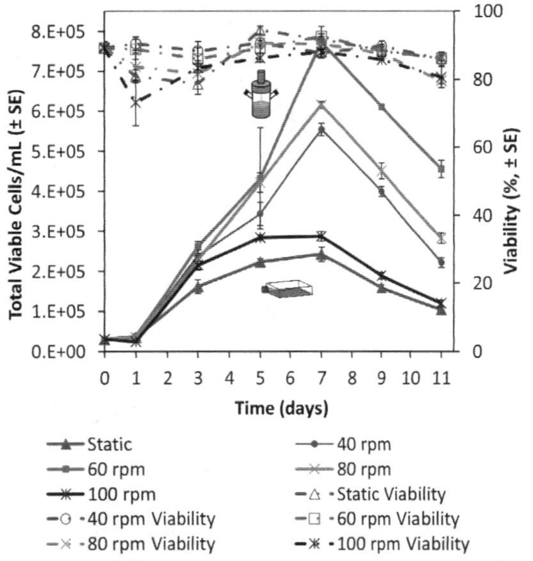

March 18, 2014 AIChE SBE Stem Cell 15

Nutrient / Waste Concentrations in Static and Suspension Batch

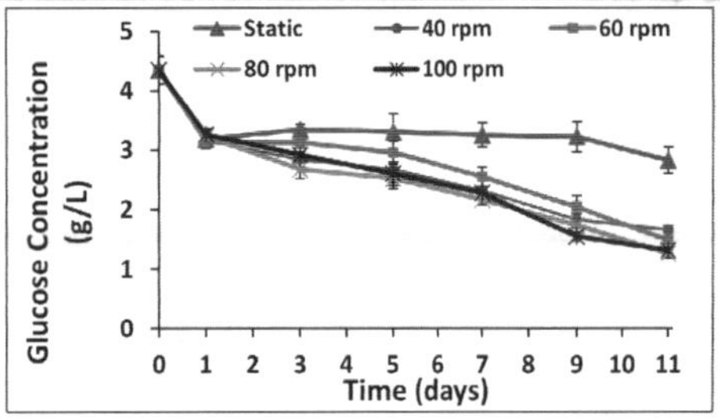

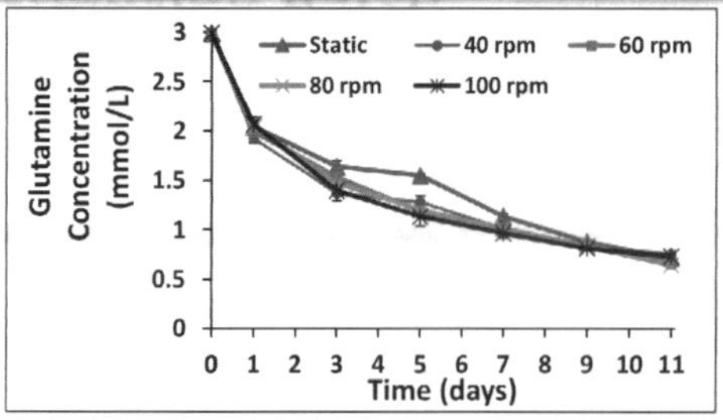

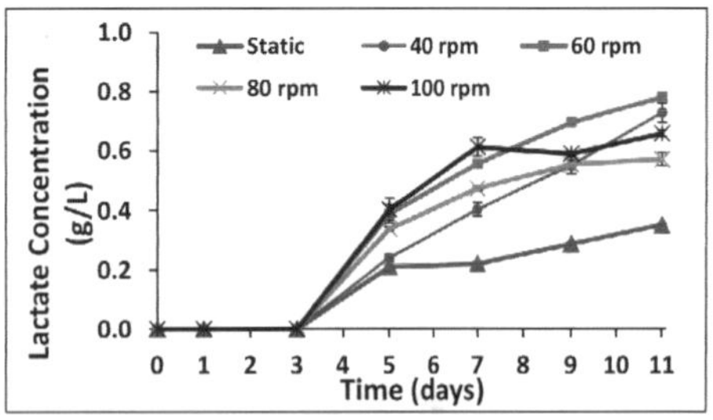

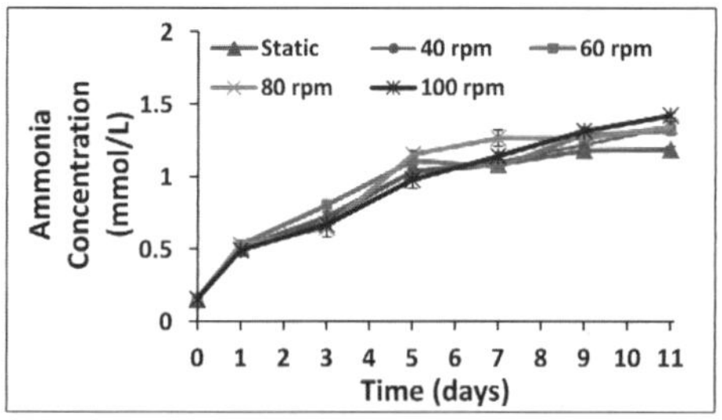

- Cell growth does not appear to be limited by glucose or glutamine consumption or buildup of lactate or ammonia

Batch Expansion of hSKPs (68yr female) - hSKPs from thigh

Aggregate Morphology

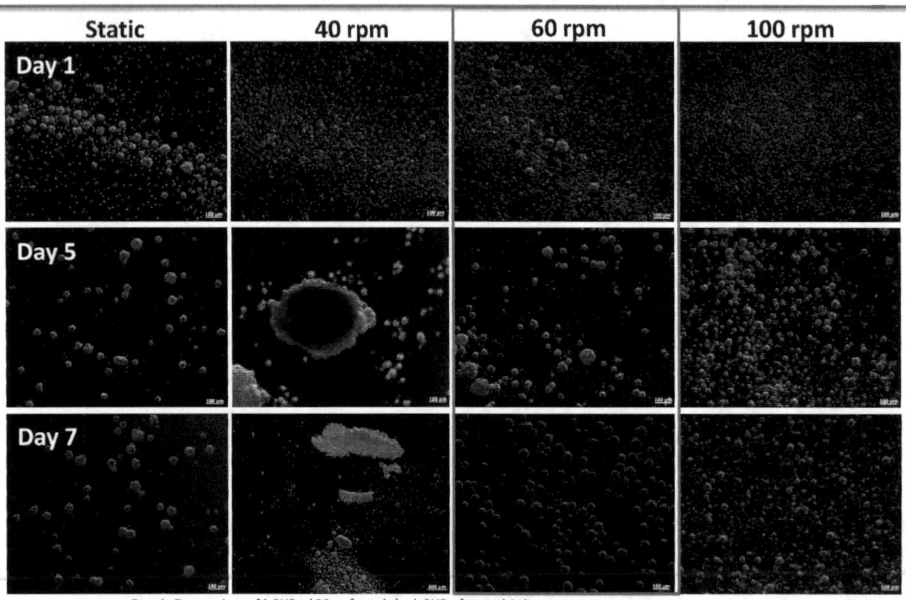

Batch Expansion of hSKPs (68yr female) - hSKPs from thigh

Aggregate Growth

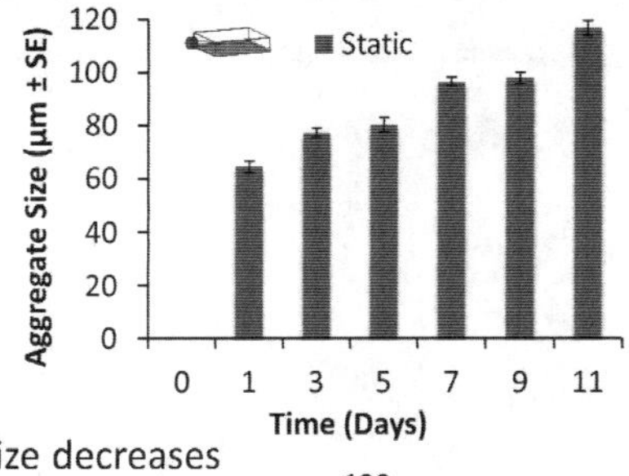

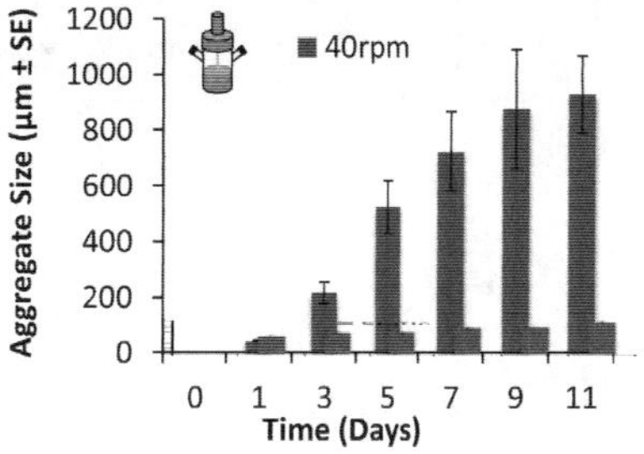

- Aggregate size decreases with increasing agitation rate
- Aggregates grown at 40 rpm are large enough to experience oxygen diffusion limitations
- Aggregates at higher agitation rates do not get to limits where necrosis would be expected

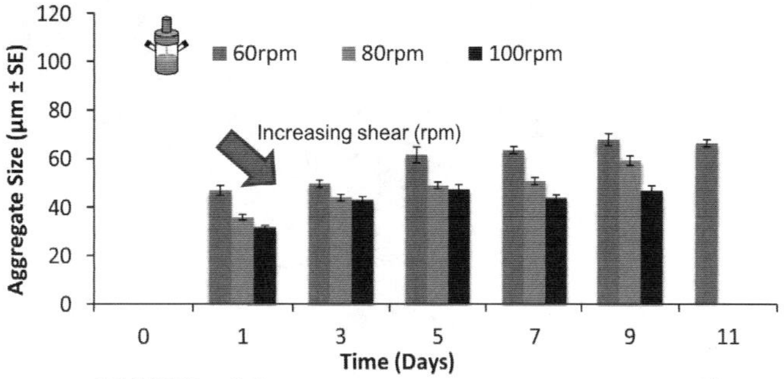

March 18, 2014 AIChE SBE Stem Cell 18

Aggregate Numbers

- The number of aggregates increases with agitation rate
- The mean aggregate diameter decreased with increasing agitation rate

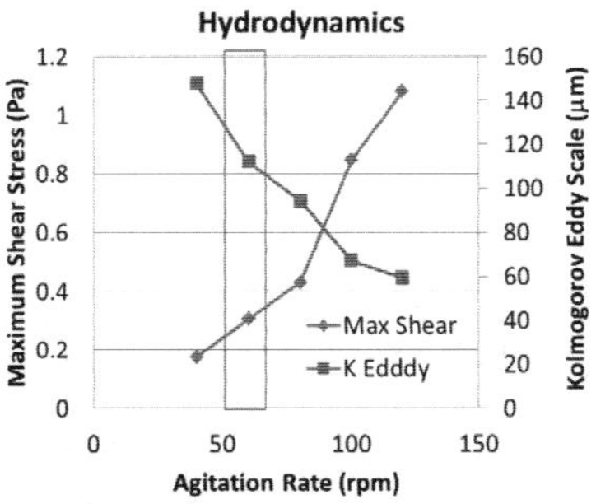

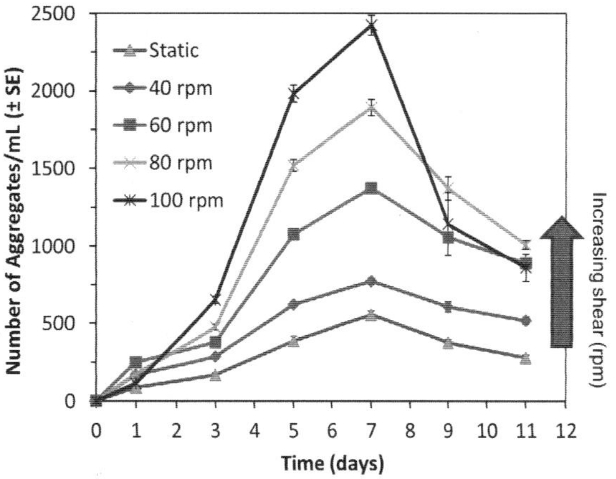

Batch Expansion of hSKPs (68yr female) - hSKPs from thigh

March 18, 2014 AIChE SBE Stem Cell 19

Oxygen Transfer / Consumption

Condition	Mass transfer of oxygen	Day 6 Photomicrograph	Mass Transfer Characteristics
Static T-flasks (75 cm^2) 	Thin liquid layer for oxygen to diffuse through (larger surface area to volume ratio)	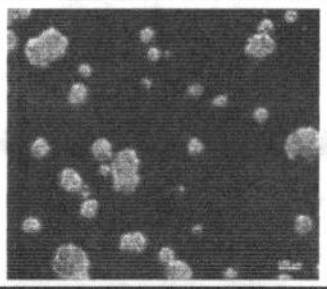	**v** = 24 mL **k_la** = 0.60 h^{-1} **SA/v** = 3.125 cm^2/cm^3
Stirred Suspension Bioreactors 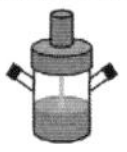	Well-mixed environment provides availability of nutrients and oxygen	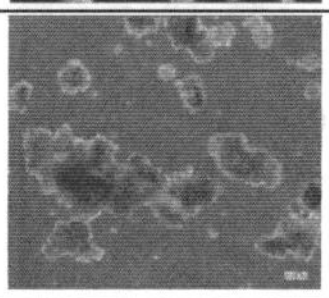	**v** = 100 mL **k_la** = 1.56 h^{-1} (60 rpm) **SA/v** = 0.252 cm^2/cm^3

Oxygen concentration profile within a spherical aggregate

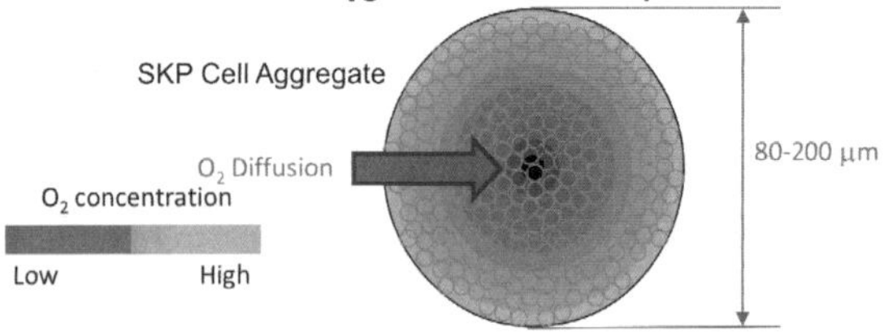

Oxygen transfer to cells within aggregates is through diffusion from the bulk media

van Winkle et al., Cells Tissues Organs 2012

Oxygen Consumption Rate

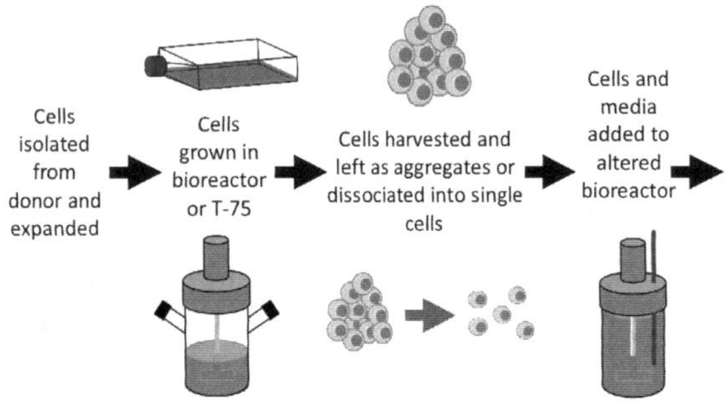

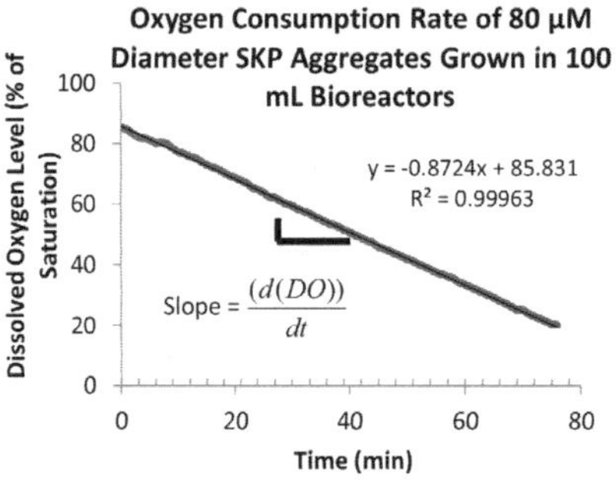

- Mass balance on oxygen in the media allows for the average per cell consumption rate to be calculated

Input	+	Generation	=	Output	+	Accumulation
0	+	Consumption by cells	=	0	+	Decrease in oxygen in media

$$(-qO_2) = \frac{(d(DO))}{dt} \times \frac{C[O_2]}{DOspan} \times \frac{1}{x}$$

qO_2 = Oxygen consumption rate

$C[O_2]$ = Equilibrium dissolved oxygen concentration

DO span = Maximum dissolved reading = 100

x = Viable cell density

$\frac{(d(DO))}{dt}$ = Change in DO concentration/change in time

Oxygen Consumption

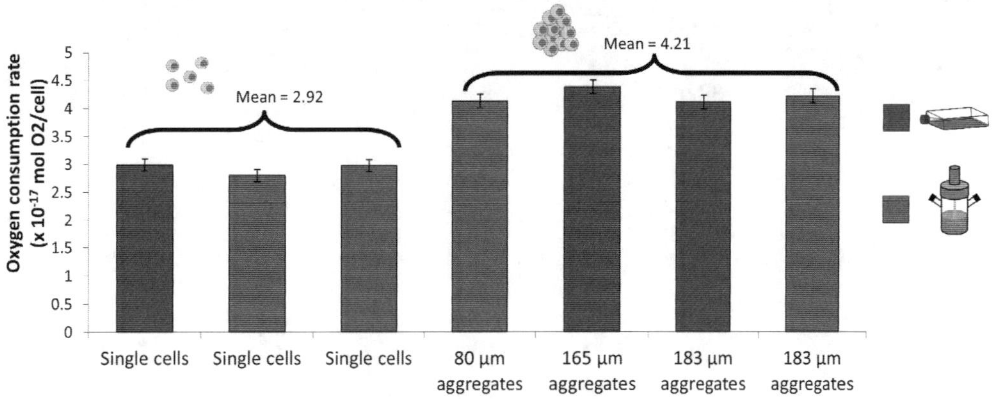

- Average cell specific oxygen consumption rate is lower for single cells than aggregates
- Rate does not depend on static or bioreactor culture
- No diffusional limitations for oxygen in aggregates grown at 60 rpm

Cell Cycle Analysis

- Examined using:
  - 5-ethynyl-2'-deoxyuridine (EdU)
    - Thymidine analog
    - Incorporated in S phase of cell division
  - Propidium iodide (PI)
    - labels DNA and RNA
    - RNAse added to samples

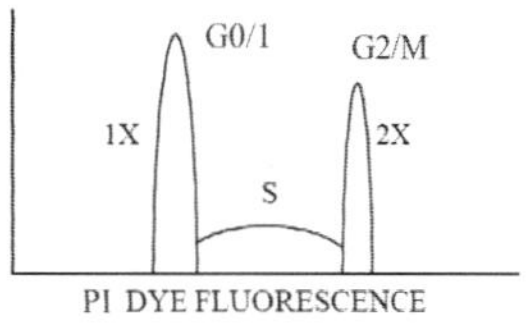

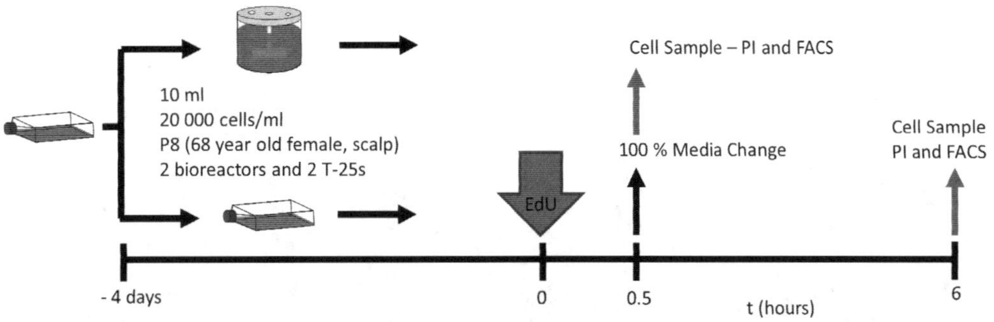

Cell Cycle

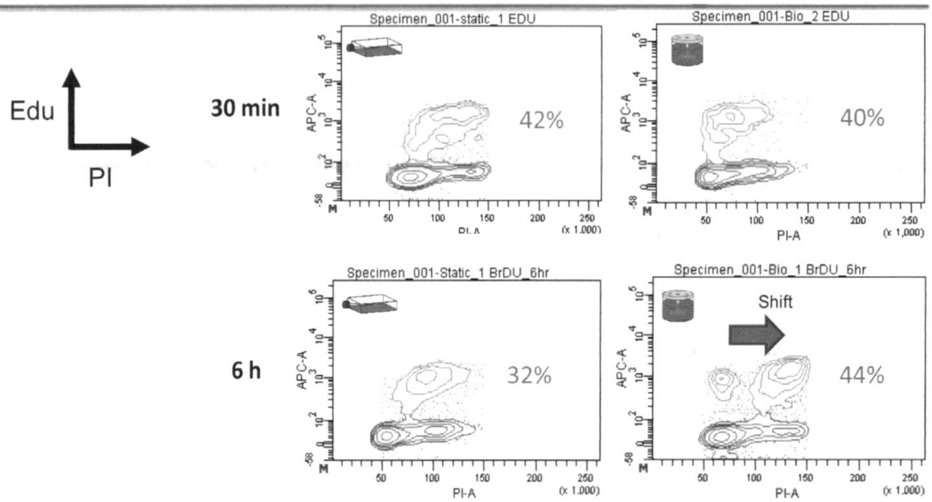

- hSKPs appear to be progressing through the cell cycle faster in stirred suspension bioreactors than in static culture

Phenotype Analysis (1/3)

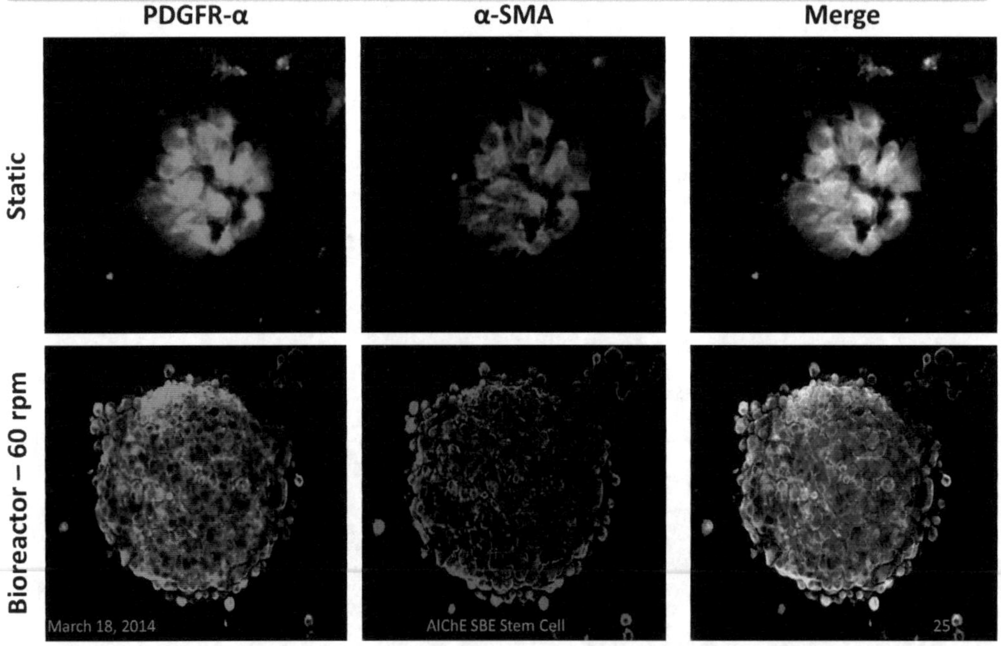

Phenotype Analysis (2/3)

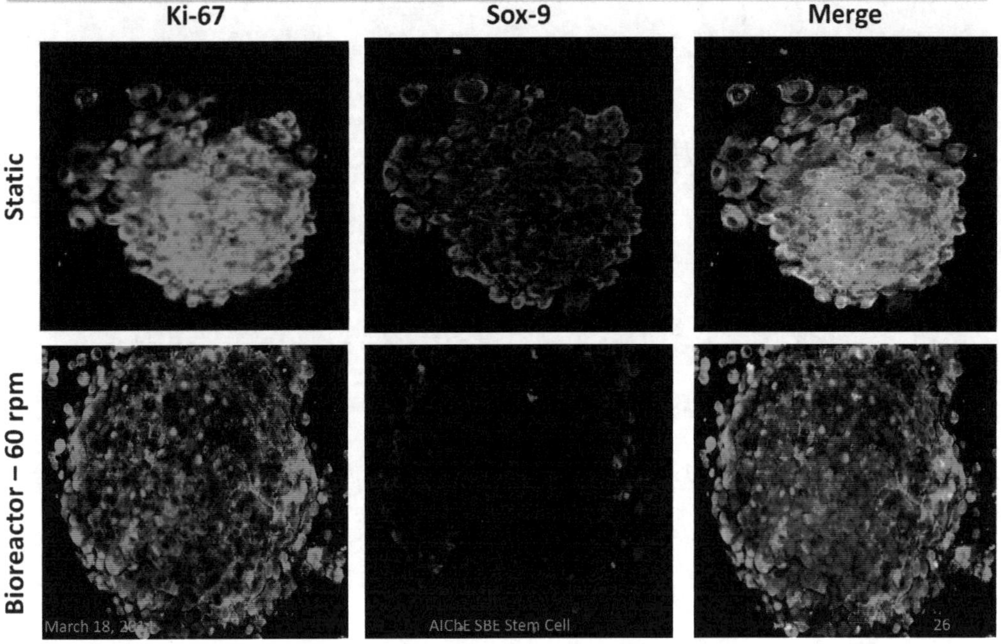

Phenotype Analysis (3/3)

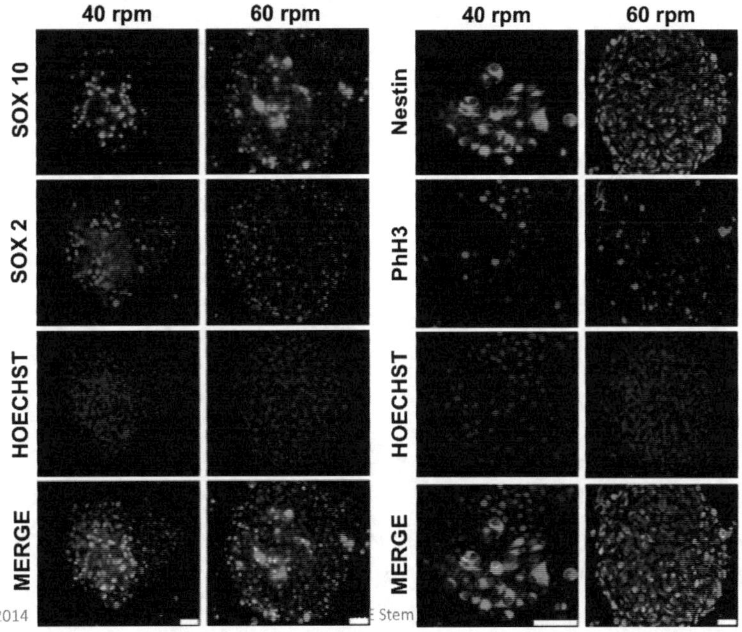

Serial Passaging Batch Bioreactor

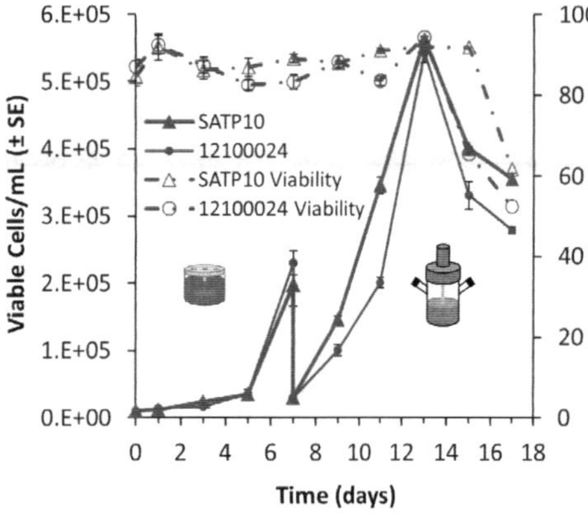

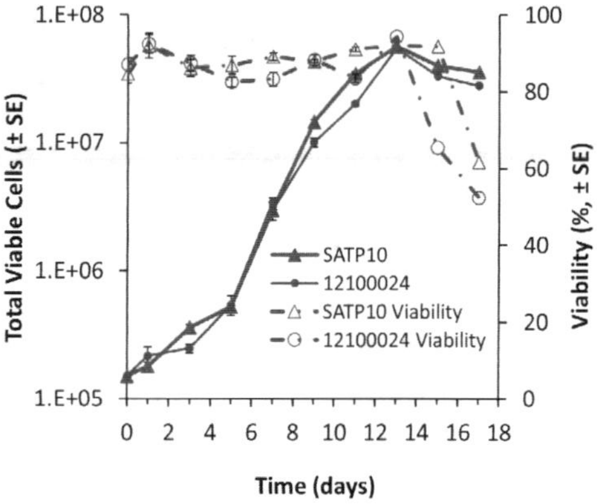

- hSKPS inoculated as single cells (10,000 cells/mL) in 15 mL
- After 7 days – passaged (single cells) into 100 mL bioreactor
- Maximum fold-expansion of 375x in 13 days
- 150k cells → 56,250k

Increasing Cell Density

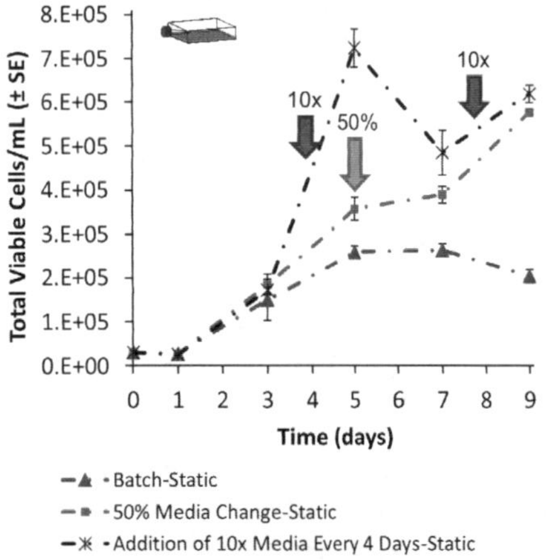

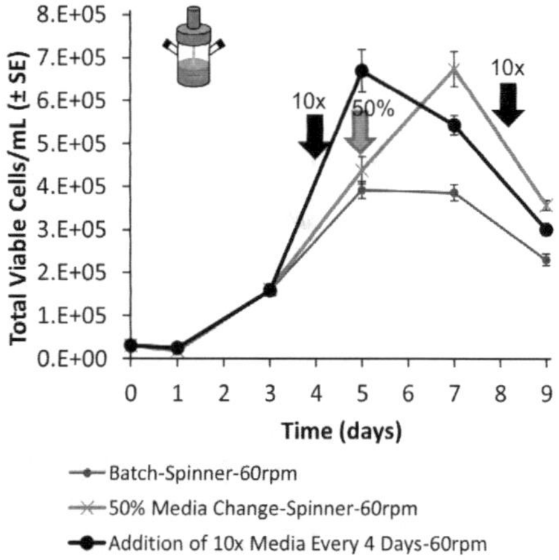

- Donor – Male (64 yrs)
- Media changes result in increased maximum cell densities in static and suspension bioreactor culture

Conclusions

- SKPs - potential to improve outcomes of split-thickness skin grafts for burn survivors
- Stirred suspension bioreactors offer an attractive culture method for scalable production
  - **Superior cell expansion compared to static culture**
  - Optimal agitation/shear for cell expansion
  - Oxygen consumption rate matches other stem cells
  - Phenotype is maintained in suspension
- Multiple donors – ages, sex – all proliferate in the bioreactors
  - Working on donor-to-donor predictability

March 18, 2014 AIChE SBE Stem Cell 30

Acknowledgements

- Dr. Vincent Gabriel, Assistant Professor of Physical Medicine & Rehabilitation, University of Calgary

Alberta Innovates Health Solutions

NSERC CRSNG

UNIVERSITY OF CALGARY

HexaScreen Culture Technologies S.L.

STEMCELL TECHNOLOGIES Scientists Helping Scientists™

March 18, 2014 AIChE SBE Stem Cell 31

Pharmaceutical Production Research Facility

McCaig Institute *for Bone and Joint Health*

UNIVERSITY OF CALGARY

Stem Cell Bioprocessing of Skin-Derived Precursor Cells (SKPs) in Stirred Suspension Bioreactors

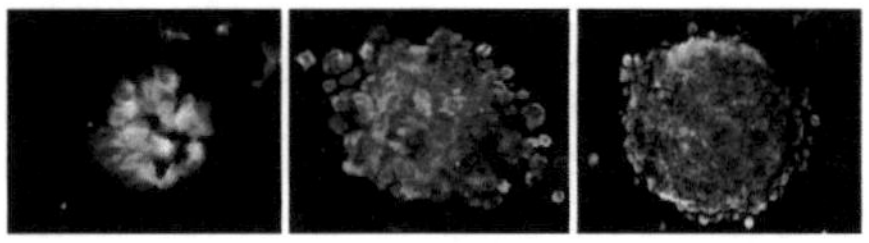

Kathryn V. Boon[1], Breanna Borys[1], Denver Surrao[1], Ranjan Kumar[2], Min Cheng[3], Eko Raharjo[3], Jeff Biernaskie[4], Michael S. Kallos[1]

(1) Pharmaceutical Production Research Facility, Schulich School of Engineering, University of Calgary, Calgary, AB, Canada,
(2) Hotchkiss Brain Institute, University of Calgary, Calgary, AB, Canada,
(3) Faculty of Veterinary Medicine, University of Calgary, Calgary, AB, Canada,
(4) Dept of Comparative Biology and Experimental Medicine, Faculty of Veterinary Medicine, University of Calgary, Calgary, AB, Canada,

AIChE SBE, 4th International Conference on Stem Cell Engineering, Coronado, CA, March 18, 2014

@kallosm mskallos@ucalgary.ca

Evaluating the effects of standard cryopreservation protocols on human mesenchymal stem cell survival and quality

Dr. Karen Coopman

Centre for Biological Engineering

Loughborough University

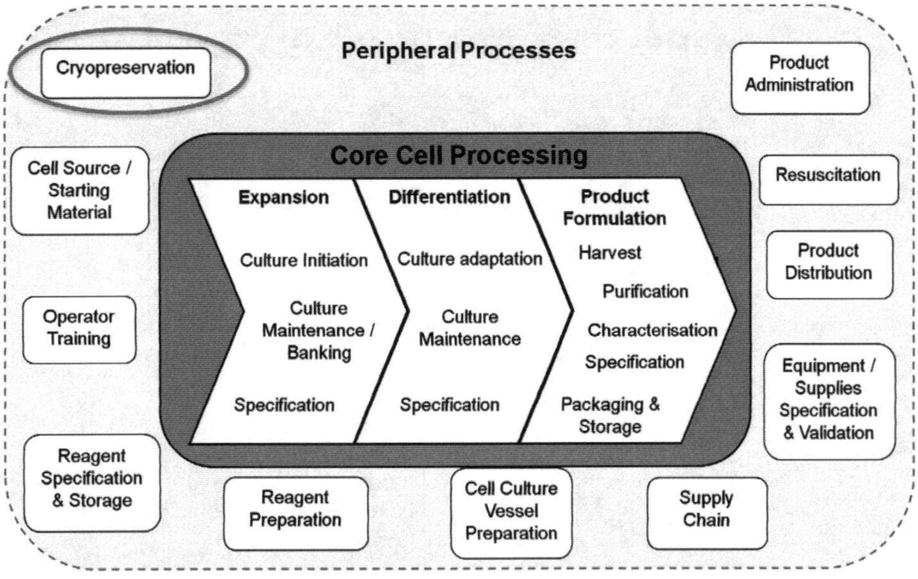

Ratcliffe *et al*, 2011, *Br Med Bull*, 100:137-155

Centre for Biological Engineering

Loughborough University

Content

- Context
  - When and why do we cryopreserve cells?
  - What damage can cryopreservation cause?
- DMSO cytotoxicity
  - Defining a processing window
  - Can we replace/reduce DMSO
- Summary

Centre for Biological Engineering

Loughborough University

Cryopreservation: when & why

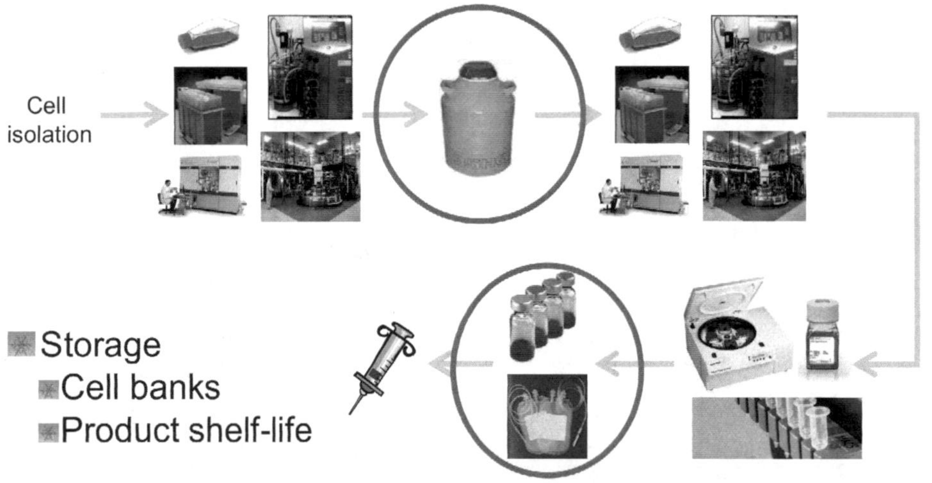

- Storage
  - Cell banks
  - Product shelf-life

Centre for Biological Engineering

Loughborough University

Slow Freezing

Typically...

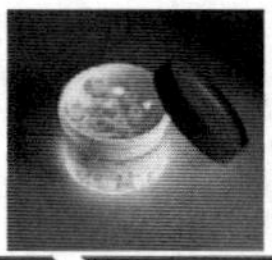

Harvest Cells → Suspend Cells in Freezing Solution → Vials or bags → Freeze at ~1°C/min

← Thaw rapidly ← (Wash cells)

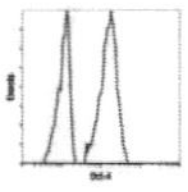

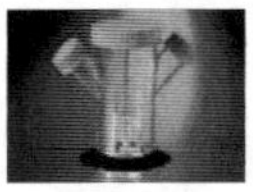

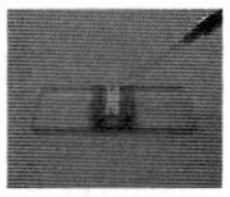

Centre for Biological Engineering   Loughborough University

Cell damage during cryopreservation

Thermal

- Cold shock, metabolic arrest

Mechanical

- ice crystal formation, osmotic shock

Chemical

- ↑ [solute], pH excursion, free radical formation

∴ use cryoprotective agents

Penetrating CPAs – e.g. DMSO

- Replace intracellular water
  - ↓ cell shrinkage & intracellular ice crystal formation, maintain [solute]

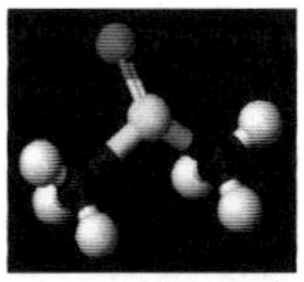

Centre for Biological Engineering   Loughborough University

Issue: DMSO cytotoxicity

- Has been reported to induce differentiation
- Widely believed to be cytotoxic at temperatures > 0ºC but mechanism poorly understood.
  - ➔ protocols are often vague and open to interpretation
  - working "quickly" and placing cells on ice is achievable at lab scale – **not** at larger scale.
    - Sizes of lots
    - Particulate tests
    - What about post-thaw exposure?

Is DMSO cytotoxicity really an issue?

Our approach: Defining a processing window for DMSO-based cryopreservation

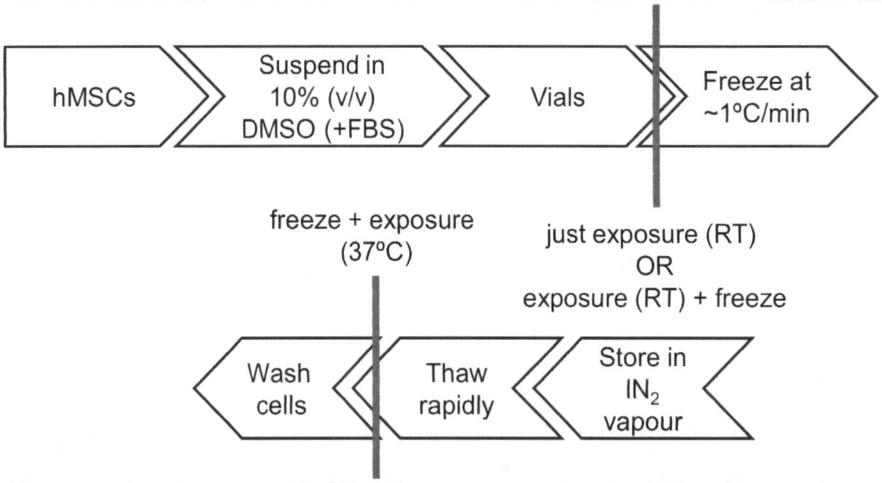

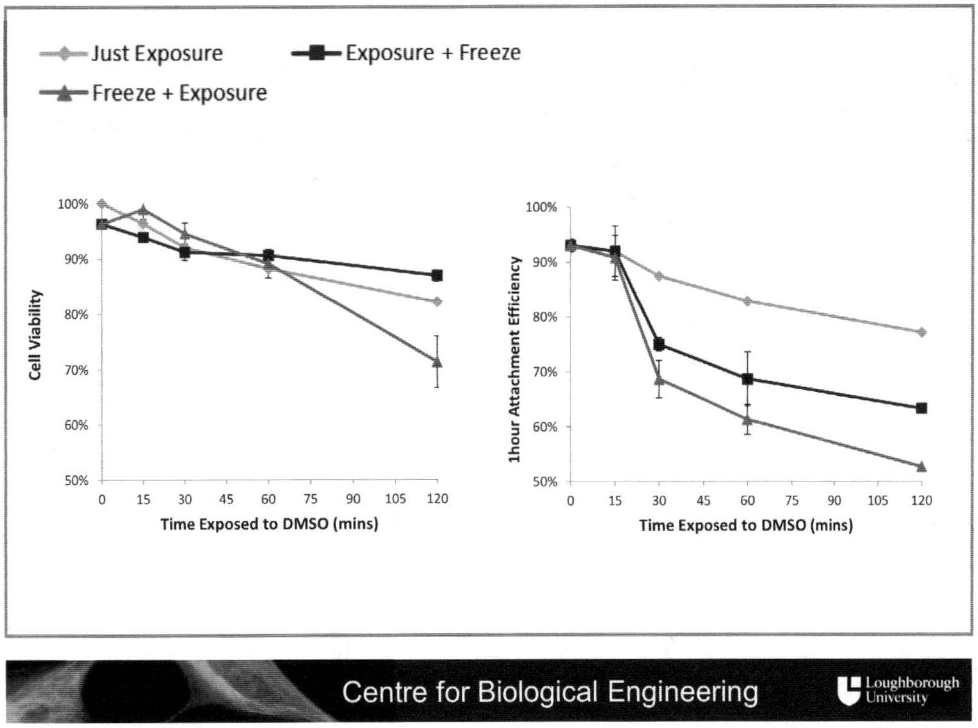
Just Exposure
Exposure + Freeze
Freeze + Exposure
Cell Viability
100%
90%
80%
70%
60%
50%
0 15 30 45 60 75 90 105 120
Time Exposed to DMSO (mins)
1hour Attachment Efficiency
100%
90%
80%
70%
60%
50%
0 15 30 45 60 75 90 105 120
Time Exposed to DMSO (mins)
Centre for Biological Engineering
Loughborough University

Multi-parameter flow cytometry

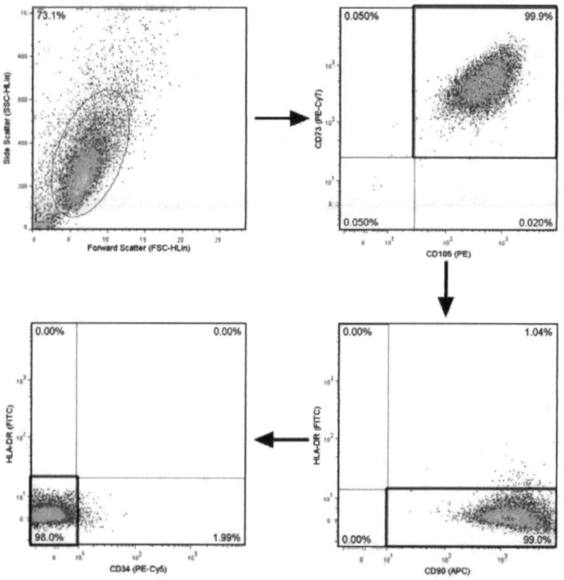

Note: as few as 91.32±2.04% of cells had the desired $CD73^{+}/CD105^{+}/CD90^{+}$, HLA-$DR^{-}/CD34^{-}$ phenotype.

Chan et al, 2013, Biotechnology Letters

Centre for Biological Engineering

Loughborough University

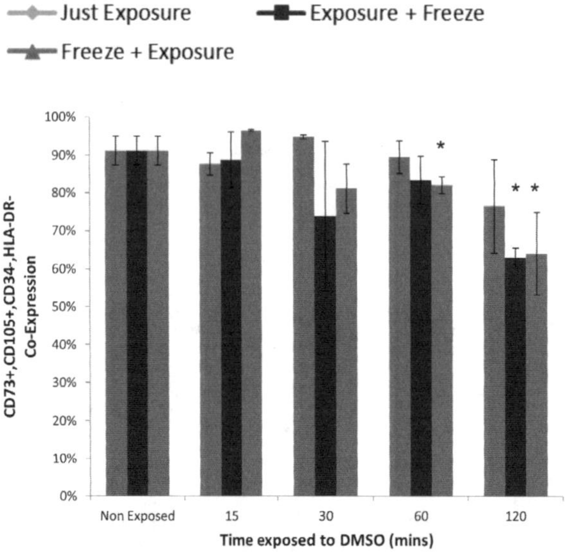

Prolonged exposure to DMSO (before or after freezing) changes hMSC phenotype

Centre for Biological Engineering

Loughborough University

Our approach: replacing DSMO

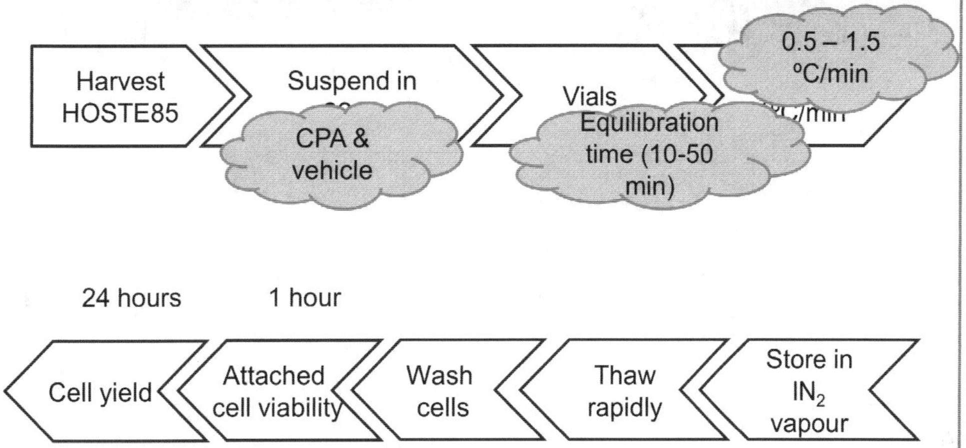

Centre for Biological Engineering
Loughborough University

Mixture blending: Simplex-Lattice designs

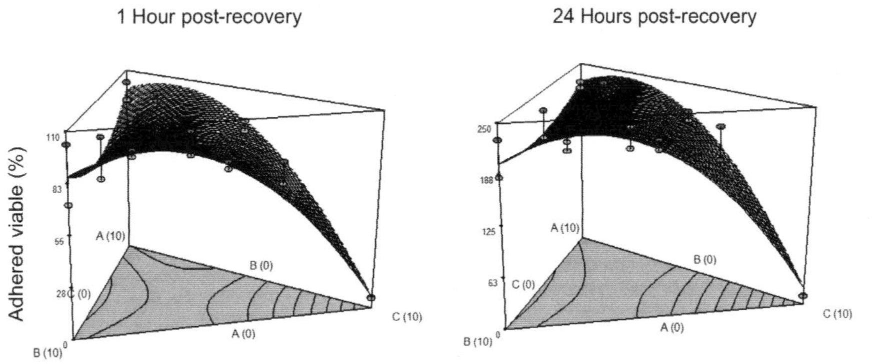

A: DMSO
B: Propylene glycol
C: Trehalose

-1°C/min controlled freezing, vehicle FBS

Centre for Biological Engineering
Loughborough University

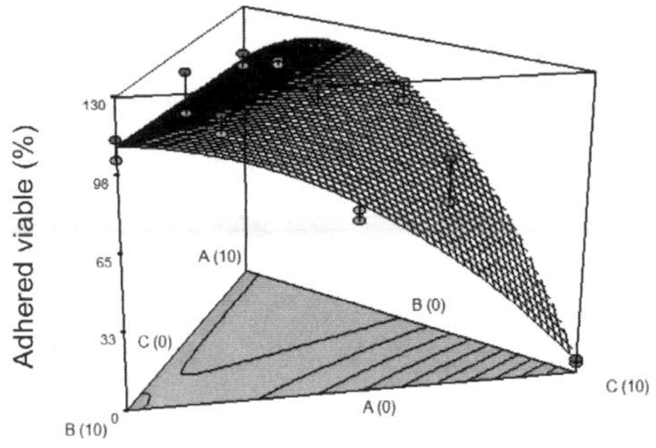

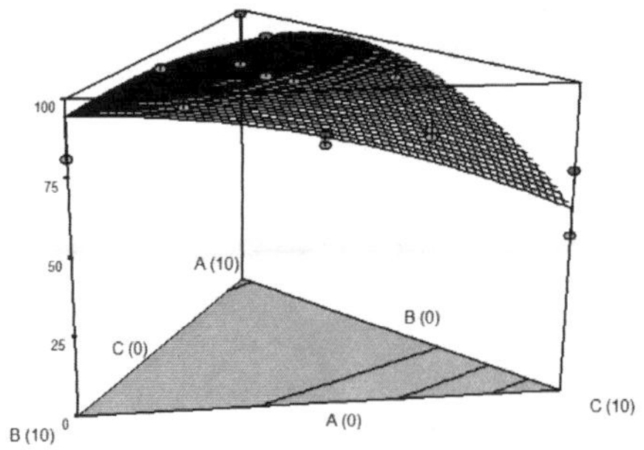

vehicle PlasmaLyte-A + 5 % HSA

vehicle hypothermic storage solution

A: DMSO
B: Propylene glycol
C: Trehalose

-1°C/min controlled freezing, 1 hour post-recovery data

Centre for Biological Engineering — Loughborough University

Factorial designs: minimize cryoprotectant

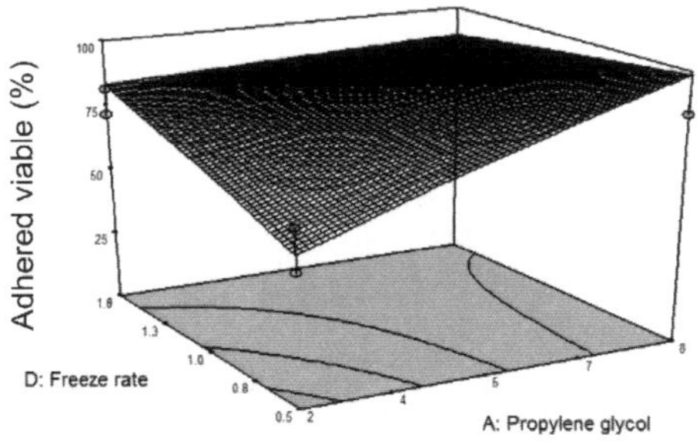

Factors
A: Propylene glycol
B: Trehalose
C: Equilibration time
D: Freeze rate
Vehicle: FBS

1 Hour post-recovery data

Centre for Biological Engineering — Loughborough University

Is DMSO cytotoxicity really an issue? YES

- Data so far indicates exposure to DMSO during processing pre/post-thaw can affect hMSC quality
  - → **Need to control process during manufacture but also at point of care to ≤ 1 hour**
  - Further define quality & understand the role of DMSO vs. freezing/thawing
  - Donor variability?
- Tools such as Simplex-lattice and factorial designs enable optimisation of protocols
  - → **Identified hypothermic storage solutions as a potential vehicle solution which enables lower [penetrating CPA] to be used**
  - Extend to hMSCs
  - Assess impact of prolonged exposure before/after freezing

Thanks to....

- Tim Morris
- Tom Heathman
- Alex Chan
- Kirsty Marrow
- Dr. Qasim Rafiq
- Dr. Andrew Picken
- Dr. Duncan Sharp
- Prof Chris Hewitt
- Prof Nigel Slater

EPSRC Doctoral Training Centre Regenerative Medicine

EPSRC Pioneering research and skills

BBSRC bioscience for the future

UNIVERSITY OF CAMBRIDGE

BRIC · BIOPROCESSING RESEARCH INDUSTRY CLUB

Centre for Biological Engineering Loughborough University

Suspension Culture and Cardiomyogenic Differentiation of Human Pluripotent Stem Cells in Stirred Bioreactor Systems

Zweigerdt, R., Hannover Medical School (MHH)
Olmer, R., Hannover Medical School (MHH)
Kropp, C.,
Kepf, H.,
Martin, U.,
Originally presented on: 3/18/2014 0:40:00 - 1:00:00

Human pluripotent stem cells (hPSC) present an attractive source to generate large amounts of lineage specific progenies for innovative cell therapies, tissue engineering, *in vitro* disease modelling and drug screening assays. The therapeutic and industrial applications of hPSCs will require large cell quantities generated in defined conditions. We have recently established single cell inoculated suspension cultures of hPSCs (Zweigerdt et al., Nature Prot. 2011) which form aggregates in stirred tank reactors (Olmer et al., Tissue Eng. 2012) in the defined medium mTeSR™ (STEMCELL Technologies). This work enabled the translation of conventional, adherence-dependent "2D" culture of hPSC to "3D" suspension culture. Since stirred tank bioreactors allow straightforward up scaling and comprehensive monitoring of process parameters these systems are widely used for the mass culture of conventional mammalian cell lines. Application of stirred reactors to hPSC culture, however, is in its infancy. Aiming at low medium consumption but integration of all probes relevant for process monitoring (pO_2, pH, biomass) a parallel "mini bioreactor system" consisting of individually controlled vessels (DASGIP / Eppendorf) was utilized. After establishing stirring-controlled aggregate formation up to $2x10^8$ hiPSCs were generated per run in 100 ml culture scale applying batch-feeding. Yet, only linear cell growth was achieved suggesting suboptimal process conditions. Here we will present how perfusion-feeding results in substantially improved process characteristics and hPS cell yields. Expanded cells were directly used for the efficient differentiation into cardiomyocytes in stirred tank reactors in a chemically defined medium resulting >80% cardiomyocyte purity without additional lineage enrichment. Technical modifications of the bioreactor system will be highlighted including: impeller design, online biomass sensor integration, establishing a cell retention system and utilization of disposable bioreactor vessel (BioBLU® 0.3) combined with the DASbox® culture control system.

Funding sources: REBIRTH Cluster of Excellence (DFG EXC62/3), BMBF (VDI grant no. 13N12606), BIOSCENT (FP7/2007-2013, grant no. 214539) and StemBANCC (Support from the Innovative Medicines Initiative joint undertaking under grant agreement n° 115439-2, resources of which are composed of financial contribution from the European Union (FP7/2007-2013) and EFPIA companies' in kind contribution). STEMCELL Technologies (Vancouver, Canada), DASGIP / Eppendorf (Germany)

Nutrient Regulation by Continuous Feeding Removes Limitations on Cell Yield in the Large-scale Expansion of Mammalian Cell Spheroids

Meri Firpo, Ph.D.
Stem Cell Institute
Schulze Diabetes Institute
University of Minnesota

Schulze Diabetes Institute

UNIVERSITY OF MINNESOTA

Stem Cell INSTITUTE

UNIVERSITY OF MINNESOTA Driven to Discover

Stem cells for transplantation in diabetes: Scale Up Cultures of Human iPS-Derived Endoderm for Transplantation

Meri Firpo, Ph.D.
Stem Cell Institute
Schulze Diabetes Institute
University of Minnesota

Schulze Diabetes Institute

UNIVERSITY OF MINNESOTA

Stem Cell INSTITUTE

UNIVERSITY OF MINNESOTA Driven to Discover

How Do You Cure Diabetes with Stem Cells?

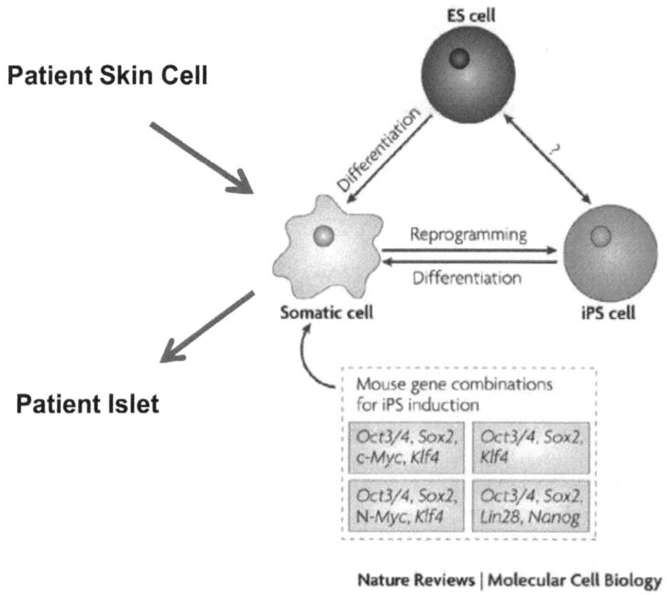

We follow differentiation by monitoring expression of specific transcription factors

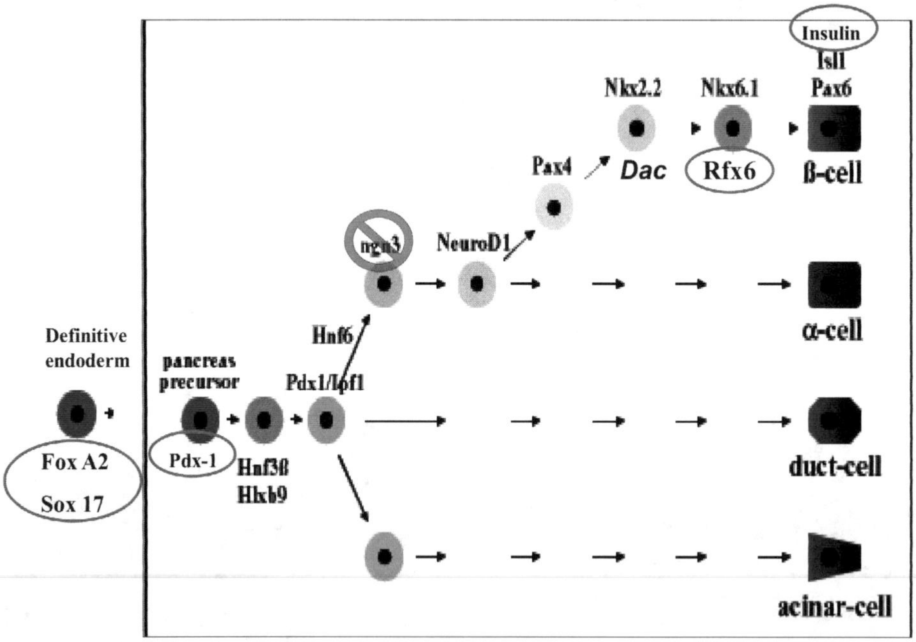

Evaluation of iPS differentiation into pancreatic cells

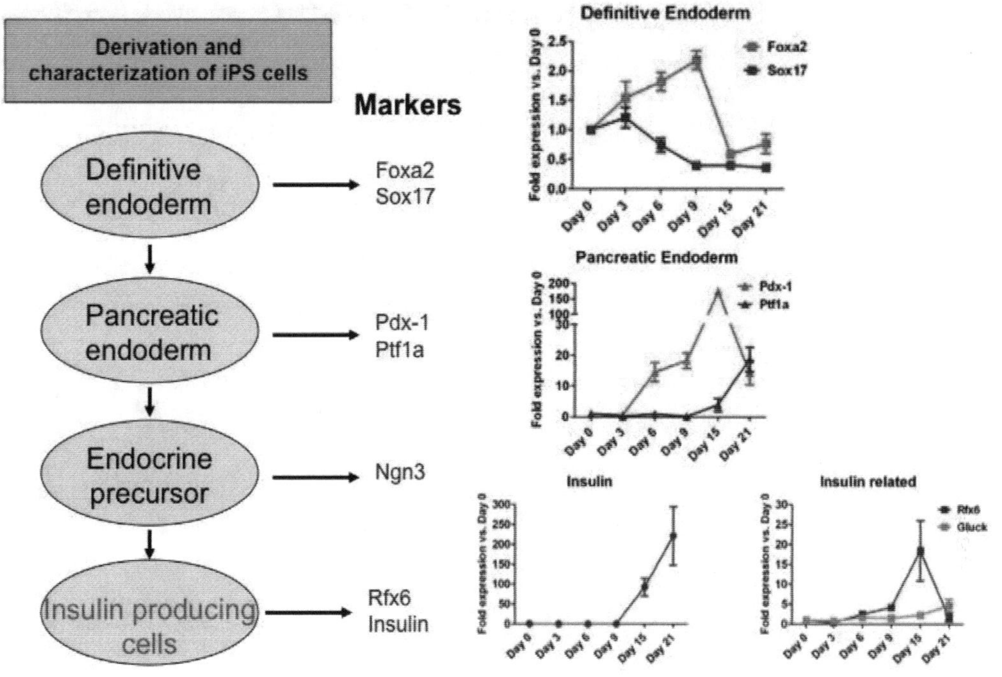

Functional analysis of differentiated iPS cells in vitro

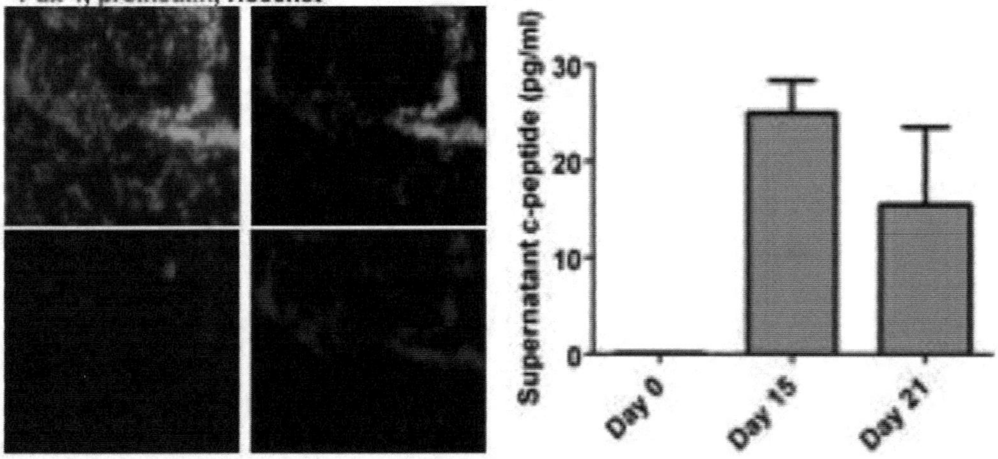

Functional analysis of differentiated iPS cells in vivo

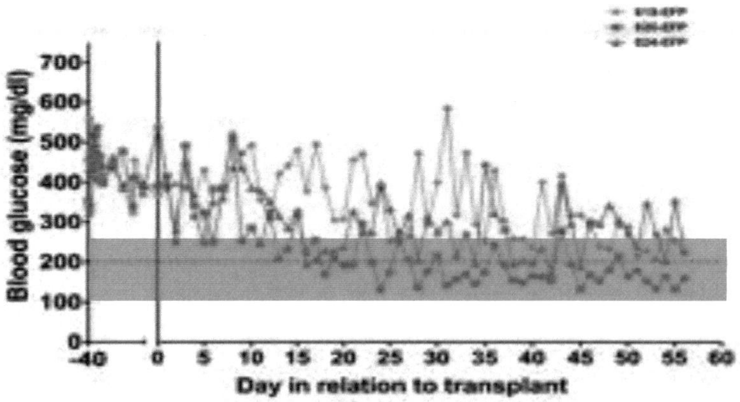

Summary of Initial Transplantation Experiments

In vitro insulin levels often peak at day 15

Glucose levels were best controlled in transplanted mice from day 15 differentiation

- Are beta cells maturing in vitro?
- Are the beta cells dying as they develop from precursors?

Can we design the culture conditions to better support beta cell maturation, survival and function?

Changing Culture System

Static Culture

Stirred bioreactor

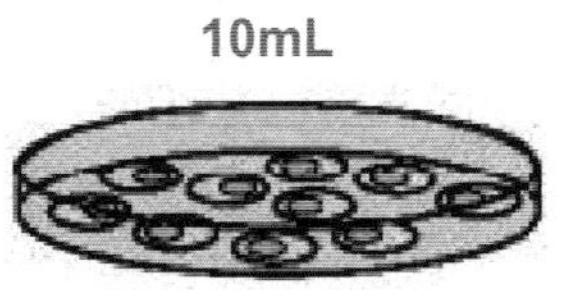

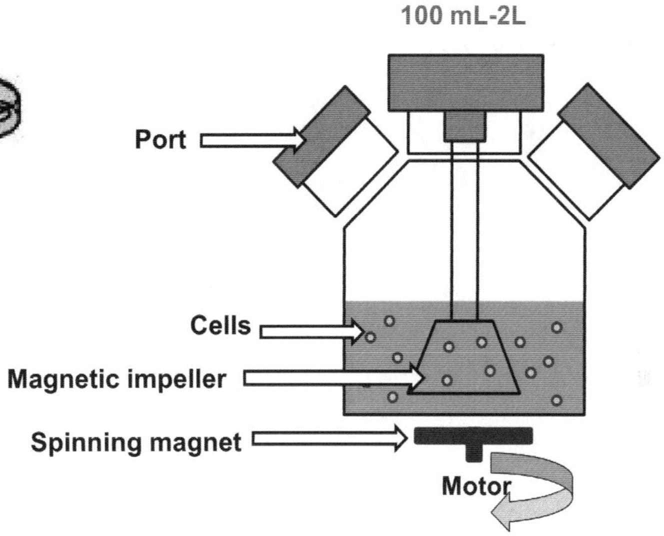

Comparison of Static and Stirred Bioreactor Cultures: Increased cell yield

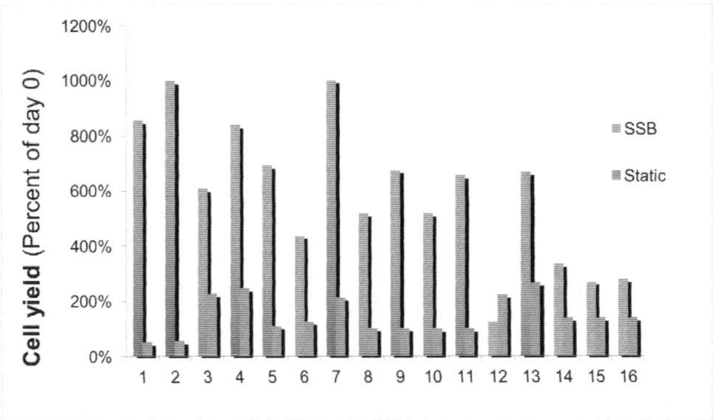

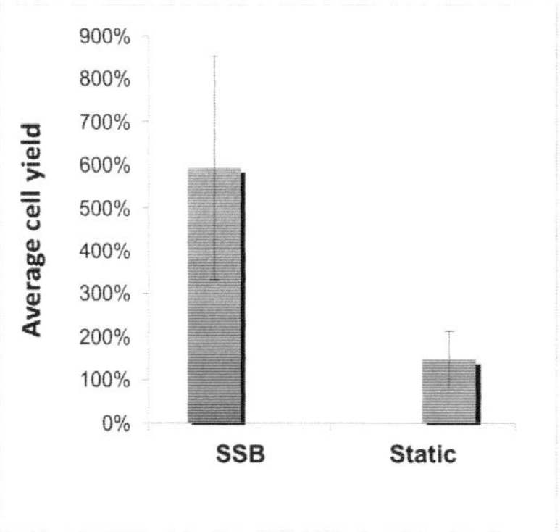

Medium changed every 3 days

Stirred Bioreactor Cultures: Limiting nutrient levels fluctuate because of batch feeding

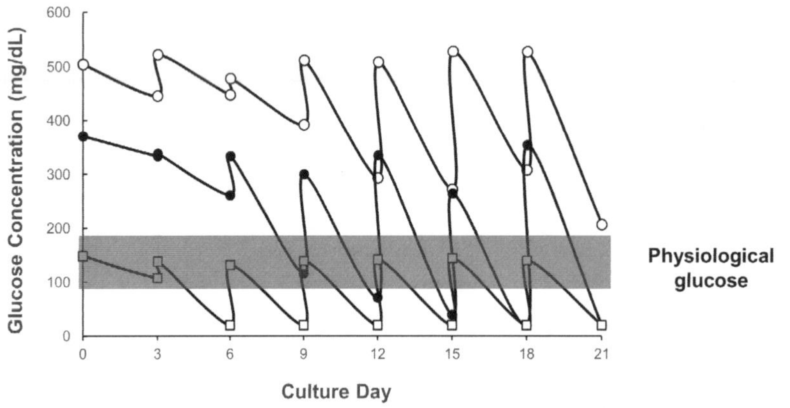

Medium changed every 3 days
βTC6 cells

Design of continuous feed apparatus

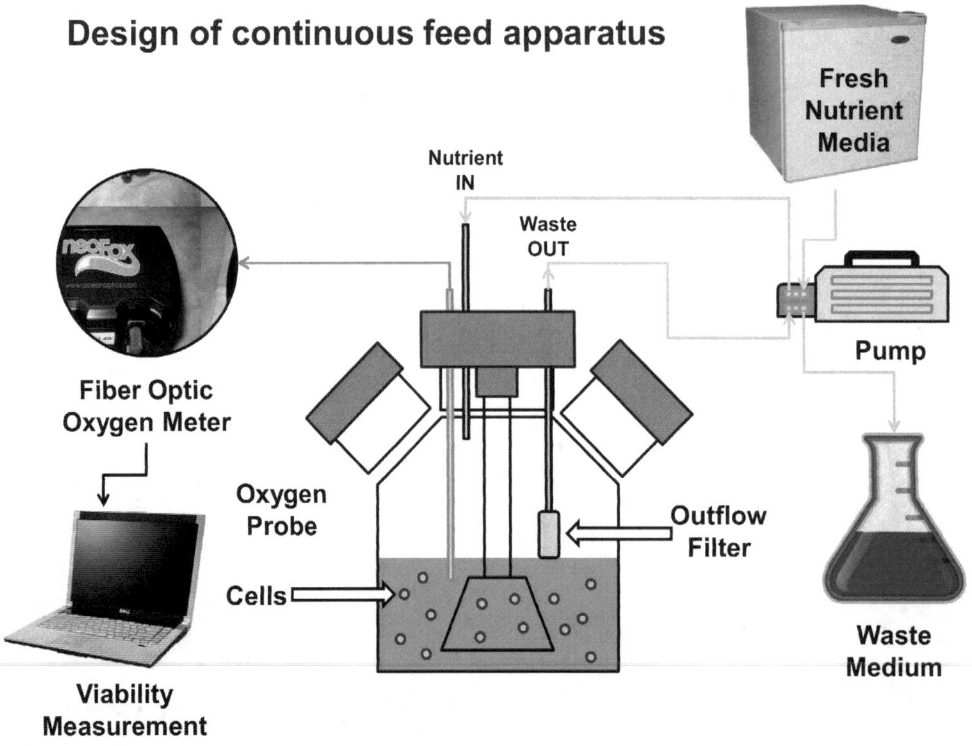

Outflow Filter Design using fritted glass

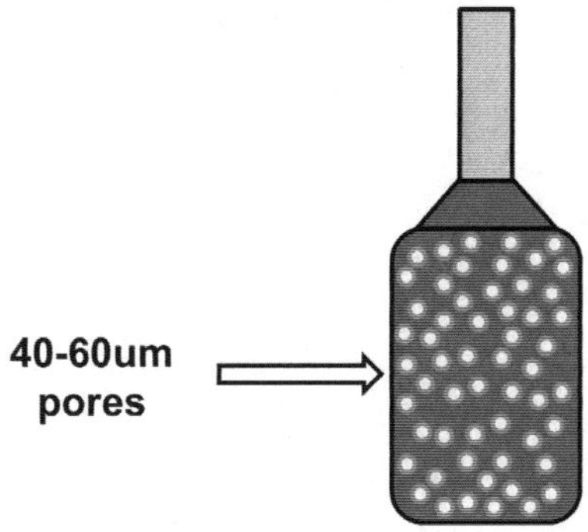

Stirred Bioreactor Cultures: Glucose levels fluctuate less with continuous feeding

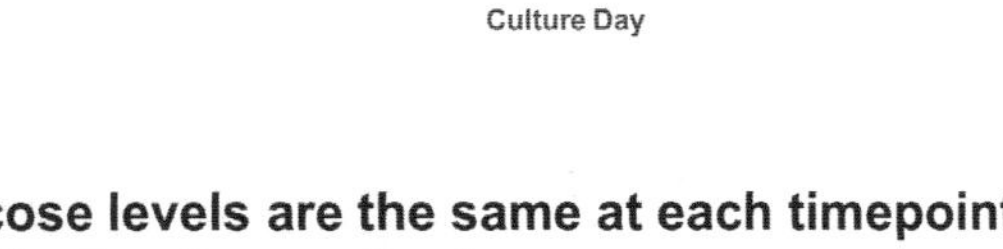

...But the glucose levels are the same at each timepoint for bioreactors regardless of the feeding method

Static and batch fed had medium changed every 3 days
Feed rate set at total medium change every 3 days

Comparison of Static, Stirred and Continuously Fed Bioreactor Cultures:

Simply removing fluctuations reversed the limits on cell yield

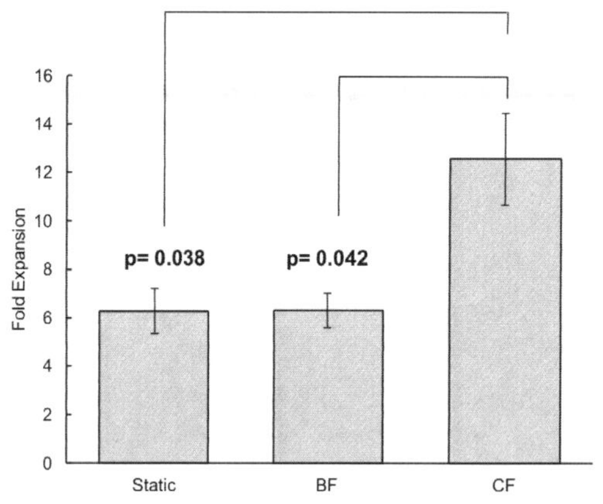

Feed rate set at total medium change every 3 days

Continuously Fed Stirred Bioreactor Cultures: Glucose is consumed with stable feed rates

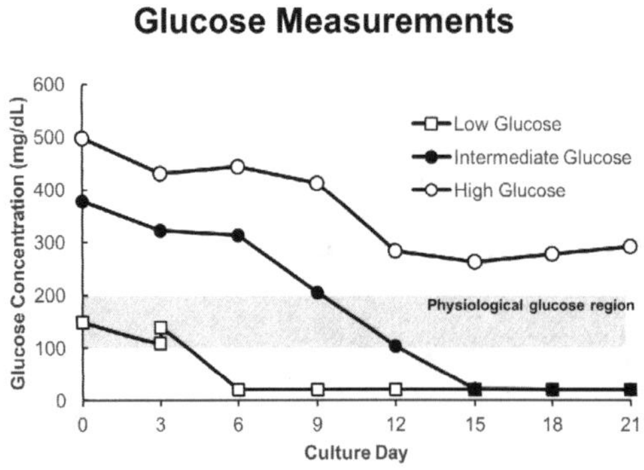

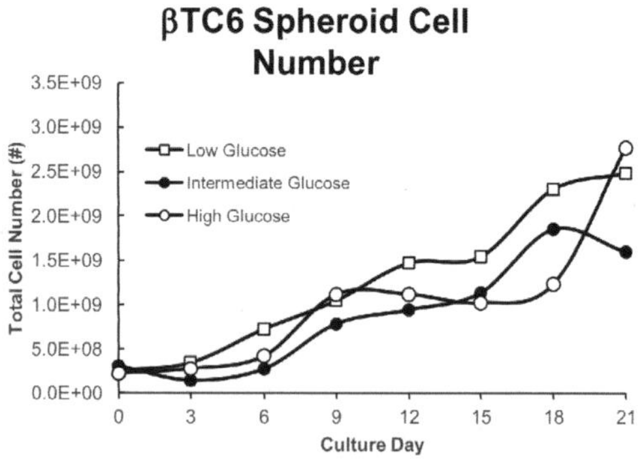

Feed rate set at total medium change every 3 days

Developing an Adjusted Feed Rate to Maintain glucose levels

Continuous fed cultures with a constant feed rate were not able to maintain a physiological glucose level during the 21 day culture period.

An algorithm was developed to adjust the feed every three days based on the sampling information. The goal was to predict the glucose consumption rate for the culture.

The algorithm incorporated:

- **-The current day cell count**
- **-A predicted growth rate based on previous data**
- **-Current medium glucose measurements**

Developing an Adjusted Feed Rate to Maintain glucose levels

Linear growth rate prediction ***(eq 1)***

N2=predicted cell count
N1=current cell count
RG=linear approximated cellular growth rate (cells/day)
t1=current culture day
t2=culture day of prediction (day 3)

$$N2 = N1 + R_G \times (t_2 - t_1)$$

Cellular consumption rate equation ***(eq 2)***

RF=Feed Rate (ml/min)
R1=cellular glucose consumption rate (mg/min/cell)
N1=current cell count
N2=predicted cell count (3 days later defined by Eq 1)
CF=glucose concentration in feed medium (mg/ml)

$$\text{Consumption} = R_F = \frac{R_1 \times \frac{2N_1+N_2}{3}}{C_F}$$

Glucose adjusted feed rate equation ***(eq 3)***

GAFR=glucose adjusted feed rate (ml/min)
X=unit less feedback control power constant
C1=medium glucose concentration on current culture day (mg/dl)
CD=desired medium glucose concentration (mg/dL)
RF=feed rate (ml/min, from eq 2)

$$\text{GAFR} = \left[1 - \frac{X\text{-}(C_1 - C_D)}{C_D} \right] R_F$$

Continuously Fed Stirred Bioreactor Cultures: Physiological glucose is maintained with adjusted feed rate

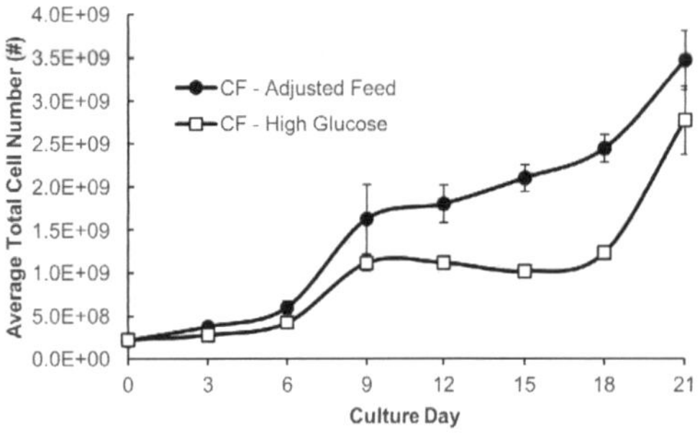

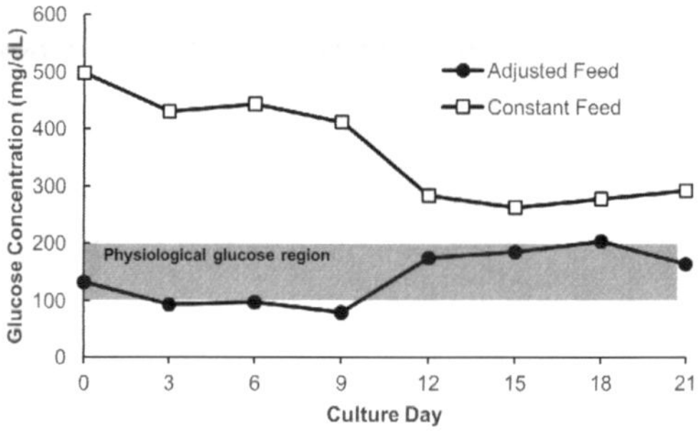

Adjusting the feed rate based on glucose use improved the cell expansion by 25% compared to constant feeding method and the glucose level in the medium was maintained near physiological levels for the duration of the 21 day culture.

Changing Bioreactor System

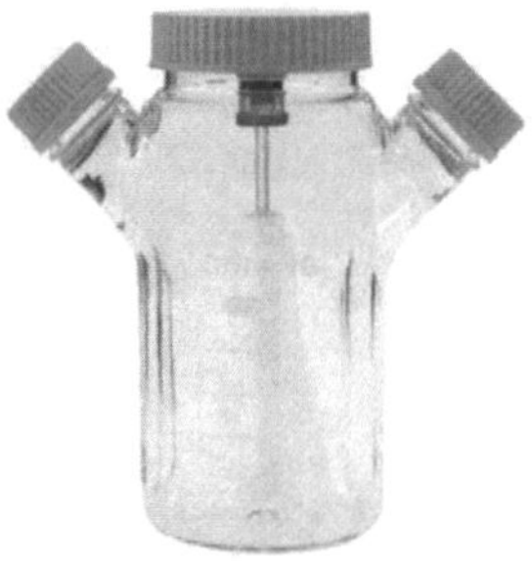

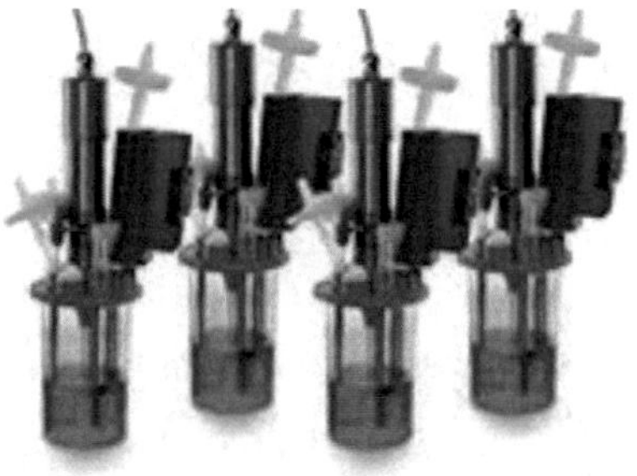

Impact of Feeding Strategies on Stem Cell Products

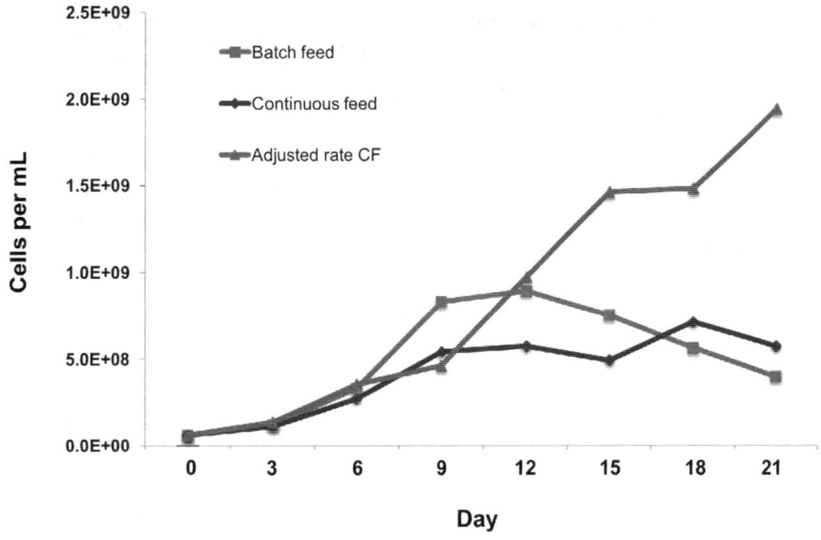

Batch fed had medium changed every 3 days
Continuous feed rate set at total medium change every 3 days

Impact of Feeding Strategies on Stem Cell Products

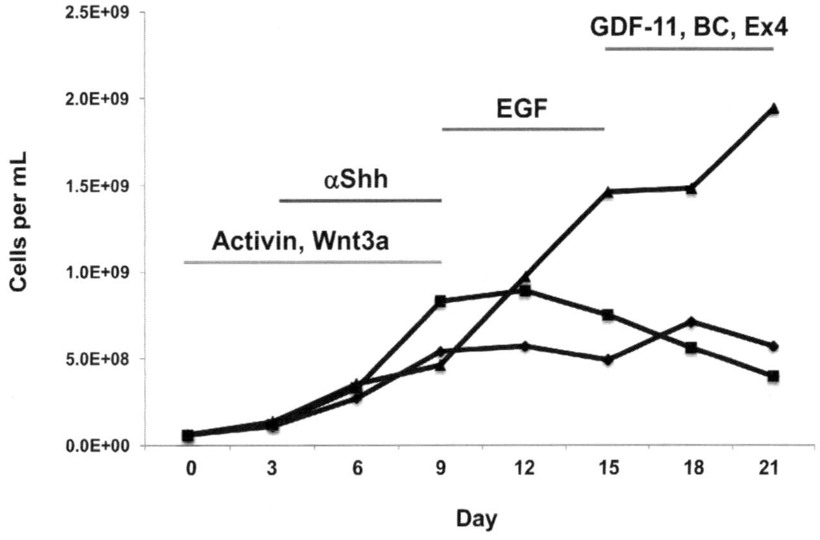

Batch fed had medium changed every 3 days
Continuous feed rate set at total medium change every 3 days

Conclusions

- iPS differentiation cultures can be conducted in large scale and at high density by using stirred bioreactors
- Continuous feeding increases cell yield
- Fluctuations in glucose levels reduce cell yield even when absolute levels are the same.
- Fluctuations in glucose levels can be eliminated by varying the feed rate based on cell counts/glucose usage, resulting in even higher cell yields.
- Adjusting the feed rate on stem cells in culture removes restrictions in response to cytokines.

Acknowledgements

Zhaohui Geng

Marjan Jahani

Ben Becker

Kristin Voltzke

Kimberly Stelzig

Oladipo Fagbemi

Erica Dobbs

Christopher Chapman

Taylor Millikan

Elise Mead

Kristen Maynard

Brad Weegman

Peter Nash

Chunlan Jiang

Schulze Diabetes Institute:

Bernhard Hering

Tim O'Brien

Melanie Graham

Jody Janecek

Klearchos Papas

Stem Cell Institute

Nobuaki Kikyo

Hiroyuki Hirai

Wei-Shou Hu

Bob Sorenson
Cliff Steer
Catherine Verfaillie

Directed differentiation of ES cells into pancreatic cells

		Cytokine	Induced cell type
Stage 1	Day 0-3	100 ng/ml Activin A 50 ng/ml Wnt 3a	Endoderm
Stage 2	Day 3-9	100 ng/ml Activin A 50 ng/ml Wnt 3a 2.5 ug/ml anti-Shh	Definitive endoderm Pancreatic endoderm
Stage 3	Day 9-15	50 ng/ml EGF	Expansion of Pancreatic endoderm
Stage 4	Day 15-21	50 ng/ml GDF-11 50 ng/ml Betacellulin 10 nM Exendin4	Pancreatic cells Maturation of β cells

Lentiviral Arrays for High-Throughput, Live Monitoring Gene and Pathway Activation during Stem Cell Differentiation

Padmashali, R., SUNY at Buffalo
Mistriotis, P., University at Buffalo, State University of New York
Liang, M. S., State University of New York at Buffalo
Andreadis, S. T., University at Buffalo, The State University of New York
Originally presented on: 3/19/2014 0:00:00 - 0:20:00

Lentiviral arrays for high-throughput, live monitoring gene and pathway activation during stem cell differentiation

Roshan Padmashali, Panagiotis Mistriotis1, Maoshih Liang1, Stelios T. Andreadis1, 2
1. Bioengineering Laboratory, 908 Furnas Hall, Department of Chemical and Biological Engineering, University at
Buffalo, State University of New York, Amherst, NY 14260-4200, USA
2. Center of Excellence in Bioinformatics and Life Sciences, Buffalo, NY 14203, USA

Introduction: Mesenchymal stem cells (MSCs) are abundant and multipotent cells with great potential for regenerative medicine and tissue engineering. The conventional tools for investigating gene expression such as qPCR are laborious, cost intensive and they cannot provide functional information on the pathways involved in biological processes. On the other hand, LentiViral microArrays (LVA) provide live-cell, high-throughput monitoring of activation of biological pathways during biological processes such as MSC differentiation. In this paper, we employed LVA with a library of lentiviral vectors to monitor pathway activation during MSC differentiation into bone, fat or cartilage. In addition, we hypothesized that use of LVA in combination with a
small molecular chemical library might identify pathways that control MSC fate commitment.

Materials and Methods: To address this hypothesis, first we engineered a dual-promoter lentiviral vectorâ??LVDP containing a sp(ae)cific transcriptional regulatory element encoding for green fluorescence protein for monitoring gene regulation. Additionally this vector contains a constitutive promoter driving red fluorescence protein providing an internal normalization strategy, thereby allowing measurements of gene or pathway dynamics independent of the gene transfer efficiency. We developed a library of lineage specific reporters containing promoters such as adiponectin, SOX9 and Runx2 for adipogenesis, chondrogenesis and osteogenesis, respectively as well as transcription factor response motifs e.g. Vitamin D, SMAD, C/EBP etc. By

Figure 1 (a) In the experiment setup, MSCs were transduced with various LVDP viruses and the
dynamics of lineage-specific reporters were
monitored real-time with fluoresunce microscopy during differentiation. (b) Heat map summarizing the effects of chemical inhibitors on lineage-specific reporters during differentiation.

employing automated fluorescence microscopy, we determined the kinetics of lineage-specific pathway activation in MSC derived from two different anatomic locations i.e. hair follicle and bone marrow, (Fig 1(a)). Furthermore, we screened a small array of chemical inhibitors to monitor the pathways that may control
differentiation into one or more lineages.

(b)

Results and Discussion: Our results demonstrate differential reporter activation between the two MSC sources, suggesting that the regulatory pathways controlling differentiation may be dependent on the source of MSC. To identify key molecules affecting MSC differentiation, several inhibitors were selected to illustrate the effects on lineage specific reporters. We found that inhibition of TGF-Î² receptor promoted adipogenesis as evidenced by decreased activity of the adiponectin promoter. Similarly, inhibition of p38 increased the activity of both chondrogenic (SMAD2/3) and osteogenic (OCBOX) reporters, while inhibition of TGF-Î² or FGF receptors down-regulated chondrogenic or osteogenic reporters, respectively. Inhibition of the PI3 kinase and glucocorticoids readily suppressed the activity of adipogenic pathways. Also, inhibition of JNK and ERK blocked the expression of osteo-specific reporters such as Vitamin-D and OCBOX. Finally, we identified novel pathways differentially regulating MSC differentiation along different lineages. The complete results were summarized in Fig 1(b).

Conclusion: We developed a LVA system that enabled efficient monitoring of gene and pathway activation in MSC differentiation. This approach also allowed us to identify differences in the kinetics of pathway activation between different MSC lines, despite their similar differentiation potential. Notably, the LVA enabled screening of small molecule libraries that identified pathways that were necessary for differentiation along particular lineages. Our results suggest that the LVA is a tool with great potential to enhance understanding of stem cell fate specification as well as pathway regulation and drug discovery.

SINGLE CELL WESTERN BLOTTING TO PROBE STEM CELL SIGNALING AND DIFFERENTIATION

Dawn Spelke
Schaffer Lab, Department of Bioengineering
University of California, Berkeley
AICHE SBE Stem Cell Engineering Conference
March 19, 2014

Stem Cell Heterogeneity

Heterogeneity hallmark of many biological processes

Adult hippocampal neural stem cells

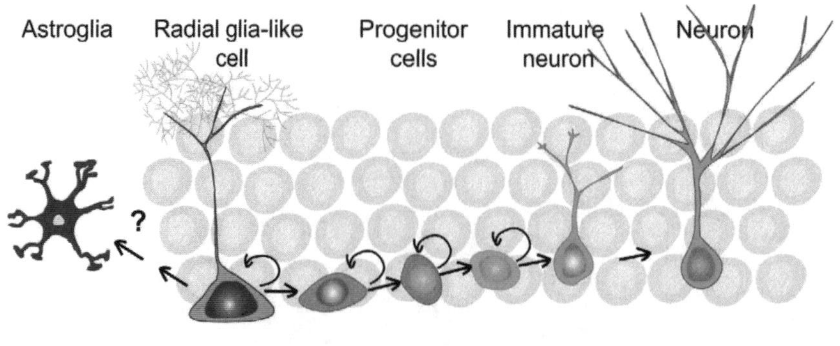

Heterogeneous base population?
Stochastic response?

Bonaguidi, et al., *Cell* 2011

Current Single Cell Analytical Tools

- Genomic and transcriptomic
  - May not correlate with protein levels
  - Inability to probe post-translational modifications
- Immunocytochemistry and flow cytometry
  - Multiplexing difficult
  - Cannot interrogate protein-protein interactions, isoforms
  - Poor antibody specificity can obscure findings

<u>Need for novel single cell proteomic tool</u>

➔ **Single cell Western Blotting:** Separate proteins by MW

Single Cell Western Blotting Workflow

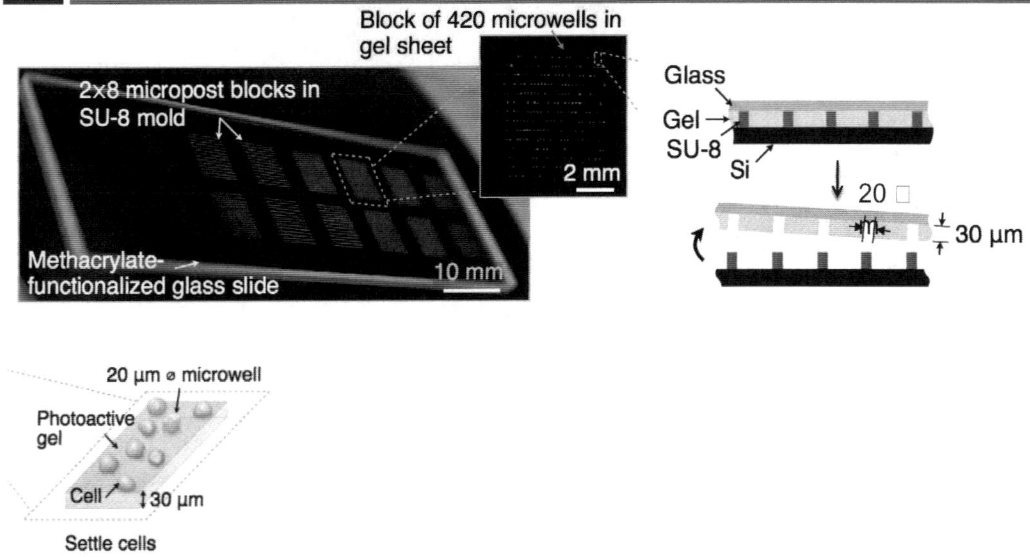

<u>~ 4 hours to complete assay</u>

Hughes*, Spelke*, et al., submitted

Single NSCs Settle into Microwells

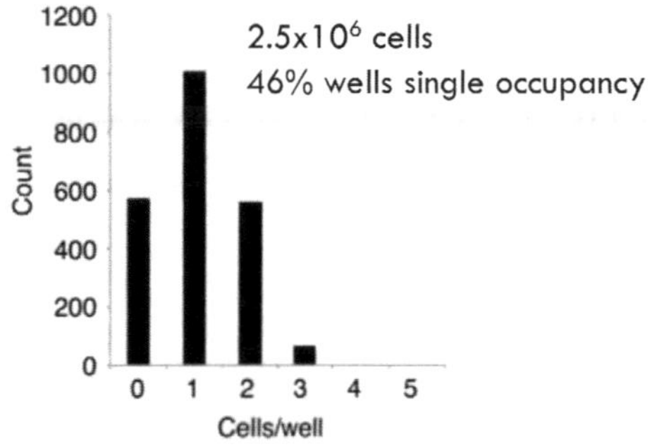

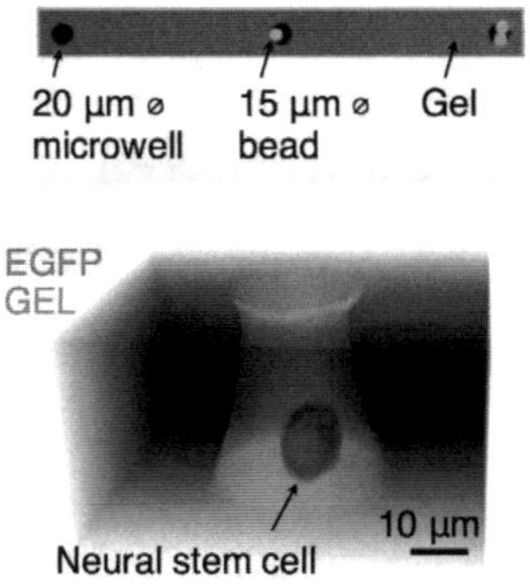

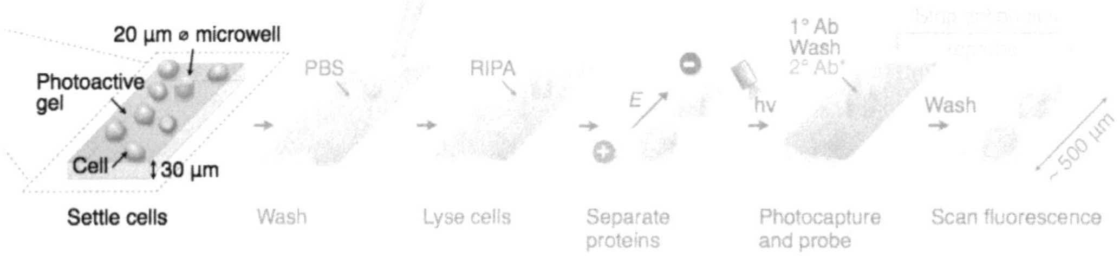

NSCs Quickly Lyse & Electrophorese into Gel

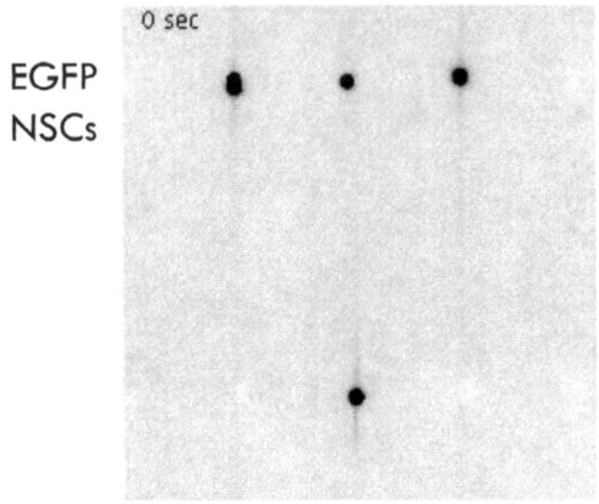

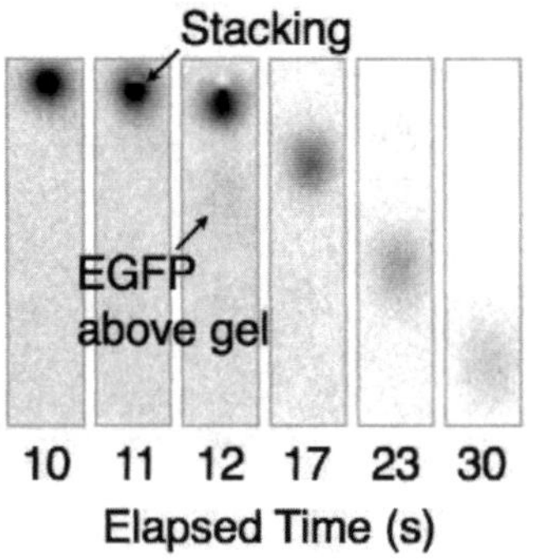

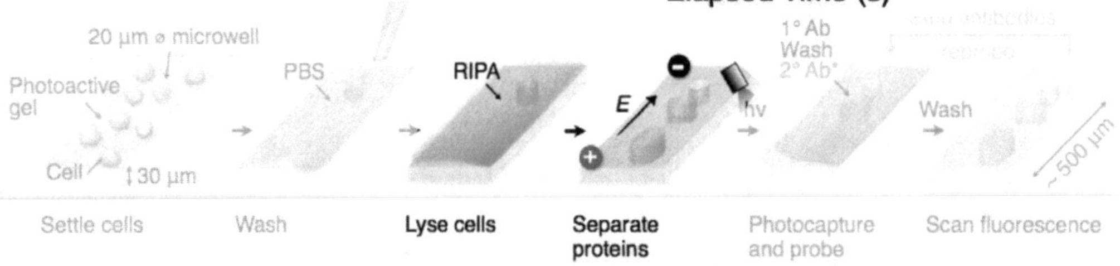

Proteins are Photocaptured and Probed

Capture 27.5% EGFP

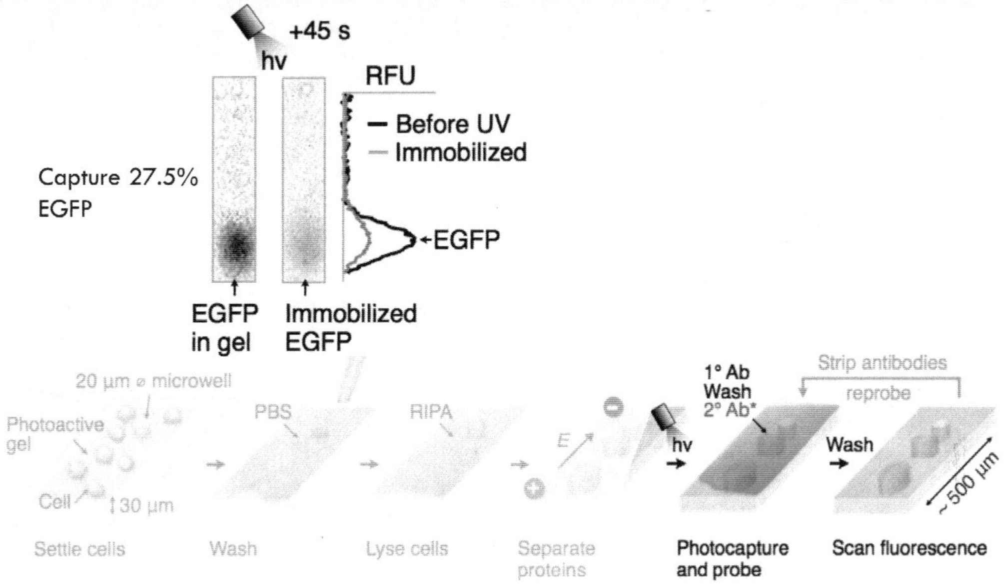

Representative Data Set

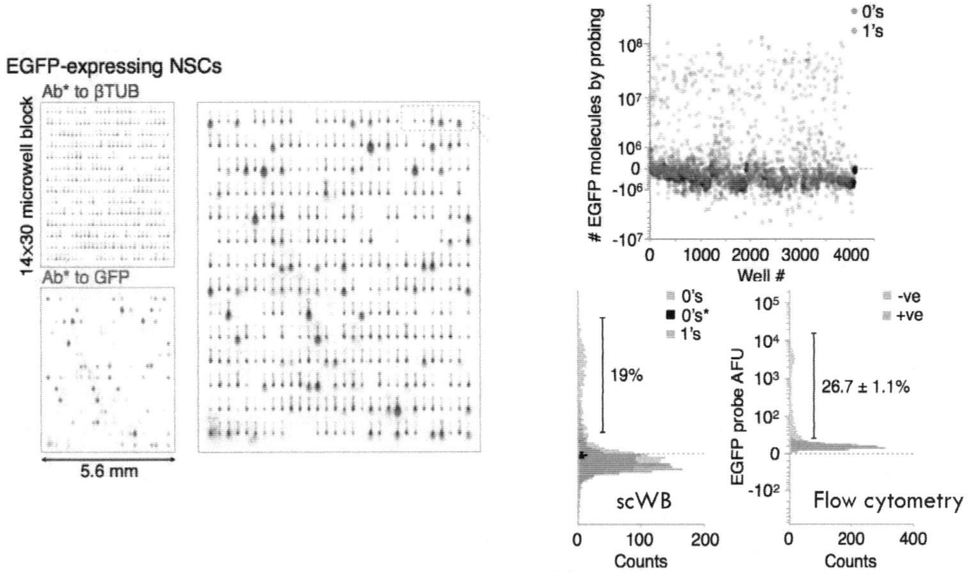

EGFP heterogeneity observable

Stripping Allows for Multiplexed Probing

11-plex assay, 9 stripping/reprobing rounds

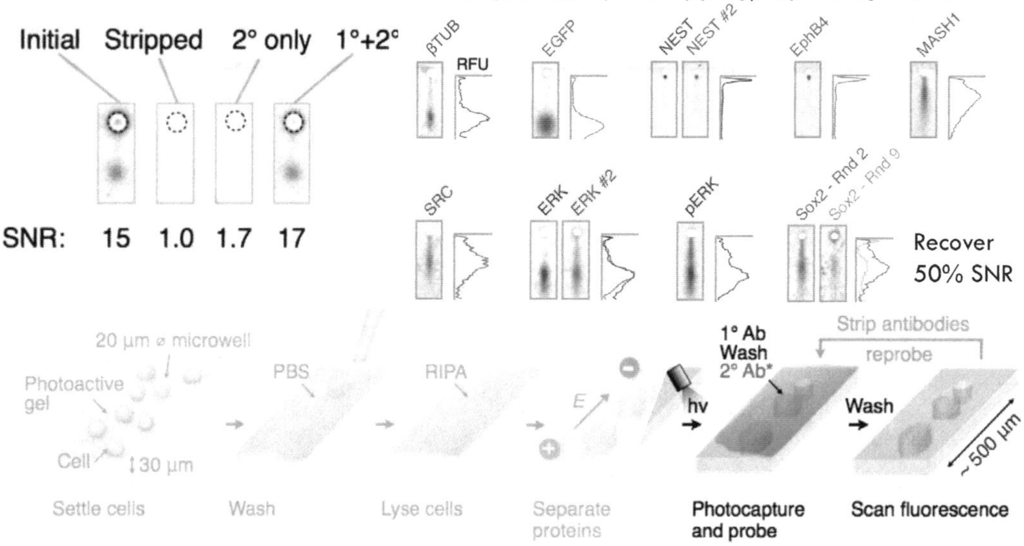

scWB Limit of Detection 27,000 Molecules

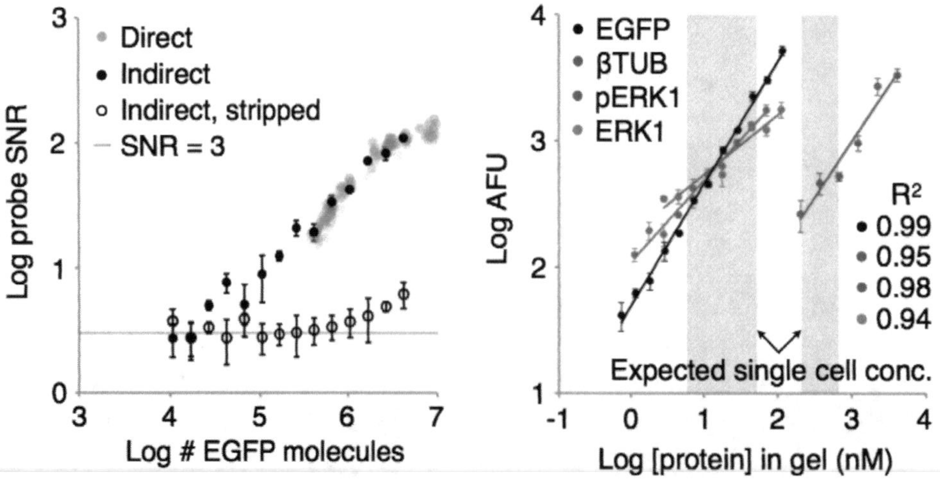

Can probe endogenous target proteins in single cells

Analysis of MAPK Signaling Dynamics

FGF stimulation time course, probe for pERK/ERK and pMEK/MEK

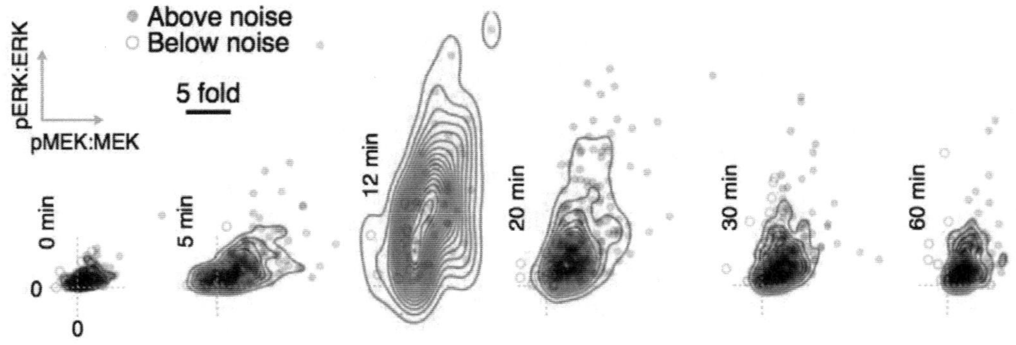

pERK and pMEK trends correlate well with conventional WB

scWB reveals **cell-cell variability** in signaling response

scWB Addresses Non-Specifc Antibody Effects

pERK antibody shows off-target band at ~103 KDa in WB

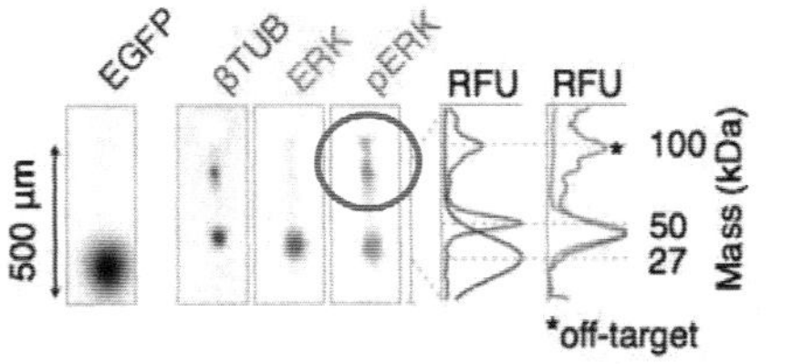

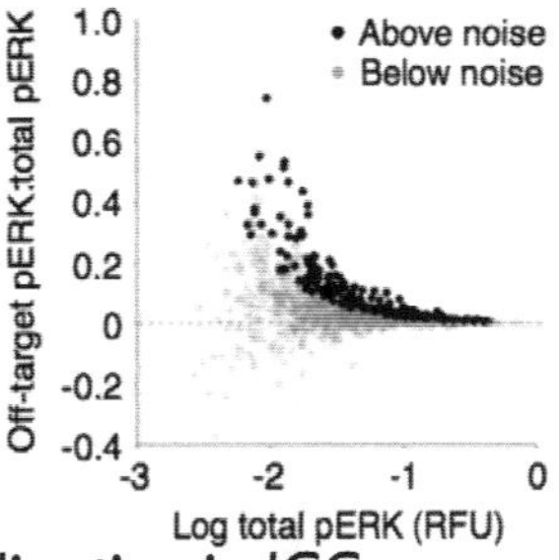

pMEK antibody shows improper nuclear localization in ICC

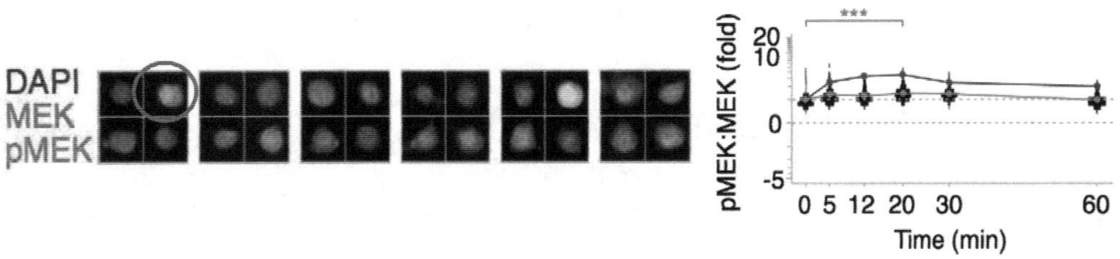

Analysis of NSC Differentiation Dynamics

NSC immature state: Sox2+/Nestin+

NSC lineages: Neurons – IIITub+, Astrocytes – GFAP+

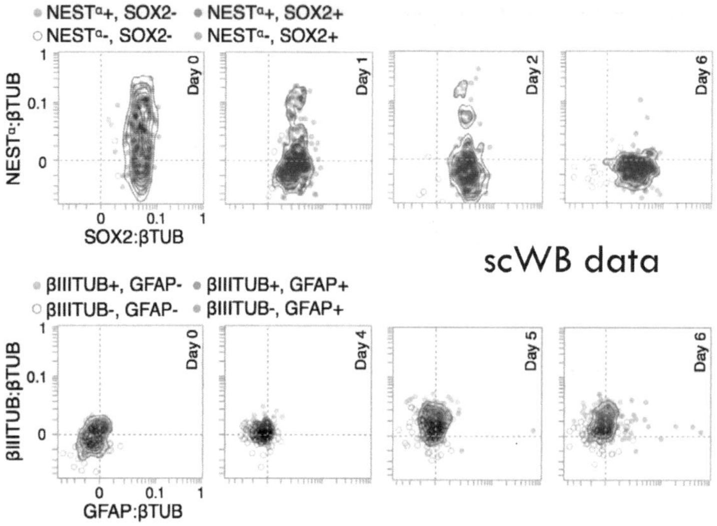

scWB data

scWB Allows for Isoform Analysis

Nestin splice variants exhibit differential expression dynamics

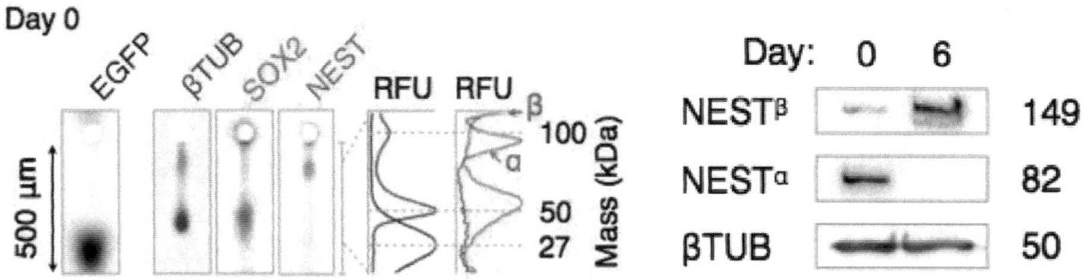

NEST **novel** neurogenic differentiation marker

Summary

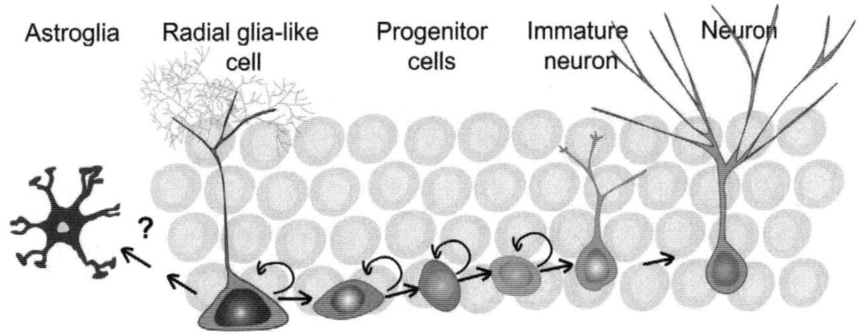

Acknowledgements

Schaffer Lab
Professor David Schaffer
Emily Connelly
Albert Keung, PhD
Anthony Conway, PhD
Ashley Fritz, PhD
Lukasz Bugaj
Sisi Chen, PhD

Herr Lab
Professor Amy Herr
Alex Hughes, PhD
Zhuchen Xu
Ginny Kang, PhD

Funding Sources
National Science Foundation Graduate Research Fellowship
California Institute for Regenerative Medicine Training Grant Fellowship
Siebel Scholars Program

Extra Slides

Direct Calibration

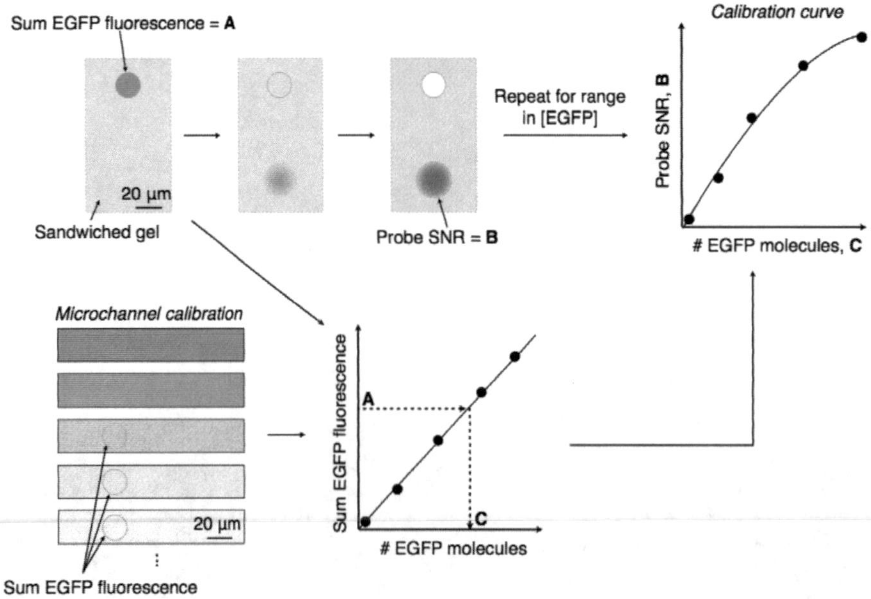

Indirect Calibration

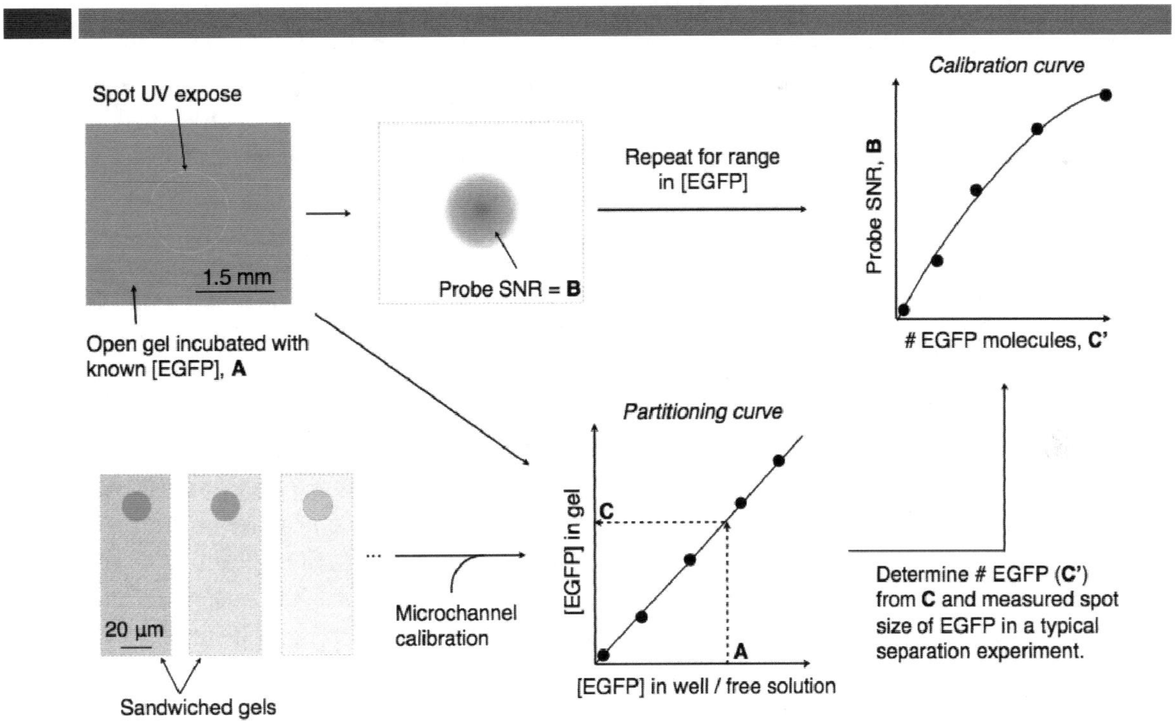

Accelerating Development of Stem Cell-Derived Therapies Using Multiplexed Microbioreactor Arrays

Titmarsh, D., The University of Queensland
Hudson, J., The University of Queensland
Frith, J. E., The University of Queensland
Padmanabhan, H.,
Hidalgo-Gonzalez, A.,
Glass, N.,
Ovchinnikov, D., The University of Queensland
Porrello, E., The University of Queensland
Wolvetang, E., The University of Queensland
Cooper-White, J., The University of Queensland
Originally presented on: 3/19/2014 0:40:00 - 1:00:00

Introduction: Effective use of human stem cells in regenerative medicine and drug screening depends on our ability to effectively direct both their undifferentiated expansion and differentiation into desired lineages. Exquisite control over stem cell fate is needed to efficiently produce sufficient, defined cell populations for such applications, yet this is substantially hindered by undefined culture components, signal crosstalk between multiple exogenous and endogenous factors, and spatiotemporal variations in microenvironmental composition inherent to conventional culture formats.

Experimental: We have developed scalable, valveless, continuous-flow microbioreactor arrays that both provide a full-factorial set of exogenous factor compositions, and also allow controlled accumulation of paracrine factors. These arrays have been used to survey up to 8100 individual, perfused cellular microenvironments in parallel.

Results: Through screens of pluripotency maintenance and mesendodermal differentiation of human pluripotent stem cells (hPSCs) as examples, we demonstrate the unique ability of this platform to separate, visualise, identify and modulate paracrine effects that are not otherwise readily accessible. In addition, we have applied this same technology platform to the optimisation of maintenance (and improvement) of potency of human mesenchymal stem cells, and the impact of small molecule agonists and antagonists on osteogenic differentiation outcomes. Most recently, with our 8100-chamber, high density microbioreactor array, we have assessed the impacts and interplay of small molecule agonists and growth factors on human cardiomyocyte proliferation. Importantly, optimization of these culture conditions with the arrays was readily translatable to improving differentiation in conventional static culture protocols, exemplifying the immediate practicality of the microbioreactor array platform in a range of stem cell screens.

Conclusion: This multiplexed microbioreactor platform deciphers factor interplay and signalling hierarchies that control stem cell fate, and is applicable as a universal microenvironmental screening platform for bioprocess optimisation, media formulation design, quality control for cellular therapeutics and cell-based drug toxicity and discovery.

Metabolic Impact of Culture Media on Pluripotent Stem Cell Growth and Differentiation

Metallo, C. M., University of California San Diego
Zhang, H.,
Originally presented on: 3/19/2014 0:00:00 - 0:20:00

Since the discovery of culture conditions that allow for the maintenance and proliferation of human pluripotent stem cells (hPSCs) researchers have attempted to develop chemically defined media which can be used to safely expand hPSCs. These efforts have included the engineering of functional matrices that mimic or replace undefined substrates (e.g. Matrigel) and the addition of signaling molecules which maintain hPSCs in a highly proliferative state. Early studies focused on the design of culture media that does not require conditioning by embryonic fibroblasts, and more recently researchers have successfully developed chemically defined liquid media containing as few as 8 components. These latter efforts have been motivated, in part, by experiences in the bioprocess industry, as undefined components added to clinical biologics processes can be a source of contamination by adventitious agents. Furthermore, such additives may also cause lot-to-lot variability which complicates production processes and/or and interpretation of experimental outcomes. These concerns are well-founded and of critical importance as we attempt to scale-up pluripotent stem cell and adult progenitor cell cultures for clinical applications in regenerative medicine. However, unlike common industrial bioprocesses that produce therapeutic protein products in Chinese Hamster Ovary (CHO) cells, cell-based therapies are themselves the clinical product and must therefore match or replace the function of tissues in the body. Importantly, these functions almost always involve highly regulated metabolic activities. Researchers must therefore pay increasingly close attention to the metabolic performance of stem cells and their derivatives as we attempt to develop and scale-up cell-based therapies for more functionalized tissues.

To better characterize the metabolic behavior of hPSCs we have applied 13C metabolic flux analysis (MFA) to human embryonic stem cells and their derivatives. These studies involve application of stable isotope-labeled substrates such as [13C]glucose or [13C]glutamine, mass spectrometry-based metabolomics, and systems-level analysis of isotope enrichment in metabolic networks to estimate intracellular fluxes. Our investigations of metabolic performance in standard (conditioned) versus several commonly used defined media formulations identified profound differences in the metabolism of hESCs, though all media maintained OCT4 expression for over 20 passages. Defined media induce hESCs to increase glycolytic flux while decreasing oxidative TCA metabolism, with a concomitant decrease in oxygen consumption. Steady state metabolomics analysis indicated that TCA intermediates were more abundant in defined culture conditions, while lipogenic metabolites such as glycerol-3-phosphate (G3P) and citrate became depleted. Reductive flux of glutamine carbon in the TCA cycle was increased to support lipid biosynthesis. In fact, while cells in standard hESC medium exhibit minimal de novo lipogenesis, hESCs in defined media synthesize significant amounts of fatty acids and cholesterol. To support these processes hESCs increase NADPH production in the oxidative pentose phosphate pathway but also exhibit increases in total reactive oxygen species (ROS). These changes are further manifested by the activation of upstream signaling pathways that are commonly upregulated in transformed cells. By supplementing these defined media with specific additives we can normalize the metabolic behavior of hESCs. Intriguingly, the nutritional microenvironment also influences differentiation efficiency. Current studies are now addressing the extent that these culture conditions affect the metabolic performance of hPSC-derivatives, including cardiomyocytes which have significant bioenergetic needs. These results suggest that caution should be used in employing minimal media for the maintenance of hPSCs and highlight the importance of metabolic function in characterizing cell function.

Linking Proliferation Dynamics and Culture Density to Human Pluripotent Stem Cell Differentiation Proclivity

Wu, J., State University of New York at Buffalo
Tzanakakis, E. S., State University of New York at Buffalo
Originally presented on: 3/19/2014 0:25:00 - 0:45:00

Human pluripotent stem cells (hPSCs) exhibit a shortened G_1 phase and rapid proliferation compared to somatic cells. Latest findings have suggested that hPSCs are permissive to differentiation during the G_1 phase and that commitment is linked to a lengthened cell cycle. We also demonstrated that stochastic partitioning at hPSC mitosis contributes to the heterogeneous response of stem cells in populations to differentiation stimuli. Therefore, hPSC proliferation and cell cycle regulation are critical for fate determination. Yet, how the proliferation dynamics influence commitment at the population level is unclear.

An approach was employed combining experiments and mathematical modeling. The hPSC cycle distribution under different culture conditions was analyzed by flow cytometry (FC) via multiple methods to ensure data accuracy and consistency. Proliferation was monitored both by traditional cell enumeration methods and automated image analysis developed in the laboratory. States of hPSC self-renewal and differentiation were assessed by FC and real-time PCR based on pertinent marker expression. A multiscale population balance model (PBM) was developed comprising integro-differential equations capturing the proliferation and differentiation kinetics of heterogeneous hPSC ensembles.

Human PSC proliferation was markedly reduced with an order-of-magnitude increase in doubling time as the population density surpassed $50x10^4$ cells/cm^2. This transition was accompanied by a rise in the fraction of G_1 phase hPSCs. Compared to low-density cultures exhibiting 80-90% $NANOG^+$ hPSCs, only 50% $NANOG^+$ cells were observed in dense cultures after 4 days. The loss of pluripotency was concomitant with the emergence of 10% of the cells entering quiescence. Four-day differentiation of hPSCs (H1, H9 and IMR90) with unconditioned medium (UM) under low or high initial density resulted in significant differences in the $NANOG^-$ cell fractions.

Further analysis was performed by combining these experimental data with the PBM. Because of the observed loss of pluripotency at higher densities, the model was employed to quantify the extent of induction of differentiation or differentiation 'strength', k_{diff}, in media routinely used for hPSC expansion. Upon leave-one-out cross validation of the model, k_{diff} was predicted for media containing differentiation factors (e.g. BMP4, activin A). The fractions of primed/committed ($NANOG^-$) cells in various media over a range of initial hPSC densities and differentiation times were calculated by solving the PBM via parallel computing. Simulation results were verified by running experiments for sets of the corresponding conditions. Moreover, the model was applied to hPSC cultures in microcarrier bioreactors. For both static and microcarrier cultures, the estimated k_{diff} was lower for media used for hPSC propagation than for those utilized in guiding hPSC commitment.

Our results suggest that the short G_1 phase acts as a safeguard of hPSC pluripotency. We demonstrated that culture density, which is easily adjustable, influences cell cycle length thereby affecting hPSC fate decisions. The approach presented is generally applicable as it allows the prediction of differentiation rates for multiple hPSC lines and culture conditions. Beyond deepening our understanding of mechanisms controlling hPSC self-renewal and differentiation, these findings will contribute to the rational design of niches and differentiation strategies for the generation of stem cell therapeutics.

Quantitative Spatial and Temporal Pattern Classification of Embryonic Stem Cell Aggregate Differentiation

White, D., Georgia Institute of Technology
Kemp, M., Georgia Institute of Technology
McDevitt, T. C., Georgia Institute of Technology
Originally presented on: 3/19/2014 0:40:00 - 1:00:00

Pluripotent embryonic stem cells (ESCs) can differentiate into the three germ lineages, making them an intriguing cell source for regenerative medicine therapies and drug screening. However, in order to precisely control differentiation it is still necessary to investigate the mechanisms governing cell fate decisions. One method of studying differentiation is to use aggregates of ESCs called embryoid bodies (EBs) which are capable of spontaneously differentiating into all three germ layers. Several groups have reported the spontaneous formation of spatially organized, complex tissue structures [1-3] in EB systems highlighting the importance of understanding the inherent emergent behavior capable of generating such complexity. We have recently described the creation of a computational model to simulate the 3D differentiation of EBs [4], but quantitative descriptions of the multicellular pattern formation in biological systems remain lacking.

Others have attempted to quantify spatial patterns using machine learning approaches [5] in combination with digital reconstruction algorithms [6]. However, such approaches are limited to specific systems of interest, or lack the necessary resolution to make claims about single cell spatial pattern dynamics. Here we describe the novel use of spatial networks for capturing single cell spatio-temporal heterogeneity associated with complex biological pattern formation. By utilizing network theory, various classifiers such as node counts, connections counts or novel user defined metrics, can be evaluated to create rich data sets for deriving predictive trends that can be used in conjunction with computational modeling approaches for model validation.

Formation and culture of murine ESCs (D3 line) into EBs was performed as previously described for two initial cell seeding densities (250 and 1000 cells/well) [4]. EB?s were fixed with formalin, counterstained with Hoechst and Phalloidin Alexflour 546 and immunostained for the pluripotency transcription factor Oct4. Stained EB?s were imaged by confocal microscopy and analyzed via Cell Profiler software coupled with a network reconstruction algorithm. Computational modeling was performed as previously described [4].

During EB differentiation, four types of multicellular patterns were observed: random, inside-out, outside-in, and connected (Fig 1A). To classify these patterns quantitatively, various metrics were validated by generating *in silico* patterns (Fig 1B). Using principal component analysis (PCA), we showed that the patterns could be correctly segregated with these metrics (Fig 1C). Furthermore, classification using machine learning approaches indicated patterns could be correctly classified with >98% accuracy (Fig 1D) suggesting that these metrics were capable of capturing pattern heterogeneity. Next we applied these metrics to our differentiating EB system [4]. PCA depicted a distinct differentiation trajectory that was independent of EB size (Fig 1E), which was further confirmed by pattern classification performed on these trajectories (Fig 1F). Additionally these patterns matched spatial patterns extracted from computational modeling traces as well (Fig 1G).

Subtle, yet distinct, spatial patterns were distinguished in the *in silico* training set utilizing network metrics which subsequently classified spatial patterns in both computational and experimental systems. These results highlight the power of our approach for quantitatively describing pattern trajectories during morphogenesis in a manner which allows direct comparison with complex computational modeling strategies.

References:

Self-organizing optic cup morphogenesis in three dimensional culture. Eiraku M, Takata N, Ishibashi H, Kawada M, Sakakura E, Okuda S, Sekiguchi K, Adachi T, Sasai Y. Nature. 2011 Apr;472(7341):51-6.

Generation of functional thyroid from embryonic stem cells. Antonica F, Kasprzyk DF, Opitz R, Iacovine M, Liao XH, Dumitrescu AM, Refetoff S, Peremans K, Manto M, Kyba M, Costagliola S. Nature. 2012 Nov1;491(7422):66-71.

Self-formation of functional adenohypophysis in three-dimensional culture. Suga H, Kadoshima T, Minaguchi M, Ohgushi M, Soen M, Nakano T, Takata N, Wataya T, Muguruma K, Miyoshi H, Yonemura S, Oiso Y, Sasai Y. Nature 2011 Nov 9;480(7375):57-62.

Spatial pattern dynamics of 3D stem cell loss of pluripotency via rules-based computational modeling. White DE, Kinney MA, McDevitt TC, Kemp ML. PLoS Comput Biol. 2013;9(3).

Towards a digital model of zebrafish embryogenesis. Integration of cell tracking and gene expression quantification. Castro-Gonzalez, C., M. A. Luengo-Oroz, et al. Conf Proc IEEE Eng Med Biol Soc 2010: 5520-5523.

Imaging morphogenesis: technological advances and biological insights. Keller, P. J. Science 340(6137): 1234168.

AUTHOR INDEX

AUTHOR INDEX

AUTHOR INDEX

AIChE
120 Wall Street, FL 23
New York, NY 10005-4020

ISBN 978-1-5108-1762-3
9 781510 817623

CURRAN ASSOCIATES INC.
proceedings.com